T0323025

Neurocognitive Mechanisms

Neurocognitive Mechanisms

Explaining Biological Cognition

GUALTIERO PICCININI

OXFORD
UNIVERSITY PRESS

Great Clarendon Street, Oxford, OX2 6DP,
United Kingdom

Oxford University Press is a department of the University of Oxford.
It furthers the University's objective of excellence in research, scholarship,
and education by publishing worldwide. Oxford is a registered trade mark of
Oxford University Press in the UK and in certain other countries

First Edition published in 2020
Impression: 2

Published in the United States of America by Oxford University Press
198 Madison Avenue, New York, NY 10016, United States of America

British Library Cataloguing in Publication Data
Data available

Library of Congress Control Number: 2020947366

ISBN 978–0–19–886628–2

Printed and bound by
CPI Group (UK) Ltd, Croydon, CR0 4YY

Contents

Acknowledgments

My undergraduate studies at the University of Turin, during the early 1990s, exposed me to the classic debate on the foundations of cognitive science. Is the brain a computer? Does cognition involve computation over representations? Is the mind the software of the brain? There were arguments on both sides, and I wanted to sort them out.

While in graduate school at the University of Pittsburgh, during the late 1990s and early 2000s, I embarked on a research program: first, figure out what physical computation is by drawing from computer science and computer engineering; then, apply that understanding to the brain and answer the foundational questions about cognition. For good measure, I included a careful study of the origin of computational theories of cognition. That was way too much for a dissertation. My dissertation included some historical research and an account of physical computation (Piccinini 2003a). I left the foundational questions for future work.

My early mechanistic account of physical computation was, roughly, that physical computation is the manipulation of digits by a functional mechanism in accordance with a rule (Piccinini 2007a). I thought this account would allow me to refute the Computational Theory of Cognition (CTC) once and for all. I wrote a paper attempting that. My argument was that computation is digital; neural activity is not digital; therefore, neural activity is not computational. But I never published that paper because I eventually realized I was missing something.

I knew there used to be analog computers, and they had been used as an alternative model of brain activity. I also knew that analog computers are quite different from digital computers, are not program-controlled (qua analog), are not universal in Alan Turing's sense, existed before digital computers, and used to be called differential analyzers. They only started to be called analog "computers" *after* digital computers became popular. So, I argued that analog "computers" weren't computers properly so called after all (Piccinini 2008a). At least two events pushed back.

First, the late Jonathan Mills asked me to collaborate on explicating analog computation. Although he and I never wrote anything together, he introduced me to the groundbreaking research on analog computation that he was doing at Indiana University, of which I was unaware. Independently, in July 2007, Marcin Milkowski wrote to me and pointed out that some computer scientists were reviving analog computation; he recommended a "humbler approach" than mine.

I wondered whether those who study analog computation, and perhaps others who study other unconventional models of (so-called) computation that are not digital, might be picking up on something that digital computation, analog "computation," and other types of "computation" have in common. From a mathematical standpoint, they are all ways of solving mathematical problems. But I was after an account of *physical* computation. What do the disparate physical processes that implement different types of computation have in common?

The received view is that computation has to do with representation. But I had arguments that, in fact, computation does not require representation (Piccinini 2004a, 2008b). I also knew that the notions of function and multiple realizability are too weak to underwrite a robust notion of computation (Piccinini 2004b, 2007b). At some point, I remembered reading an insightful paper in which Justin Garson (2003) uses the notion of medium independence to characterize information. (John Haugeland (1985) uses a similar notion to characterize automatic formal systems, but I only found out about that much later.) I re-read Justin's paper and found that medium independence was just what I needed. After that, I generalized my account of physical computation to cover analog computation and other unconventional models of computation. The resulting account is that physical computation is the manipulation of medium independent vehicles by a functional mechanism in accordance with a rule (Piccinini and Scarantino 2011, Piccinini 2015).

Given this broader understanding of physical computation, my attitude about the brain and cognition switched. I now had the materials for a cogent argument *in favor of* CTC—with a twist. With help from neuroscientist Sonya Bahar, I rewrote my early paper against CTC to argue that, on one hand, neurocognitive processes are computations because they are medium independent and, on the other hand, neural computations are neither digital nor analog—they are sui generis (Piccinini and Bahar 2013).

That seemed important but, to answer the foundational questions, much more needed to be done. I needed a proper ontological foundation, an adequate account of functions, a way to integrate psychology and neuroscience, an assessment of other arguments pro and con CTC, and a place for representation and consciousness. After many more collaborations and publications and much additional research, this book is the result. It presents a comprehensive defense of CTC, updated for the era of cognitive neuroscience, with surprises for both defenders and critics of traditional forms of CTC.

Way more people have helped me than I can thank individually. I collectively thank my teachers as well as the many audiences to whom I presented versions of these ideas. I learned a lot from them.

My work builds on those who came before me. Besides those I already mentioned and my collaborators, those with the greatest philosophical influence on this project include Bill Bechtel, Christopher Boorse, Paul and Patricia Churchland, Jack Copeland, Carl Craver, Robert Cummins, Daniel Dennett, Fred Dretske, Frankie Egan, Jerry Fodor, Gilbert Harman, John Heil, Jaegwon

Kim, Ruth Millikan, Hilary Putnam, Jonathan Schaffer, Oron Shagrir, Wilfried Sieg, and Stephen Stich. On the science side, those from whom I learned the most include Warren McCulloch, Allen Newell, Walter Pitts, Claude Shannon, Herbert Simon, Alan Turing, and John von Neumann.

For comments on many papers and illuminating discussions over many years I owe special thanks to Neal Anderson, Ken Aizawa, Trey Boone, Carl Craver, Corey Maley, Andrea Scarantino, Oron Shagrir, Mark Sprevak, John Heil, Justin Garson, and Eric Thomson. For comments on aspects of this work, I thanks Darren Abramson, Anibal Astobiza, Brice Bantegnie, Sergio Barberis, Bill Bechtel, Peter Bradley, Giovanni Camardi, Glenn Carruthers, David Chalmers, Mazviita Chirimuuta, Michelle Ciurria, Judith Crane, Robert Cummins, Tony Dardis, Joe Dewhurst, Frankie Egan, Chris Eliasmith, Ilke Ercan, Nir Fresco, Rachel Francon, Erik Funkhouser, Carl Gillett, Stuart Glennan, Mahi Hardalupas, Inman Harvey, Eric Hochstein, Phyllis Illari, Anne Jacobson, David M. Kaplan, Daniel Kramer, Arnon Levy, Bill Lycan, Peter Machamer, Jack Mallah, Corey Maley, Diego Marconi, Marcin Milkowski, Kevin Morris, Alyssa Ney, Abel Wajnerman Paz, Alessio Plebe, Russ Poldrack, Tom Polger, Charles Rathkopf, Michael Rescorla, Sarah Robins, Waldemar Rohloff, Robert Rupert, Anna-Mari Rusanen, Kevin Ryan, Matthias Scheutz, Susan Schneider, Whit Schonbein, Alex Schumm, Paul Schweizer, Sam Scott, Oron Shagrir, Larry Shapiro, Kent Staley, Terrence Stewart, Jackie Sullivan, Brandon Towl, Charles Wallis, Dan Weiskopf, Gene Witmer, Mark Zeise, many anonymous referees, and others I can't recall. Special thanks to Michael Barkasi, Mark Couch, and the anonymous referees for OUP for insightful comments on the book manuscript.

Thanks to Crystal Brown, Frank Faries, Mirinda James, Elliott Risch, and James Virtel for editorial assistance.

The Introduction, Chapter 1, and Chapter 3 are new.

Chapter 2 includes a substantially revised and expanded descendant of Piccinini and Maley 2014. Sections 2.1, 2.3, and 2.5 are new.

Chapter 4 includes a substantially revised and expanded descendant of Maley and Piccinini 2013. Sections 4.1.1 and 4.1.2 are new.

Chapter 5 is a revised and expanded descendant of parts of Piccinini 2004c.

Sections 6.2–6.8 of Chapter 6 are a revised and expanded descendant of Piccinini 2017a. Section 6.9 is a revised descendant of a section of Piccinini 2009. Sections 6.1 and 6.10–6.12 are new.

Chapter 7 includes a revised and expanded descendant of Boone and Piccinini 2016a. Sections 7.1–7.3 and 7.7 are new.

Chapter 8 is a revised descendant of Boone and Piccinini 2016b.

Sections 9.1 and 9.2 of Chapter 9 include revised versions of portions of Piccinini 2016. Most of Section 9.2 is a revised and expanded descendant of a section of Piccinini and Bahar 2013. Sections 9.3–9.5 are a revised descendant of parts of Piccinini 2009.

Chapter 10 is a substantially revised descendant of Piccinini 2007c.

Chapter 11 is a substantially revised and expanded descendant of Piccinini 2010a.

Chapter 12 is a substantially revised descendant of Thomson and Piccinini 2018.

Chapter 13 is a revised and expanded descendant of much of Piccinini and Bahar 2013.

Chapter 14 includes a substantially revised descendant of Piccinini 2010b. Sections 14.1 and 14.9 are new.

It was a collective effort. I am deeply grateful to my co-authors on some of the articles from which some of the chapters derive—Sonya Bahar, Trey Boone, Corey Maley, and Eric Thomson—two philosophers and two neuroscientists. They graciously agreed to let me use our joint work in this book and deserve credit for much of what is correct in the chapters derived from it. My research program benefited immensely from each collaboration.

This material is based upon work supported by the National Science Foundation under Grants No. SES-0216981, SES-0924527, and especially SES-1654982. I also received research support from a University of Missouri Research Board Award, a 2006 NEH Summer Seminar at Washington University in St. Louis, a University of Missouri Research Grant, a grant from the International Studies and Programs at the University of Missouri—St. Louis, an Adelle and Erwin Tomash Fellowship, an Andrew Mellon Predoctoral Fellowship, and a Regione Sardegna Doctoral Scholarship. Any opinions, findings, conclusions, and recommendations expressed in this work are those of the author and do not necessarily reflect the views of these funding institutions.

Thanks to Peter Momtchiloff and the whole team at OUP for shepherding my manuscript through the publication process.

Deep thanks to my friends and family for their love and support during all these years. I'm grateful to my partner Michelle and daughters Violet, Brie, and Martine for bringing so much meaning and joy to my life.

List of Figures

Introduction

This book defends a neurocomputational theory of cognition grounded in a mechanistic, functionalist, egalitarian ontology. I argue that *biological cognitive capacities are constitutively explained by multilevel neurocognitive mechanisms*, which *perform neural computations over neural representations*. Providing a scientific explanation of cognition requires understanding how neurocognitive mechanisms work. Therefore, the science of cognition ought to include neuroscience to a degree that traditional cognitive science was not expected to. Scientists on the ground have been working on this for a while. Psychology is becoming more and more integrated with neuroscience.

The picture I defend stands opposite to both traditional reductionism (type-identity theory) and anti-reductionism (autonomy of psychology). Contrary to traditional reductionism, neural computations and representations are not identical to their lower-level realizers. Contrary to traditional anti-reductionism, neural computations and representations are not entirely distinct from their realizers. Instead, neural computations and representations—both types and tokens—are *aspects* of their realizers.

The picture I defend stands opposite to both computational chauvinism (computation and representation are proprietary psychological notions) and anti-realism (computations and representations are mere manners of speaking). Contrary to computational chauvinism, computations and representations are properties of the nervous system. Contrary to anti-realism, computations and representations are real, causally efficacious properties—as real as any other properties of the nervous system.

I will begin by providing accounts of levels of composition and realization, mechanisms, functions, computation, and multilevel neurocognitive mechanisms. This ontological foundation will allow me to improve on existing versions of functionalism and make clear what the computational theory of cognition does and does not say. After that, I will address some fallacies and objections to the computational theory of cognition. Finally, I will provide empirical evidence that neurocognitive systems perform sui generis computations over neural representations. I will conclude by clarifying the relation between computation and consciousness and offering a noncomputational functionalism about phenomenal consciousness.

By *capacity*, I mean a causal power subject to normative evaluation—a power that can be manifested correctly or incorrectly.

Neurocognitive Mechanisms: Explaining Biological Cognition. Gualtiero Piccinini, Oxford University Press (2020).

DOI: 10.1093/oso/9780198866282.001.0001

By *cognitive* capacities, I mean capacities such as perception, memory, reasoning, emotion, language, planning, and motor control. Cognitive capacities explain some of the most interesting behaviors exhibited by physical systems.

By *biological* cognition, I mean cognition carried out by biological organisms—more specifically, organisms with a specialized control organ. The specialized control organ of earthly organisms is the nervous system. Some philosophers argue that plants have cognitive capacities (Calvo 2016). Plants, fungi, and even bacteria are exquisitely adapted to their environments. Yet they lack a specialized organ for control of their organismic functions based on processing information that results from integrating inputs from multiple, physically distinct sources. My topic is the cognitive capacities of organisms with a specialized control organ that integrates multiple sources of information.

There are also artifacts such as robots and digital assistants that possess computational and information processing capacities, and there are analogies between such artificial systems and biological cognizers. This book focuses on *biological* cognition exhibited by organisms with a nervous system.

Some argue that the realizers of cognitive states and processes in biological organisms include not only the nervous system but also some things outside it (e.g., Wilson 1994). I will mostly ignore this possibility to simplify the exposition; this does not affect my arguments.

By *constitutive explanation*, I mean explanation of a capacity of a system in terms of the system's causal structure. I will argue that constitutive explanation is provided by *mechanisms*—that is, pluralities of components, component functions, and organizational relations that, collectively, possess the capacity and produce its manifestations (Chapters 3, 7).

Typically, components of mechanisms are themselves mechanisms whose capacities are explained mechanistically. Ditto for the components' components. This multilevel mechanistic structure requires an ontological foundation. I argue that all *levels* are equally real: neither higher levels nor lower levels are more fundamental than one another. Higher-level objects are invariants under the addition and subtraction of some parts, while higher-level properties are aspects of their lower-level realizers. Adequate constitutive explanation requires identifying the higher-level properties and organizational relations that produce a capacity. Therefore, adequate constitutive explanation requires identifying appropriate higher-level objects and their relevant properties (Chapter 1).

Capacities such as cognition are causal roles—that is, they are higher-level properties defined solely by their regular effects under certain conditions. Causal roles are often *multiply realizable*, meaning that there are different kinds of mechanism that can perform the same causal role. In addition, some multiply realizable causal roles are *medium independent*, meaning that even their inputs and outputs are multiply realizable (Chapter 2).

Causal roles are one, broad notion of function. In Chapter 3, I argue that the notion of causal role is too broad to do justice to mechanistic explanation of organisms and artifacts. The capacities of organisms and artifacts are explained by causal roles that make regular contributions to the goals of organisms. These contributions are *teleological functions* (Chapter 3).

On this basis, I propose an improved formulation of *functionalism*: the view that the mind is the functional organization of the brain. I argue that functional organization should be understood mechanistically, as encompassing not only causal relations between internal states but also the components bearing the states, their functions, and their organizational relations (Chapter 4).

Functionalism is a close ally of the computational theory of cognition, which was first proposed by Warren McCulloch and Walter Pitts in 1943. They created a neural network formalism for representing neuronal activity in a simplified and idealized way, and they argued that neural networks perform digital computations. Their idea had an immense impact on the field (Chapter 5).

Some mechanisms, such as the neural networks devised by McCulloch and Pitts, have the special teleological functions of *computing* and *processing information*. Computing is processing medium-independent vehicles in accordance with a rule, while information processing is the processing of vehicles that carry information. Computational vehicles may or may not carry information, and information processing may or may not be done by computing (Chapter 6).

Many authors have argued that there are constitutive explanations—whether computational or not—that are not mechanistic. I argue that such putatively nonmechanistic constitutive explanations are *aspects* of mechanisms. Adequate constitutive explanation—including computational explanation—involves mechanisms (Chapter 7).

The scientific study of neurocognitive mechanisms is cognitive neuroscience. To a large extent, cognitive science has already turned into cognitive neuroscience. This is the integrated study of how multilevel neurocomputational mechanisms that process neural representations explain cognition (Chapter 8).

The *Computational Theory of Cognition* (CTC) is the theory that cognitive processes are computations, or cognition is explained computationally. Since computation is a mechanistic process, CTC is a mechanistic hypothesis. Since the organ of biological cognition is the nervous system, CTC for biological organisms is the claim that neurocognitive processes are computations. I argue that CTC in its generic formulation is correct for a couple of reasons. First, the main vehicles of neural processing—spike trains—are functionally significant thanks primarily to firing rates and spike timing, which are medium independent. Second, neural signals carry, integrate, and process information from physically different sources, which requires transducing them into shared, medium-independent vehicles (Chapter 9).

Assessing CTC requires understanding not only the reasons for it, but also what is *not* a reason for it. A range of putative arguments for CTC are based on the Church–Turing thesis, the thesis that the functions that are computable in an intuitive sense are precisely those functions that are computable by Turing machines. I argue that these arguments are fallacious: the Church–Turing thesis does not help establish CTC (Chapter 10).

Assessing CTC also requires understanding putative objections. There are two classes of objections. Insufficiency objections maintain that cognition involves X (for some X) and computation is insufficient for X. Candidate Xs include consciousness, intentionality, embodiment, embeddedness, dynamics, and mathematical insight. I reply that insufficiency objections do not undermine CTC; at most, they show that something else is needed, in addition to computation, to explain cognition. I emphasize that an adequate computational explanation of biological cognition involves computations that are embodied and embedded. Objections from neural realization argue that neural processes involve Y (for some Y) and computation does not involve Y; therefore, neural processes are not computations. I reply that none of the putative Y's undermines CTC (Chapter 11).

Although computation can occur in the absence of information processing, this is not what happens in the nervous system. Neural processes carry, integrate, and process information in the service of teleological control functions. Informational Teleosemantics is the view that representations are precisely states that carry information in the service of teleological functions. Thus, Informational Teleosemantics applied to the neurocognitive system entails that the neural states that carry information in the service of control functions are *neural representations*. I argue that neural representations are not only real but routinely observed by neuroscientists (Chapter 12).

One long-standing dispute about neural computation is whether it is digital or analog. I argue that it is neither, at least in the general case. Instead, neural computation is sui generis. Therefore, theories of cognition should take into account what is known about neural computation (Chapter 13).

The *Computational Theory of Mind* (CTM), as I use this term, is a stronger view than the Computational Theory of *Cognition*. CTM covers the whole mind—both cognition *and* consciousness. Thus, CTM says that the whole mind has a computational nature, or that all there is to the nature of mental states, including conscious states, is being computational states. Since computation is a mechanistic process, CTM is a mechanistic hypothesis. It is the computational version of mechanistic functionalism. Contrary to a common assumption, the alternative to CTM is not just the type-identity theory of mind. There is also a noncomputational version of functionalism about consciousness that deserves to be explored (Chapter 14).

Biological cognition, or at least biological cognition in organisms with a centralized control system that integrates multiple sources of information, turns out to be neural computation over neural representations. Neural computation is a sui generis kind of computation—neither digital nor analog—that neurocognitive mechanisms perform. Neural representations are simulations of the organism, its environment, and their interaction that neurocognitive mechanisms construct. Neural computations process neural representations. The result of this process is biological cognition.

Neurocognitive mechanisms span multiple levels of organization, from single neurons up to the whole nervous system. At each neurocognitive level, neural computations process neural representations. Except for the lowest level, each neurocognitive level is realized by the level below it. Except for the highest level, each neurocognitive level realizes the level above it. Thus, lower-level neural representations and computations realize higher-level neural computations and representations.

Needless to say, explaining how cognitively endowed organisms behave requires considering the dynamical coupling among nervous system, body, and environment. In other words, explaining how organisms behave requires considering how nervous systems are embodied and embedded in their environment. Considering all of this is what cognitive neuroscience does. Nevertheless, nervous systems contribute something distinctively cognitive to the explanation: the performance of neural computations over neural representations. That is the focus of this book.

Although the chapters are deeply intertwined, I outline the main point of each chapter in its first section—except for Chapter 1. A reader who wants the gist of the book quickly can read Chapter 1 plus the first section of the other chapters.

1
Levels of Being

1.1 Parts, Wholes, and their Properties

This book argues that cognition is explained by multilevel neurocognitive mechanisms. The first step towards this view—or any view involving levels of composition and realization, for that matter—is clarifying what levels are and how they are related. This chapter lays out an account of levels on which the rest of the book will build.

The way we talk about things and their *proper*[1] parts gives rise to puzzles. For example, a hammer has two parts: a handle and a head. If you have a hammer, how many objects do you have? If you answer one, you are not counting the head and the handle. If you answer two, you are not counting the whole hammer. If you answer three, you seem to be counting the same portion of reality twice. None of the answers are entirely satisfying. And then there are the hammer's molecules, atoms, and subatomic particles. How many things are there?

Puzzles like this make it difficult to understand how our discourse about wholes fits with our discourse about parts. Both lay people and scientists talk about different sorts of objects, some of which are part of others. We talk about people, nervous systems, brains, neural networks, neurons, neurotransmitters, and so on. We describe properties that these objects have and their manifestations: people walk, brains develop, and neurons send action potentials.

We also describe *relations* between objects. For convenience, in this chapter I will often follow the common convention of using the term "property" to include relations. For instance, I will treat *A being inside B* as a spatial property possessed by A and B collectively. So, I will often talk simply of objects and properties, with the understanding that relations are also included. Nothing substantive hinges on this.

Some relations occur between objects that are wholly distinct from one another—objects *at the same level*: people talk to one another, brains are encased in skulls, and neurons attach to one another (via synapses). Other relations occur between objects that are part of one another. For example, synapses are part of

[1] In mereology—the formal study of the part whole relation—it is convenient to define "part" so that everything is part of itself, and "proper part" as a part that is not identical to the whole. By contrast, in ordinary language, we usually presuppose that a part is not identical to the whole. From now on, I will follow ordinary usage and use "part" to mean proper part.

Neurocognitive Mechanisms: Explaining Biological Cognition. Gualtiero Piccinini, Oxford University Press (2020).

DOI: 10.1093/oso/9780198866282.001.0001

neurons, which together form neural networks; nervous systems are partially made out of brains, which are partially made out of networks, among other structures. *Being part of* something and *being made out of* something are relations that occur between objects *at different levels* of being.

Sorting out what there is, what properties objects have, and how they relate is the business of science. Sorting out what *sorts* of objects there are as well as what *sorts* of properties and relations they have is the business of metaphysics. Scientists rely on experimenting and theorizing. Metaphysicians rely on scientific results and conceptual considerations. By using these tools, metaphysicians attempt to organize our concepts so we don't get confused or say things that make little sense. This book begins with metaphysics and progresses from there towards the philosophy of mind and cognitive neuroscience.

I will use the term "composition" for the relation between wholes and their parts: parts compose a whole; a whole decomposes into its parts. I will use the term "realization" for the relation between properties of a whole and properties of its parts taken collectively, or between its parts taken collectively and all of the parts' parts taken collectively: higher-level properties are realized by lower-level properties; lower-level properties realize higher-level properties.[2]

Properties come in types and tokens (instances). For example, each planet in the solar system is an oblate spheroid. That is, each planet instantiates the property *type* of oblate spheroidicity. Equivalently, each planet has its own *token*, or instance, of oblate spheroidicity. We can talk about type realization or token realization. I will mostly focus on type realization, without making it explicit every time. Much of what I say applies to token realization as well.

As I use these terms, both composition and realization are asymmetric, irreflexive, and transitive.[3] Both are relations of synchronic metaphysical necessitation: once you have the parts with their properties and relations, you necessarily have the whole with its properties and relations. Necessitation does *not* entail *dependence*! Both composition and realization are often *assumed* to be relations of ontological dependence, meaning that wholes and their properties ontologically depend on (are grounded in, are posterior to) their parts and their properties. Not only am I *not* making this dependence assumption; I will soon argue *against* it.

There are a number of puzzles that arise from our talk about things at different levels of being. To keep the discussion manageable, let's focus on one especially difficult and theoretically important puzzle: the puzzle of causal exclusion. I'll

[2] There is a dispute as to whether, when property P is realized by property Q (where Q could also be a plurality of properties), P and Q belong to the same object X or Q belongs to X's parts. I will discuss this in Chapter 2, where I argue that, in the sense of realization that matters most to the metaphysics of science, Q belongs to X's parts. In this chapter, I'm just *defining* "realization" as the relation that holds between the properties of a whole and the properties of its parts, or between its parts taken collectively and all of the parts' parts taken collectively.

[3] Asymmetry: xRy → not yRx. Irreflexivity: not xRx. Transitivity: (xRy and yRz) → xRz. On the features of realization, cf. Baysan 2015.

propose a framework for thinking about levels of being that solves the causal exclusion puzzle and explains how different levels fit together. Beware: a proper treatment of these topics would require a book of its own; in this chapter I only have room for a brief sketch that will help us through the rest of *this* book.

1.2 Causal Exclusion

Suppose that brain B is composed of a bunch of neurons, glia, other cells, and extracellular fluid, which I'll collectively label NN. For simplicity, I will refer to NN simply as neurons. Different people disagree about the relationship between wholes and their parts—e.g., between B and NN. Some think they are identical, others that they are distinct, yet others that one of the two is grounded in the other. We'll suspend judgment for now.

Consider a voluntary action E performed by an organism, and consider two causal claims that are entailed by what is found in any neuroscience textbook:

(1) B causes E.
(2) NN cause E.

When we make this kind of causal claim, we usually assume that a system causes an effect in virtue of being in a particular state or having a particular property. I assume that *being in a state* is the same as *possessing a property*, so we can talk about states and properties interchangeably.

To take the role of properties into account, let's reformulate (1) and (2) in terms of brain state S_B and neural state S_{NN}, such that S_{NN} realizes S_B:

(3) S_B causes E.
(4) S_{NN} causes E.

Again, different people disagree about the relationship between properties of wholes and properties of their parts—e.g., between S_B and S_{NN}. Some think they are identical, others that they are distinct, yet others that one of the two is grounded in the other. Again, we'll suspend judgment for now.

When we make this kind of causal claim, we generally assume that each cause is sufficient for the effect under the relevant background conditions. That is to say, if S_B causes E, then—given relevant background conditions—the occurrence of S_B is enough, all by itself, to bring about E. By the same token, if S_{NN} causes E, then—given relevant background conditions—the occurrence of S_{NN} is enough, all by itself, to bring about E.

The puzzle of causal exclusion arises because if S_{NN} is already enough to cause E, there seems to be nothing left for S_B to do. Suppose that we establish (4)—that a neural state causes a voluntary action. Why would we say that a brain state also causes that action, if we already established that a neural state is enough to cause that very same action? The problem is entirely symmetrical. Suppose we establish (3)—that a certain brain state causes a certain action. Why add that a neural state also causes that action, if we already established that a brain state is enough to cause that very same action? If one of the two is enough to produce the action, it seems that the other is either identical to it or should be dispensed with. But which is it?

The problem generalizes. I picked neurons and brains among many other levels that I could have picked. The same puzzle can be run with neural systems, neural networks, molecules, atoms, and so on—not to mention whole organisms. This is the puzzle of causal exclusion: if any level is enough, all by itself, to cause an effect such as E, then other levels seem either dispensable or identical to the one causally efficacious level (Kim 1998, 2005; Merricks 2001). Yet both laypeople and scientists attribute causal efficacy to multiple levels of being, and we don't seem to take all levels to be identical to one another. What gives?

1.3 Overdetermination

One answer to the causal exclusion puzzle is that distinct levels cause the same effect at the same time. This theoretical option is called overdetermination. To understand overdetermination, consider the classic example of a firing squad, each of whose members fires a deadly bullet. Given the operative background conditions, each bullet is sufficient, all by itself, to kill a person. In this case, the presence of a sufficient cause is compatible with many other sufficient causes for the same effect. By analogy, levels of being might be like soldiers in a firing squad: each level produces the same effect independently of the others (Loewer 2007; Pereboom 2002; Schaffer 2003; and Sider 2003). Might this be a solution to the causal exclusion puzzle?

There does seem to be a similarity between the two situations. In both cases, multiple things occurring at the same time are said to be causally sufficient for the same effect. Yet there are also profound dissimilarities.

In cases of ordinary causal overdetermination, such as the firing squad, the different causes can be independently observed and tracked. For example, we can check the different rifles before the squad shoots and see if they contain live ammunition. After the shooting, we can search for distinct bullets and see where they ended up. This sort of empirical investigation is not an option when different levels of being are in play. We can't investigate the effects of a brain independently of those of the neurons that make up that very brain. We can't observe or manipulate the state of the brain, separately observe or manipulate the state of the neurons that make up that same brain, and see if they have distinct causal powers analogous to the powers of distinct bullets. Therefore, we cannot collect any evidence that there are distinct causal powers at distinct levels that are sufficient for the same effect. Without evidence, it is both ontologically profligate and epistemically irresponsible to posit them.

A defender of overdetermination may reply that the disanalogy between multi-level overdetermination and ordinary—firing-squad-type—overdetermination is immaterial. Distinct levels of being (wholes and their parts, higher-level properties and their realizers) are metaphysically connected to one another in a way that distinct bullets are not. Given the tight connection between levels, a defender of overdetermination might insist that distinct levels may well be causally sufficient for the same effect—at least in a loose-enough sense of "cause."

This reply misunderstands the stakes of the problem. The question is not whether a suitably watered-down notion of causation can underwrite claims to the effect that distinct levels cause the same effect. The question is whether it is legitimate to posit distinct causal powers that are causally sufficient for the same effect in situations where it is impossible, as a matter of principle, to collect empirical evidence of their distinct existence.

Even if we set aside epistemic considerations, it makes no sense to posit that things that stand in a relation of composition or realization have distinct causal powers for the same effect. The objects themselves—a whole and its parts—are not even (wholly) distinct. The parts make up the whole; the whole consists of the parts. By the same token, higher-level properties and their realizers are too intimately connected to have distinct causal powers for the same effect. Once you have the realizer, you also have the realized property; and once you have the realized property (token), you also have its realizer. How could things that are as intimately related as a whole and its parts, or a higher-level property and its realizer, come to have distinct causal powers, each sufficient for the same effect (cf. Bernstein 2016)? Later I will elucidate the metaphysical connections between wholes and their parts as well as higher-level properties and their realizers, arguing that such connections are inconsistent with overdetermination.

In summary, the overdetermination solution to the puzzle of causal exclusion posits multiple distinct causes for the same effect that are possessed by a whole and its parts, or by a higher-level property and its realizers. These are more causes than are necessary to explain phenomena, without any evidence that they are there or any way to investigate them separately. Because of this, overdetermination is a desperate option. Perhaps we should consider it as a last resort if all other options fail. Let's ponder more appealing alternatives.

1.4 Reduction

The simplest solution to the puzzle of causal exclusion would be to identify the one *genuine* cause of an effect and reduce all the other putative causes to *that* one.

For example, suppose we established that neural states are the genuine cause of voluntary actions. If we could reduce brains to collections of neurons, and brain states to neural states, then we would conclude that there is no real conflict between (1) and (2), or between (3) and (4), because (1) reduces to (2), and (3) reduces to (4). There is only one genuine cause of E—neural states.

If we are going to consider this sort of *reductionism*, we need to say what we mean by reduction. This is a vexed question in both metaphysics and the philosophy of science (van Riel and van Gulick 2019). For present purposes, I will adopt the simplest and clearest notion of *ontological* reduction: identity with a direction. That is to say, X reduces to Y if and only if X is strictly identical to Y and Y is more fundamental than X. Specifically, B reduces to NN if and only if B = NN and NN is more fundamental than B, and S_B reduces to S_{NN} if and only if $S_B = S_{NN}$ and S_{NN} is more fundamental than S_B. The difficulties with reduction arise for any reasonable notion of reduction; we might as well keep things manageable by focusing on reduction as identity plus direction.[4]

It should go without saying that reduction in the relevant sense is not just a matter of what is part of what. Everyone agrees that neurons are parts of brains, molecules are parts of neurons (as well as brains), and every physical object is part of the whole universe. Yet reasonable people disagree about whether large things reduce to small things, small things reduce to the universe (e.g., Schaffer 2010), or neither reduces to the other.

Assume reductionism holds. The first question is, *which* direction does reduction go? Do brains reduce to neurons or vice versa? Do neural states reduce to brain states or vice versa? We've also seen that there aren't just two levels—there are many. We need to say where neural systems and their properties, people and their properties, molecules and their properties, and so forth fit within the reductionist picture. For any one of these levels, either it is the reduction base for all the others, or it reduces to whichever level all others reduce to. For reduction to work, we need to find the one *fundamental* level to which all other levels reduce. I doubt we can.

In recent decades, perhaps the most popular solution is to pick the level of elementary fermions and bosons as the fundamental ontological level, to which everything else supposedly reduces. This does have some appeal—at least to those, like me, with a taste for particle physics. Physicists themselves call elementary fermions and bosons the *fundamental particles*. Maybe the particles that are *physically* fundamental are also *ontologically* fundamental?[5]

[4] A different proposal by Hemmo and Shenker (2015; Shenker unpublished) is that X reduces to Y if and only if X is an *aspect* of Y. Their flat physicalism, which was developed independently of the present work, goes in the direction of the egalitarian ontology I defend here and is more compatible with it than traditional reductionism.

[5] The precise ontological status of fermions and bosons, and of quantum mechanical systems in general, is controversial. Debates include whether fermions and bosons are individual objects (French

Before answering that question, let me point out that physical fundamentality does *not* entail ontological fundamentality. Something is physically fundamental, roughly, just in case it has no physical parts—no parts that are distinguishable via physical operations or theories. Being physically fundamental is awesome and important: theories that are true of physically fundamental particles are true of everything physical. Nevertheless, physical fundamentality is not the same as ontological fundamentality.

Something is *ontologically* fundamental just in case everything else synchronically and asymmetrically *depends* on it. The ontologically fundamental stuff *grounds* all the other stuff. The ontologically fundamental stuff is ontologically *prior* to the other stuff. These are the metaphors used by metaphysicians; they are not easy to pin down.

Ontological dependence is easy to grasp when time flows and new objects and properties come into existence. For instance, if a group of people forms a new club, the club ontologically depends on the temporally prior existence of the group of people. The reason is obvious: without people, no club could be formed. The same idea is also easy to grasp when we deal with arbitrary fusions of objects or derivative properties. For example, the object formed by combining the Statue of Liberty and the Taj Mahal depends ontologically on the existence of the Statue of Liberty and the Taj Mahal; the property of being a mile away from the Statue of Liberty depends ontologically on the existence and location of the Statue of Liberty.

It is less clear what being ontologically more fundamental means when we consider, at one and the same time, a whole versus its parts, or a whole's properties versus its parts' properties, when the whole and its properties are individuated independently of the parts and their properties.

In spite of this lack of clarity about what it means to be ontologically fundamental, lots of philosophers talk about one level being more fundamental than another, or one level being absolutely fundamental. Ultimately the lack of clarity about ontological fundamentality will not matter to us because I will argue that, given any relevant notion of fundamentality, no level of being is more ontologically fundamental than any other. For now, though, let's pretend we have at least a faint grasp of what ontological fundamentality means and proceed to examine the prospects of this idea. Could the ontologically fundamental level be the physically fundamental level?

Elementary fermions and bosons are the building blocks of the universe: they have no physical parts and they constitute every physical object. In addition,

2015) and whether quantum mechanical systems admit of unique decompositions (Healey 2013). Addressing these debates falls outside the scope of this chapter. For present purposes, I assume that fermions and bosons can be individuated by a combination of invariants and observables (Castellani 1998).

according to the Standard Model of particle physics, the properties and interactions of fermions and bosons constitutively explain all physical phenomena. That includes, presumably, all higher-level phenomena such as brain states causing voluntary action. The Standard Model is by far the most successful and well-confirmed scientific theory ever devised. That's impressive. Thus, *metaphysical atomism*—reducing higher levels of being to fundamental physical particles and their properties—seems like a good reductionist option. But again, ontological fundamentality does not *follow* from physical fundamentality.

Another popular reductionist option is to pick the whole universe as the fundamental ontological level, to which everything else reduces (Schaffer 2010). If we look at the history of philosophy, this *monistic* solution may be more popular than atomism. One advantage of monism is that, by definition, the universe includes everything within itself—all objects and properties. If we take the universe to be ontologically fundamental and reduce everything else to *it*, surely we are not leaving anything out.

Another metaphysical advantage of monism is that it is less subject to the vagaries of physical theory than metaphysical atomism. Some physicists hope to replace the Standard Model with something even more physically fundamental, which will unify all physical forces. To perform this unification, theoretical physicists explore the possibility that there are entities even more physically fundamental than fermions and bosons. If there are such fundamental physical entities, then perhaps everything reduces to *them*. But what if there are no true atoms—no indivisible physical entities? What if every physical entity has smaller and smaller parts all the way down? This seems to make reduction impossible, for lack of a reduction base. Monism—the view that everything reduces to the whole universe—does not seem to face this sort of risk (cf. Schaffer 2010).

Still, monism shares with atomism the problem that picking any level as a reduction base seems arbitrary. Aside from the puzzle of causal exclusion, it's unclear why any level should be more ontologically fundamental than any other. And the puzzle of causal exclusion is entirely neutral on which level is more fundamental. It could be the whole universe, the level of fermions and bosons, the level of brains and other mid-size objects, the level of molecules (cf. Bickle 2003), or any level in between. In addition to this arbitrariness, reductionism seems to be stuck with lack of testability: even if there is a fundamental level, it's unclear how to find it.

Another drawback is that reductionism gives the impression that some sciences are more important than others. Specifically, whichever science investigates the fundamental level—the genuine causes—seems to have an edge over the other sciences. After all, other sciences are just studying stuff whose causal powers depend on reducing it to the fundamental stuff. Human nature being what it is, many people don't like that their favorite subject matter, to which they devote their working hours, is just derivative on something more fundamental, which is studied by someone else. Plus, if everything reduces to the fundamental level, so do people. Many find it personally insulting to be reduced to something else—it makes them feel deprived of their personal agency. This is not a substantive objection; nonetheless, it makes reductionism unpopular.

Most of the people who like reductionism are those who claim to be studying the fundamental stuff. That includes some neuroscientists, to be sure—most commonly molecular and cellular neuroscientists. The reason they like reductionism is that they focus more on reducing higher levels to cells and molecules than on reducing cells and molecules to fermions and bosons. Be that as it may, a lot of philosophers and scientists who study higher-level entities hate reductionism. If they could find a viable alternative, they'd jump on it.

1.5 Anti-Reductionism

Many advocates of the special sciences—especially psychology—defend some form of anti-reductionism (e.g., Fodor 1974; Pereboom and Kornblith 1991; Pereboom 2002; Gillett 2002, 2010; List and Menzies 2009). According to standard anti-reductionism, higher-level entities and properties are both causally efficacious and distinct from the lower level entities and properties that compose and realize them. For instance, minds are distinct from collections of neurons; mental states are distinct from neural states. Talking about minds and mental states raises special questions that would be distracting here, so let's stick with unequivocally physical systems and properties. In the next few chapters I will argue that this restriction makes no difference for our purposes. According to standard anti-reductionism, then, brains are distinct from collections of neurons; brain states are distinct from neural states. Each type of entity and property has its own distinct causal powers. This is the kind of anti-reductionism I discuss in this section.

If anti-reductionism is correct, each scientific discipline can study its subject matter in the hope of finding something real in its own right, including real causal relations. Scientific explanations are *autonomous* from one another: no science is hegemonic and each scientific community can do their empirical investigations at their own level without getting in each other's way. That might explain why anti-reductionism is popular among those who study higher levels.

One crucial question for anti-reductionists is whether a whole's properties endow it with causal powers that go beyond the causal powers possessed by the organized parts that make up that very whole. In other words, suppose that a whole object O is made out of parts $O_1, O_2, \ldots O_n$ standing in appropriate relations. Does O have causal powers that $O_1, O_2, \ldots O_n$, when they are organized to form O, collectively lack? A positive answer to this question yields a version of anti-reductionism known as *strong synchronic emergence*.

Strong synchronic emergence is sometimes confused with the uncontroversial claim that, when a set of objects enter new organizational relations among them, their properties change. Objects that organize together in specific physical configurations possess causal powers that they lack when they are isolated from one another. For instance, Quantum Mechanics teaches us that particles can become entangled with each other; when they are so entangled, their properties change; as a result, their behavior is different than when they are untangled. Novel physical ties lead to novel properties. No one disputes this sort of *diachronic* emergence, because there is nothing especially mysterious about it. Here, we are not concerned with diachronic emergence at all.

Genuine strong synchronic emergence is not a diachronic relation between unorganized parts and organized wholes—it is a *synchronic* relation between organized parts and the wholes they compose. More precisely, strong synchronic emergence is a relation that holds between the properties of an object and the properties of its parts at one and the same time. In our example, strong synchronic emergence does not say that entangled particles have causal powers that untangled particles lack—that's diachronic emergence and it's uncontroversial. Strong synchronic emergence says that, at one and the same time, the entangled particle system—the whole—has causal powers that the entangled particles—the parts of that very same whole, entangled as they are and considered collectively—lack.

It is unbearably mysterious how, *after* a plurality of objects has become organized, the whole could possess causal powers that are lacked by the organized parts taken collectively. Yet this is what strong emergentists claim.[6] Needless to say, no one has been able to make any sense of how this could be—how a higher-level property could possibly endow an object with causal powers that its lower-level realizers—considered collectively at the same time—lack. Or, if a strong emergentist were to deny that higher-level properties are wholly realized by lower-level ones, no one has been able to make sense of where the unrealized portion of the property comes from, or how it relates to the properties of the object's parts.

Extraordinary claims require extraordinary evidence. Yet when strong emergentists purport to give evidence, they fall short. They offer two sorts of examples. On one hand are examples of entities that acquire new powers and hence new behaviors by entering into new organizational relations (e.g., untangled particles that become entangled). This is not evidence of strong synchronic emergence but of good old diachronic emergence, which virtually no one disputes because it poses no ontological puzzles. On the other hand are higher-level capacities or generalizations for which we lack lower-level mechanistic explanations. This is not evidence of strong emergence either—it is evidence of our epistemic limitations and it is consistent with the denial of strong synchronic emergence. In fact, it is unclear how anyone could possibly show that, at one and the same time, a whole has causal powers that its organized parts, considered collectively, lack. On the

[6] Recent defenses of strong synchronic emergentism include Mitchell 2009; Thalos 2013; Gillett 2016. By contrast, Humphreys 2016 focuses primarily on diachronic emergence.

contrary, the historical track record indicates that higher-level phenomena can be explained without positing novel higher-level causal powers (e.g., McLaughlin 1992; Papineau 2001). In light of these deficiencies, I set strong synchronic emergence aside and return to more sensible forms of anti-reductionism.[7] According to ordinary anti-reductionism, wholes and their properties are distinct from parts and their properties, although their causal powers do not surpass those of their parts.

The most popular argument for anti-reductionism is based on multiple realizability (Fodor 1974; Polger and Shapiro 2016). Multiple realizability is the claim that at least some higher-level properties—the kind of property putatively studied by the special sciences—can be realized in different ways. For example, there are different ways of making a mousetrap, an engine, or a computer. All the different realizers share higher-level properties: they catch mice, generate motive power, or perform computations. Furthermore, according to anti-reductionism, the different realizers of a higher-level property share no lower-level property. Because of this, says anti-reductionism, higher-level properties such as catching mice, generating motive power, or computing cannot reduce to any lower-level property.

This argument for anti-reductionism from multiple realizability faces a dilemma (cf. Kim 1998, 2005). To set up the dilemma, consider any putatively multiply realizable higher-level property P—such as catching mice, generating motive power, or computing—and its putative realizers R_1, R_2, ... R_n. Let's assume, along with most participants in this debate, that lower-level realizers are more fundamental than the higher-level properties they realize. Let's rule out overdetermination—the unwarranted, ontologically profligate view that both P and its realizers are independently causally sufficient for P's effects (per Section 1.3). Let's also rule out strong synchronic emergence, the mysterious and unsubstantiated claim that P's causal powers outstrip those of its realizers. From the absence of strong emergence, it follows that any realizer of a higher-level property—e.g., any R_i—is causally sufficient for P's effects.

The dilemma is this: either P is merely a disjunction of lower-level properties—something of the form *R_1 or R_2 or... R_n*, where R_1 ... R_n are its realizers—or P is a genuine property with genuine causal powers. If P is merely a disjunction of lower-level realizers, all the genuine causal work is done by the realizers, so multiple realizability is just a relation between a disjunction and its disjuncts. This form of multiple realizability is consistent with reductionism, because all the causal work is done by the realizers and there is nothing distinctive about higher levels. So anti-reductionism fails. On the other hand, if P is a genuine property, then its causal powers must be identical—and thereby reduce—to those of its realizers. For we already established that P has no additional causal powers beyond

[7] For more detailed critiques of strong emergentism, see McLaughlin 1992; Kim 2006.

those of its realizers, and we are assuming that there is no overdetermination. Therefore, its realizers already have all of P's causal powers. Ergo, there is something in P's realizers that is identical to P's causal powers. There is no genuine multiple realizability after all and anti-reductionism fails again. On either horn of the dilemma, the real causal work is done by the lower-level realizers. On either horn of the dilemma, higher-level properties, understood as pluralities of causal powers, reduce to lower-level properties and anti-reductionism fails.

Anti-reductionists have attempted to respond to this dilemma. A popular response invokes special science generalizations: different sciences (or scientific subdisciplines) establish generalizations at different levels; established generalizations at different levels are equally legitimate; therefore, higher-level generalizations should be accepted as part of our scientific understanding of the world (cf. Fodor 1997; Block 1997; Aizawa and Gillett 2011). Another popular response invokes proportionality considerations; we can intervene on higher-level variables and observe effects, and the resulting higher-level causal *statements* sound more proportional to the effect than lower-level causal statements. From this, some anti-reductionists infer that higher-level statements capture genuine causal relations involving genuine causal powers, which cannot reduce to causal relations at lower levels (Yablo 1992; List and Menzies 2009; Raatikainen 2010; Woodward 2015). As supporting evidence, they may point out that there are few successful derivations of higher-level generalizations from lower-level ones.[8]

These responses are unpersuasive because they invoke a fair epistemic point—the plausibility of generalizations and causal statements at different levels—to draw an ontological conclusion—the distinctness of levels of being and their causal efficacy—that simply does not follow. The reductionist is unmoved. Barring strong emergence, we cannot intervene on higher-level variables without also intervening on their realizers; so, any alleged evidence of higher-level causal efficacy is always also evidence of lower-level causal efficacy (cf. Baumgartner 2013, 2017; Hoffmann-Kolss 2014; McDonnell 2017). No matter how often special scientists propose higher-level generalizations and models or how rarely they derive them from lower-level generalizations, the dilemma stands. The dilemma is not about statements—it's about properties (in the sparse sense; more on this below). Either higher-level properties are just causally inert disjunctions of lower-level properties, or they are identical to lower-level properties. Either way, reductionism wins.

Except for two things that are rarely mentioned. First, the standard dialectic between reductionism and anti-reductionism neglects a third option: that higher-level properties are neither identical to nor entirely distinct from their lower-level realizers—rather, they are *an aspect of* their realizers. I'll come back to this option

[8] A third response is that higher-level properties have *fewer* causal powers than their realizers. That is on the right track so let's bracket it for now; more about it shortly.

later. Second, as I pointed out in the previous section, reductionism says that one level is fundamental and other levels reduce to it. It doesn't have any principled means of identifying the fundamental level. For all that reductionism says, the fundamental level can be arbitrarily high—as high as the whole universe, in fact. So, for all the reductionist dilemma shows, the reduction base could be the higher level—the level of P. Or it could be an even higher level. Anti-reductionists may not find this observation especially reassuring, because they have no way of showing that their preferred level is the fundamental one. But reductionists should find this observation even more disconcerting. The inability to identify the fundamental level makes reductionism unsatisfying.

1.6 A Single Level

To get out of the reductionism vs. anti-reductionism morass, it is tempting to get rid of multiple levels altogether. If we didn't have multiple levels, we wouldn't have to worry about which one is fundamental and which ones are not. To see how this might go, we need to take a step back and distinguish between concrete reality and our descriptions of it.

In our discussion so far, I've assumed that there is a real concrete universe independent of our thoughts and descriptions of it. So far, so good. I've also assumed that when we talk about things at different levels (atoms, molecules, cells, tissues, etc.), we capture something objective about the structure of the universe—a genuine articulation of concrete reality into objects that compose one another and properties that realize one another. This latter assumption may be questioned. Perhaps reality is not structured in levels of being at all. The multiple levels might be a feature of our thoughts and descriptions, not reality.

If there are no multiple levels of being—no layers of objects standing in composition relations, no layers of properties realizing one another—the next question is what our discourse refers to; what makes it true or false. The simplest answer is that there are, in fact, some objects and properties in the world. They are just not organized in multiple levels of composition and realization. They form *a single genuine level*. To make sense of that, let's go back to our earlier claims:

(1) B causes E.
(2) NN cause E.
(3) S_B causes E.
(4) S_{NN} causes E.

Previously, I assumed that for (1)–(4) to be true, there must be real, concrete brains and neurons in the world, with real, concrete states, and they must really cause E. We saw that this way of thinking led to trouble.

Here is a different way of thinking about it. All these putative things—brains and their states, neurons and their states—are not genuine things to be found within concrete reality. Rather, statements that purport to be about them, such as (1)–(4), are true if and only if *some* aspect of concrete reality—whatever exactly that amounts to—makes them true.

To illustrate more vividly, suppose that reality consists of elementary fermions and bosons and nothing else. Fermions and bosons do not genuinely compose atoms, molecules, and so forth. Rather, the terms "atom," "molecule," etc., including "neuron" and "brain," are just shorthand for unmanageably complex arrangements of fermions and bosons. Ditto for the properties of fermions and bosons—they are the only real properties, and any names of putative higher-level properties are just shorthand for unmanageably complex arrangements of properties of fermions and bosons.

According to this *single level* picture, we might say that there is exactly *one* genuine level of objects and properties. All parts of our discourse—including all of our science—refer to objects and properties at the one and only genuine level of being. Our discourse is true to the extent that it accurately describes the genuine level. All other putative levels are illusory—they are aspects of our discourse and thought, not reality.[9]

This single level picture sounds a bit like reductionism, but there is a subtle difference. Reductionism retains many levels as real, whereas the single level picture eliminates them. Reductionism claims that there are multiple genuine levels of reality: our talk of neurons, brains, etc., and their properties does refer to real things—it's just that they all reduce to the one *fundamental* level. By contrast, the single level view claims that there are no multiple levels at all—no genuine neurons, brains, and higher-level properties: our talk of such things is just shorthand for talk of things at the one and only *genuine* level.

If this single level picture is right, then the causal exclusion puzzle doesn't even arise. For the causal exclusion puzzle presupposes multiple levels of reality and questions how they can all be causally efficacious without redundancies. If there is only one genuine level, then that is the one and only locus of genuine causation.

Elegant as the single level view is, it leaves much to be desired. Specifically, it calls for an account of three things. First, which is the genuine level of being? Second, in virtue of what do we make true statements about the one genuine level of being by using shorthand statements that purport to refer to other levels? And third, what's the relation between shorthand statements that appear to capture the

[9] There are at several routes to the single-level view. One route is nihilism—the denial that the many fundamental objects compose (Cameron 2010; Dorr and Rosen 2002; Sider 2013). Another route is a sufficiently sparse view of properties (Heil 2012). A third route is *existence monism* (as defined, not endorsed, in Schaffer 2010): the view that there is only one object—the universe—that does not decompose into parts.

same portion of reality—such as (1) and (2), or (3) and (4)? Let's look at each question in turn.

Finding the one *genuine* level of being among the alleged fakes is just as difficult as finding the one *fundamental* level among the many allegedly nonfundamental ones. The genuine level could be elementary fermions and bosons, the universe as a whole, or any level in between. There is no obvious principled way of deciding. This is disappointing enough. The other questions lead to even deeper trouble.

The second question is, in virtue of what are special science generalizations and explanations true of the world? This is a difficult question. Part of the answer has to do with what the one genuine level is. Since we don't know, let's pick an arbitrary level as a working hypothesis. Let's hypothesize that the one genuine level is the elementary fermions and bosons.

Under this hypothesis, the single level view says that when we use terms like "fermion" and "boson," we genuinely refer to real fermions and bosons and say things that are genuinely true or false of them. By contrast, when we use terms like "neuron," "brain," and "person," what we are really talking about is unmanageably complex arrangements of fermions and bosons. What makes our special science statements true or false are not neurons, brains, or people—strictly speaking, there are no such things. What makes our special science statements true or false are unmanageably complex arrangements of fermions and bosons.

But what are these *arrangements*? There are two options. Either "arrangement" is the kind of term that refers to some genuine aspect of reality (like "fermion" and "boson"), or it isn't. If it isn't, then we have no genuine answer to our question—no answer in terms of genuine aspects of reality. Is there *nothing* in concrete reality in virtue of which special science statements are true or false? Are they just true or false simpliciter? Is their truth or falsehood just a primitive fact about them? This makes the success of the special sciences mysterious—it makes it a mystery what, exactly, scientists are latching onto when they say true things about the world.

If "arrangement" does refer to some genuine aspect of reality, however, we want to know what aspect that is. The obvious answer is: the objective and specific ways in which fermions and bosons are physically arranged. But if there is an objective, genuine way that fermions and bosons are arranged and we can refer to such arrangements, there is no reason why we couldn't name those objective arrangements and say true things using their names. For instance, some of those arrangements could be called "atoms," more complex ones "molecules," and so forth. Ditto for the properties of fermions and bosons vis à vis the properties of their objective arrangements. In sum, if we allow that "arrangement" refers to a genuine aspect of reality, we recover multiple levels of being. We've just renamed composition and realization. We have abandoned the single level view.

The same thing happens if we try to answer our third question: what's the relation between distinct shorthand statements that appear to capture the same portion of reality, such as (1) and (2), or (3) and (4)? The traditional answer begins with the claim that brains are made of neurons, and brain states are realized by neural states. That leads to the difficult question of whether brains reduce to (collections of) neurons, and whether brain states reduce to neural states. If there are no genuine composition and realization, the question of reduction does not arise. But there remains the question of whether different statements from different special sciences capture the same portion of reality. If two statements capture the same portion of reality, I'll call them *ontologically equivalent*. Is (1) ontologically equivalent to (2)? Is (3) ontologically equivalent to (4)?

Again, there are two options. Either there is some aspect of reality that answers this question—presumably at the one and only genuine level—or there isn't. If there isn't, there's just no fact of the matter as to whether statements like (1) and (2), or (3) and (4), ultimately say the same thing. This is hard to live with. It's not just that metaphysicians have spent centuries debating whether different special science statements are ontologically equivalent. If metaphysicians turned out to be misguided, we could live with that.

It's also that, as we shall see in more detail in later chapters, scientists devote enormous energy to integrating statements about brains and their properties with statements about neurons and their properties, as well as statements about all kinds of levels in between. Scientists construct multilevel mechanistic explanations of phenomena, which integrate statements at different levels. In doing so, they make specific claims about which objects are part of others and what properties realize others.

What are they doing? What makes their statements true? If nothing in reality tells us how statements at different levels relate to one another, if nothing in reality makes statements about composition and realization true, how can we make sense of multilevel model building and explanation? If the single level view entails that nothing in reality helps us make sense of scientific practices, the single level view is an inadequate foundation for science.

As I mentioned, there is another way to answer our third question. According to this alternate answer there is, in fact, some aspect of reality that answers questions about ontological equivalence, composition, and realization. Something in reality determines whether one or another special science statement is ontologically equivalent to another as well as whether statements about composition and realization are true. What is this aspect of reality?

The obvious answer is analogous to the second answer to the second question: what answers questions about ontological equivalence, composition, and realization is the objective arrangement of the elementary fermions and bosons. If two special science statements describe the same facts about how fermions and bosons are arranged, then they are ontologically equivalent to one another. If, instead, two statements describe different facts about how fermions and bosons are arranged, then they are not ontologically equivalent to one another. A special science statement about composition or realization is true if and only if some fermions, bosons, and their properties are objectively arranged in a certain way.

(Again, elementary fermions and bosons need not be the one genuine level—they ought to be replaced by talk of whatever the genuine level turns out to be. I am just using fermions and bosons as placeholders to illustrate the point.)

We've already seen the cost of this second option: by positing objective arrangements of fermions and bosons, it allows us to recover multiple levels of being. Here is another way to put the point. Call elementary fermions and bosons level 0. If there are objective arrangements of fermions and bosons and their properties (level 1), this raises the question of how such arrangements relate to arrangements of arrangements of fermions and bosons and their properties (level 2), and then to the next level of arrangements (level 3), and so forth across the multifarious levels of being. Again, we have abandoned the single level view in favor of good old multiple levels. What are we going to do? What is a level of being anyway?

1.7 Against Ontological Hierarchy

All the views we've discussed have something in common: they posit an ontological hierarchy between levels. As metaphysicians like to say, they posit that some levels are more *ontologically fundamental* than others. Sometimes they express this by saying that some levels are ontologically *prior* to others, or that some levels *ground* others.[10]

According to reductionism, the reduction base is fundamental, while all other levels are derivative. According to the single level view, there is only one genuine level, while other levels are illusory. In principle, anti-reductionism may seem to stand against ontological hierarchy between levels, because it rejects reductionism. In practice, most anti-reductionists still subscribe to ontological hierarchy. They maintain that higher levels are ontologically dependent on (or grounded in) their lower-level realizers, even though higher levels don't reduce to their realizers.[11]

This ontological hierarchy gives rise to a number of difficult puzzles. We've encountered two: First, which level is fundamental (or genuine)? Second, what happens if everything has parts all the way down, without ever reaching an ultimate lower level? There are no easy answers to these questions. We'd be better

[10] Two exceptions are Hütteman (2004) and Hütteman and Papineau (2005), who argue that higher levels and lower levels mutually determine one another and therefore there is no asymmetric determination of some levels by others. My discussion is indebted to theirs.

[11] According to overdetermination, each level has its own causal powers, which duplicate other levels' causal powers. This does not require a hierarchy between levels; nevertheless, usually even defenders of overdetermination endorse ontological hierarchy. At any rate, overdetermination is a desperate measure, which we are not taking seriously.

off if we didn't have to ask them at all. We'd be better off if we abandoned ontological hierarchy between levels of concrete being. If we get rid of ontological hierarchy, many problems don't even arise.

Why do metaphysicians posit an ontological hierarchy at all? Why do they posit that one level is more genuine or fundamental than others? We've seen that one goal is solving the puzzle of causal exclusion. Let's set that aside for a moment; I'll soon offer a nonhierarchical ontology that solves causal exclusion. Besides causal exclusion, is there any *other* reason?

There are probably complex historical *causes* of why metaphysicians posit an ontological hierarchy. In the old days, there was the Great Chain of Being, with all the layers of being emanating from God, which stood as the ultimate foundation of everything (Lovejoy 1936). Perhaps after God died (Nietzsche 1882, 1883), metaphysicians replaced God with the fundamental physical level. I will leave it to historians to sort out how that went. Setting history aside, I see no reason to posit an ontological hierarchy among levels of being.

The notion of ontological hierarchy between levels of being strikes me as a constructional analogy run amok. By constructional analogy, I mean an analogy between levels of being and a multi-story building. Each story, or level, in a multi-story building sits on the level below it—if a level crumbles, all higher levels fall down with it. Therefore, each level of a building asymmetrically depends on the level(s) below it. As a result, any multi-story building needs a solid foundation. Thus, there is a privileged level—the foundation—on which all others rest.

By analogy, being seems to be organized into levels: atoms and their properties, molecules and their properties, cells and their properties, organs and their properties, etc. Each level is made of things at the next lower level, and each level constitutes things at the next higher level. Furthermore, phenomena at each level are mechanistically explained by components and their organization at the next lower level, and they also contribute to mechanistically explain phenomena at the next higher level. This is how we talk and how scientists talk. Nothing wrong so far.

The problem arises when we take this talk of "higher" and "lower" levels to carry the same implications they have when we talk about buildings. Just as a building's higher levels asymmetrically depend on the lower levels and ultimately on a solid foundation, we are tempted to think that higher levels of being asymmetrically depend on the lower levels and ultimately on one fundamental (or genuine) level.

But this part of the analogy need not hold. In fact, there are many *dis*analogies here. You can add a new level to a building, or take a level apart. To add a new level to a building, you need to add more bricks, or glass and steel, or whatever the building is made of. To take a level off, you need to subtract material from the building. Levels of being are not like that: you can't add a new level of being to preexisting levels, nor can you subtract one. Levels of being are all in the same place at the same time—they can't be added to or removed from a portion of reality.[12] Our job is just to discover what they are. In addition, there is no analogue among levels of being to the spatial relation among stories—being stacked on top of one another to stand against gravity—that gives rise to the hierarchy. So, the reason that levels of buildings are structurally dependent on one another—the need to support higher levels against gravity—does not apply to levels of being, and there is nothing analogous to it either.

I don't know the degree to which either the Great Chain of Being or the constructional analogy played a role in the origin of the ontological hierarchy between levels. All I know is that I haven't encountered any good argument for an ontological hierarchy among levels of being. And given the headaches it generates, I propose to get rid of it. I conclude that levels of concrete physical being do not form an ontological hierarchy.

I will now argue that if we reject ontological hierarchy, we can reach a satisfying solution to the puzzle of causal exclusion—a solution that does not require overdetermination. We also gain resources to make sense of scientific practices such as multilevel mechanistic explanations. As a bonus, the egalitarian ontology I defend avoids raising unanswerable questions such as what is the fundamental level of being.

1.8 Levels of Properties

Since properties are involved in causal work, I must say more about them. I adopt a *sparse* conception of properties as qualities, causal powers, or some combination of both. Sparse properties are natural; they carve nature at its joints (cf. Lewis 1983, 1986, 59–60). I also count having structure—having certain parts organized in a certain way—as a property of a whole. I assume that any nonsparse properties, including "modal properties," are grounded in the sparse properties.

What matters most here is the relation between properties and causal powers—the powers that allow objects to do things. There are two simple views.

Categoricalism is the view that properties are qualities. Qualities, in turn, are ways objects are that do not, by themselves, include any causal powers (e.g., Lewis 1986; Armstrong 1997). Paradigmatic examples include being square versus round or being a certain size. In order for objects to do things, their qualities must be paired with natural laws. Objects do what they do because they obey laws in virtue of the qualities they possess. If we could change the laws, we would change objects' behavior—even if their qualities remained the same.

[12] Of course, parts can come together, over a time interval, to form novel wholes and wholes can be decomposed, over a time interval, into isolated parts. That's not what we are talking about. We are talking about the relation between parts and wholes and their properties at one and the same time.

Given categoricalism, then, causal powers are not basic—they are byproducts of qualities and laws.

By contrast, *dispositionalism* is the view that properties are dispositions, or causal powers (e.g., Shoemaker 1980; Bird 2007). Paradigmatic examples include being soluble or fragile. In order to do things, all that objects have to do is possess causal powers and encounter conditions that trigger their powers' manifestations. Laws are not needed. In fact, law statements are just descriptions of how objects behave in virtue of their causal powers. Given dispositionalism, causal powers are basic while laws are derivative.

There are also hybrid views, according to which some properties are qualitative while others are causal powers (e.g., Ellis 2010), and identity views, according to which qualities just are causal powers and vice versa (Martin 2007; Heil 2003, 2012). These additional views do not affect the rest of the argument below so I will no longer mention them explicitly.

Let me illustrate the dialectic between categoricalism and dispositionalism. Take a paradigmatic quality—being square. What could be less dispositional? How can dispositionalism account for an object being square in terms of dispositions? Well, according to dispositionalism, despite appearances, being square is just a cluster of causal powers: the power to reflect light so as to form square images (when the square object is also opaque), the power to make square impressions on soft surfaces (when the square object is also solid), etc. What appeared to be a single pure quality turns out to be a plurality of dispositions.

On the flip side, take a paradigmatic disposition, like being water-soluble. What could be less qualitative? How can categoricalism account for an object being water-soluble in terms of qualities? Well, according to categoricalism, despite appearances, being water-soluble is just a byproduct of the underlying qualities of the object: having certain chemical qualities, which interact with the qualities of water thanks to appropriate natural laws so as to cause dissolution. What appeared to be a single causal power turned out to be a byproduct of underlying qualities plus natural laws.

Since I will discuss the relations between properties at different levels, we need to see what these relations could be depending on either categoricalism or dispositionalism. Consider two properties at two different levels that putatively cause the same effect—e.g., S_B and S_{NN} (Figure 1.1).

According to dispositionalism, the two properties consist of two pluralities of causal powers. Overdetermination is the view that the two pluralities contain distinct tokens of the same causal power—the power to cause E. This is too ontologically redundant to take seriously. The other options are (i) *identity*, according to which the powers of the higher-level property are identical to the powers of the lower-level property ($S_B = S_{NN}$), (ii) *strong emergence*, according to which S_B's powers include some that are not included among S_{NN}'s powers, and (iii) *the aspect view*, according to which S_B's powers are some of S_{NN}'s powers ($S_B < S_{NN}$).

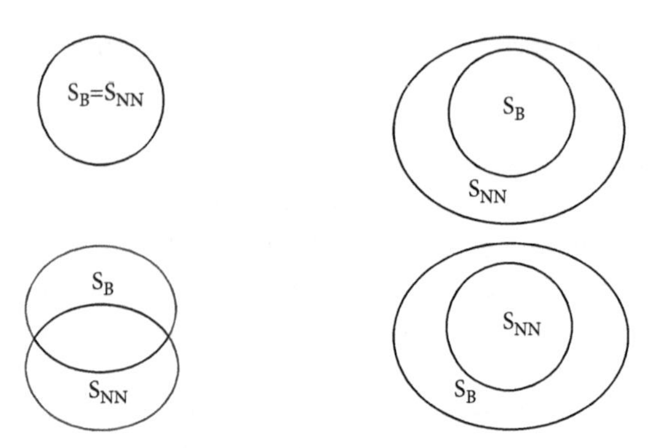

Figure 1.1 Possible relations between higher-level properties and lower-level properties. Clockwise, starting from the top left corner: identity, aspect view, and two forms of strong emergence.

By contrast, according to categoricalism, the two properties are qualities. Overdetermination is the view that the two qualities are distinct from one another yet they fall under the same law. Again, this is too ontologically redundant. The other options are that (i) the higher-level quality is identical to the lower-level quality (identity: $S_B = S_{NN}$) so that they fall under the same law, (ii) part of the higher-level quality is not included in the lower-level quality (strong emergence) so that there is at least one law governing the higher-level quality that does not govern the lower-level quality, and (iii) the higher-level quality is an aspect of the lower level quality (the aspect view: $S_B < S_{NN}$) so that there is at least one law governing the lower-level quality that does not govern the higher-level quality.

Setting aside overdetermination and strong emergence due to their glaring inadequacy, the serious options are identity and the aspect view. An aspect, as I use the term, is just a *part* of a property.[13]

As we saw above, identity with a direction is the clearest way to make sense of ontological reduction. Reductionism is hierarchical only because it posits a direction between the terms of the identity. Identity by itself is not hierarchical. If X = Y, neither X nor Y needs to be ontologically prior to, or more fundamental than, the other. If anything, X and Y should both be equally real, since they are the same thing! So much the worse for reductionism. The same is true of the aspect view: by itself, there is nothing ontologically hierarchical about it. If X is an aspect of Y, neither X nor Y needs to be more fundamental than the other. In addition, both identity and aspects explain why realization is a relation of synchronic

[13] I assume that either the parthood relation is sufficiently general to apply to any category of being, including property types (cf. Varzi 2016, section 1), or the aspect view can be cashed out in terms of property instances, which are particulars that can stand in part-whole relations to one another. The aspect view takes a slightly different form depending on whether properties are particulars (tropes), universals, or classes of particulars (Orilia and Swoyer 2020). I will stay neutral on that point.

metaphysical necessitation. So, a nonhierarchical picture of levels could appeal to either identity or aspects.

The aspect view is inspired to an extent by the so-called subset view of realization. Roughly, the subset view is that properties are (or "bestow," or "confer" on objects) sets of causal powers, and higher-level properties are subsets of the causal powers of their lower-level realizers. Versions of the subset view have been defended by several authors (e.g., Wilson 1999, 2010, 2011; Shoemaker 2007; see also Fales 1990; Clapp 2001; Polger and Shapiro 2008; Watkins 2002; Yablo 1992).

There are three main differences between the aspect view as I define it and the subset view as is usually defended. First, whereas the subset view is committed to dispositionalism (Baysan 2017), the aspect view is neutral between dispositionalism, categoricalism, and other views about the metaphysics of properties. Properties may be qualities, causal powers, or a combination of both. This means that the aspect view is more general and less affected by debates about the metaphysics of properties than the subset view.[14]

Second, defenders of the subset view typically subscribe to ontological hierarchy—the view that some levels ground others. In addition, defenders of the subset view typically maintain that higher-level properties are somehow wholly distinct from their realizers, even though they endow objects with only a subset of the causal powers that their realizers endow them with.[15] I reject ontological hierarchy, and with that any notion that higher-level properties are wholly distinct from their realizers. Higher-level properties are only partially distinct from their realizers because they are *an aspect* of their realizers.

Third, the aspect view is defined in terms of the *part of* relation, whereas the subset view is defined in terms of the *subset of* relation. To lay people, this may sound like a notational variant. To many philosophers, however, sets are abstract objects, and the subset relation is a relation between sets. To these philosophers, appealing to *sets* commits proponents of the subset view to the existence of sets qua abstract objects. By contrast, the *part of* relation holds between concrete things (as well as abstract things, if there are any), so appealing to the *part of* relation does *not* commit anyone to the existence of abstract objects. Since I don't believe in abstract objects, I'd rather not run the risk of committing to their existence.

[14] The subset view can be reconciled with categoricalism if someone accepts that whether one property realizes another is contingent. This is because, according to categoricalism, which properties confer which causal powers depends on which laws of nature hold. Therefore, which properties confer causal powers that are a subset of those conferred by another property depends on which laws of nature hold. Here I'm indebted to discussion with Daniel Pritt and Paul Powell.

[15] The subset relation itself does not entail any hierarchy between subsets and supersets, in either direction. Ironically, this is sometimes seen as a liability for the subset view, because it fails to explain ontological hierarchy (Morris 2011; Audi 2012; Pineda and Vicente 2017). Rejecting ontological hierarchy, as I do, eliminates this problem.

To recapitulate, we've identified two serious nonhierarchical options about the relation between higher-level properties and lower-level properties—identity and the aspect view. Either higher-level properties are identical to or they are aspects of lower-level properties. To choose one of these options, we need to consider what sorts of objects possess higher- versus lower-level properties, how they relate to one another, and why we need to talk about multiple levels.

Preview: it will turn out that higher-level objects are invariants under certain kinds of additions to and subtractions from their parts, and we need to talk about all these levels to (i) understand the various phenomena we encounter and (ii) give constitutive explanations of them. It will also turn out that, in general, higher-level properties are aspects of lower level ones.

1.9 Synchronic Composition

To make progress, let's take a step back. The universe unfolds over time, and science studies this unfolding. Nevertheless, it is convenient to first consider the way things are at a single time instant. Let's mentally freeze time at an arbitrary instant and consider the whole universe. There are objects and properties distributed throughout space.[16] Some objects are part of others; the properties of the parts realize the properties of the whole.

Now consider a portion of reality at a time instant—say, a portion that we normally call a brain—and consider two of its levels. That is, consider a brain with all of its properties at a time. Then consider a plurality of parts that cover that same portion of the universe without gaps and without overlaps—say, a plurality of neurons, glia, other cells, and extracellular fluid. As before, I will keep things simple by referring to the latter plurality as a bunch of "neurons." Finally, consider not only the neurons but also all of their properties at a time.

As long as we are considering such things at one instant of frozen time—without any change being possible because time is not flowing—it makes no difference whether we are considering the whole brain or its parts. Either way, we are considering the same portion of reality. When we consider the higher level, we often abstract away from lower-level details. That is, by considering a whole brain and its global properties we abstract away from the details of the billions of neurons and other components that make up the brain. Abstracting away from details is very important when time flows and we want to build a scientific model of a dynamical process, but for now we are considering a single time instant. We can consider the whole object with all of its structure—a whole brain with all of its

[16] More precisely, consider spacetime. Pick an arbitrary frame of reference. Take the 3-dimensional hyperplane of simultaneity defined by that frame of reference at an arbitrary time. Consider the objects and properties distributed throughout the hyperplane of simultaneity.

parts and their properties. By the same token, if we consider all the neurons that make up a particular brain and their properties, and we leave no gaps, we already included everything we need in order to consider the whole brain and its global properties. Thus, when we consider a portion of reality with all of its structure at a time, there is no ontological distinction or difference between the whole and its parts; there is no ontological distinction or difference between the whole's properties and the parts' properties. They are identical.

In short, synchronic composition—composition at a time—is strict identity. As long as we freeze time, a whole is identical to its parts taken collectively, which in turn are identical to their parts taken collectively, and so forth (Schumm, Rohloff, and Piccinini unpublished). Furthermore, a whole's properties are identical to its parts' properties taken collectively, which in turn are identical to their parts' parts' properties taken collectively, and so on for as many levels as there are.

As I already mentioned, we can always abstract away from some of the parts and their properties. We often do so for many ordinary and scientific purposes. In the rest of this chapter, I will discuss the aspects of the world that this kind of abstraction gets at. But if we freeze time and consider a whole with all of its structure, we already included its parts and their properties. Vice versa, if we consider a whole's parts with all of their properties, we already included the whole and its properties in our consideration.

This doctrine—that composition is identity—has recently encountered a revival (e.g., Baxter and Cotnoir 2014). As I've argued, I think it's exactly right when it comes to *synchronic* composition, and an analogous point applies to properties. Perhaps this is why the view that higher levels are identical to lower levels appeals to people. But composition as identity is radically wrong about *diachronic* composition and realization—composition of objects and realization of properties over time.

1.10 Diachronic Composition

Consider the same portion of reality—the same brain you considered before. Unfreeze time. Everything starts changing. The brain changes state over time, and so do its parts. Worse: some parts move relative to one another, others are lost, yet others are added. Old cells die and new ones are born. Nutrients flow into the brain, waste is expelled. New axonal and dendritic branches form, other branches are pruned. And so forth.

Throughout all these changes, the whole remains the same object. It's still the same brain over time. Or so we say, and so do scientists. We find it convenient to abstract away from all that change and focus on what remains stable. We identify and re-identify the same object—i.e., the same brain—at different times, in spite of the myriad changes that its parts go through. We do this at every level: each neuron remains the same neuron throughout its existence, and so does each neural system, person, and object at any other level. Or so we say.

A strict metaphysician might object to this way of talking: if the parts change, the old whole must become a new, distinct whole. The whole is made out of the parts—it's nothing over and above the parts. Therefore, if any part is lost or acquired over time, the old whole goes out of existence and a new whole comes into existence. This doctrine is called *mereological essentialism* (Chisholm 1969).

Mereological essentialism is a coherent but impractical view. No one talks like a mereological essentialist—not even mereological essentialists—for a very good reason. Most of the complex wholes that we are interested in lose old parts and acquire new parts all the time. If we had to refer to them as new objects—e.g., if we had to refer to our friends and family as new people—every time a part is replaced, we'd never been able to re-identify most objects. Our ability to communicate effectively would collapse.[17]

Re-identifying people and other complex objects over time is valuable. For, in spite of all the change, there is a lot of stability. Identifying stable aspects of the world allows us to predict and explain phenomena; explanation and prediction, in turn, are core values of both commonsense and science. Not to mention, focusing on stability allows us to maintain relationships with our friends and family. This is why mereological essentialism is a nonstarter when it comes to the metaphysics of both science and commonsense. It is also why composition as identity is a nonstarter when it comes to diachronic composition.

Composition as identity says that a whole is identical to its parts. But we typically refer to a whole as the same object across time, even though it loses some of its parts and acquires others. The parts change while the whole remains the same. If this *being the same* is supposed to be *identity*, then—by the transitivity of identity—the whole is not and cannot be identical to the parts.

Someone might argue that, precisely because the parts are no longer the same, *being the same whole through a time interval* is not the same as *being identical*. On this view, strict identity between two things requires that those two things have exactly the same properties (Leibniz's Law), including structural properties such as having certain parts. This is a fair point, and there are various ways to try to get around it (cf. Gallois 2016). Be that as it may, it makes no difference for present purposes.

Our goal is to make sense of scientific discourse and what it captures about the objective structure of the physical universe, which is continuous with what our everyday language captures. Given the way we talk, there is no problem

[17] What about wholes that are not individuated by properties but simply by enumerating their parts? For instance, what about the whole composed of Venus and Mars? That's a special case, because such wholes have their parts essentially by definition. Thus, they do not survive the addition or subtraction of their parts. But that's not the kind of thing we are usually interested in, especially in science.

identifying a whole with its parts at a time instant—but there is no way to identify a whole with its parts across time. The parts change; the whole remains the same. What objective aspect of reality does this way of talking capture?

1.11 Invariants

In mathematics and physics, an invariant is a property that remains constant when a transformation is applied to its bearer. The force of gravity is invariant under translation, the speed of light is invariant under Lorentz transformation, etc. We can use the same notion of invariance in other disciplines as well as everyday language. For example, changing an organism's environment does not alter what species it belongs to; changing its DNA (sufficiently radically) does—that's how speciation occurs. Thus, species identity is invariant under environmental change but not (radical) DNA change.

Invariance allows us to identify and reidentify objects over time. In fact, if we take the notion of invariance broadly enough, object identification and reidentification depends on individuating an object via one or more invariants. Specifically, particular objects either lose or retain their identity under certain transformations. Consider the classic example of the statue and the lump of bronze that constitutes the statue. Are they one object or two? Given that they occupy the same space, there seems to be just one object. But let's suppose the original lump was later molded into the statue. Given that the lump existed before it took the shape of the statue, there seem to be two distinct objects. What gives? If we reframe the situation in terms of invariants, the puzzlement vanishes.

Squashing a statue destroys it; breaking a tiny chip from a statue does not. That is, statues are invariant under chip removal but not under squashing. Why? Because breaking a tiny chip from a statue leaves much of the statue's shape intact, whereas squashing the statue obliterates its shape. Since one of the properties we use to identify and re-identify statues is their shape and shape is invariant (enough) under chip removal but not under squashing, we say that having (roughly) a certain shape is essential to being a statue.[18] That's why being a statue is invariant under small chip removal but not under squashing.

On the other hand, squashing a lump of bronze does not destroy it; removing a tiny chip does. Lumps of bronze are invariant under squashing but not under chip removal. Why? Because squashing a lump of bronze preserves the parts that make up the lump of bronze, whereas removing even a tiny chip does not. Since having the same parts is one of the properties we use to identify and re-identify lumps and having the same parts is invariant under squashing but not under chip removal,

[18] For more on essential properties, see Robertson and Atkins 2016.

we say that being made of those specific parts is essential to being that particular lump of bronze. That's why being a particular lump of bronze—unlike being a statue—is invariant under squashing but not under chip removal.

Once a lump of bronze becomes a statue by acquiring a new shape, it has the right properties to be individuated either as a lump or as a statue. Depending on how we individuate it—depending on which invariants we use to individuate it—it will continue to exist or cease to exist under different transformations. Individuated as a lump, it will cease to exist if a chip breaks off; individuated as a statue, it will cease to exist if it's squashed. We can individuate portions of reality in terms of any combination of their properties. Each individuation defines one or more invariants, which in turn give rise to different identity conditions.

In summary, invariants allow us to separate the objective properties of portions of reality from the way we use them to individuate and track objects. Each portion of the concrete physical universe has many properties. We can appeal to any combination of properties to individuate, type, and track a portion of reality. As soon as we choose a plurality of properties $P_1, \ldots P_n$ to individuate a portion of reality, we thereby define an object that is invariant under transformations that preserve $P_1, \ldots P_n$. We can then look at whether there is something that retains $P_1, \ldots P_n$ over time. If there is, that portion of reality continues to exist as the same object. If there isn't, that object ceases to exist.

In practice, the list of properties that we use to individuate a portion of reality may be largely implicit and vague. One thing that is common to particular objects is that for them to be preserved, they typically must retain most of their parts during short time intervals, even though all of their parts may be replaced during a longer time interval. For instance, for something to count as the same person in the ordinary sense, it must retain most of its parts during short time intervals, even though all of its parts may be replaced during a longer time interval.[19]

1.12 Subtractions of Being

When we consider a portion of reality as a whole object, we can do it in one of two ways. Wholes may be considered either together with their parts—with their structural properties and hence including their parts—or in abstraction from their parts. For example, we may consider a neural circuit with all of the distinct neurons that constitute it, or in abstraction from its neurons (Figure 1.2). As we shall see in Chapter 7, this sort of abstraction (partial consideration, inclusion of

[19] This part-preservation principle does not apply to objects with no parts, such as elementary fermions and bosons. They can be individuated by a combination of invariants and observables (Castellani 1998). Castellani builds on a rich tradition among physicists of individuating objects in terms of invariants and observables. For what it's worth, I discovered this tradition *after* developing the proposal about complex objects that I defend in the main text.

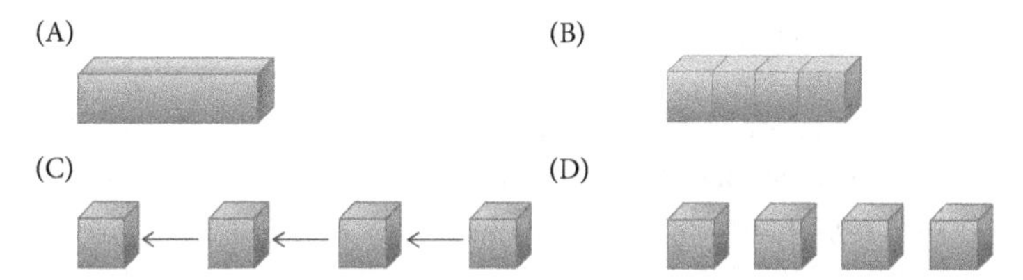

Figure 1.2 Ways of representing a whole and its parts. (A) A whole in abstraction from its parts. (B) A whole considered together with its parts. (C) Parts considered in isolation but with their organizational relations. (D) Parts considered in isolation without their organizational relations.

more or fewer properties) is an important aspect of multilevel mechanistic explanation that needs to be reconciled with an appropriate ontology of levels.

Depending on how we individuate a whole, the object that results from such individuation stands in different ontological relations to its parts. If a whole is considered together with all of its (organized) parts at a time instant, what is thought of is identical to those parts. Any addition or subtraction of any part would replace that whole with a distinct whole.

In many contexts including most scientific ones, however, we consider wholes in abstraction from their parts, or at least in abstraction from most of their individual parts (as opposed to part *types*). By the same token, we often study wholes as instances of a type—that is, as possessing certain property types but not others. For instance, when neurophysiologists study a neuron, they don't study all or even most of its parts at the same time. When a whole is considered in abstraction from its parts, as it normally is, what is being considered is an *invariant*—a particular that retains certain properties—under part addition, subtraction, and rearrangement, within limits. In other words, and for most scientific purposes, wholes are particulars that can be reidentified through certain kinds of part addition, subtraction, and rearrangement.

In yet other words, wholes are *subtractions of being* from their parts. This subtraction is an epistemic operation—something that we do in our mind by abstracting away from the parts—but it does have an ontological consequence. The ontological consequence is that, by abstracting away from the parts, we identify and become able to track a stable aspect of the world that we miss when we consider a whole with all of its parts.

The boundary between changes that preserve a whole and changes that destroy a whole lies in the global properties that individuate the whole. Global properties include qualities, causal powers, and structural types. Structural types are properties such as *being composed of parts with certain property types* or *composing wholes with certain property types*. In other words, structural types abstract away from the particulars that an object is composed of or composes, while preserving the types of part it has or type of whole it composes. Belonging to a structural type means either possessing that type of structure or being part of that type of structure.

For instance, a neuron is individuated by certain higher-level properties (say, a certain shape, size, types of component, and producing action potentials). As long as a particular object has those properties, it remains a neuron. As long as it remains a neuron, it remains that particular neuron. As soon as it loses enough of those properties, it ceases to be a neuron, and a fortiori it ceases to be that particular neuron.

Since this account relies on the notion of properties that individuate a whole, it requires an account of properties at different levels (properties of wholes vs. properties of their parts) that does not make them redundant. If there are no higher-level properties, then higher-level properties cannot individuate wholes and cannot be used to tell when a whole is preserved or destroyed by a change of properties. Therefore, we need an account of properties at different levels.

Fortunately, we've already encountered the account we need in Section 1.8. I called it the aspect view: higher-level properties are aspects of their lower-level realizers. That is to say, if an object is made out of a plurality of parts, each global property of the whole is an aspect of some properties of its parts. If properties are causal powers, then the causal powers of the whole are among the causal powers of the parts. If properties are qualities, then the qualities of the whole are among the qualities of the parts, and there is at least one law governing the parts that does not govern the whole. Either way, higher-level properties—properties of wholes—are *subtractions of being* from lower level realizers—properties of parts (Figure 1.3). This account must be interpreted *dynamically*: what matters are properties as producers of change, which manifest themselves in the activities of objects.

Why aspects and not identity? Why are higher-level properties *aspects of* lower-level properties rather than *identical* to them? In principle, they can certainly be identical. If we freeze time and consider all the properties of a whole including its structural ones, then higher-level properties are identical to lower-level properties. As soon as we consider wholes in abstraction from their parts or unfreeze time, however, higher-level properties cease to be identical to lower level ones. Let's see why.

When we abstract away from an object's parts, or even from some of them, we are no longer including the particular structural properties of the whole among the properties we are considering—except for structural types. For example, we may be considering a whole brain and its global properties without bothering to consider all of the neurons that make up that brain and their properties. Therefore, by definition, we are omitting both the parts and their specific properties—properties that attach to them qua parts—from our consideration. Thus, the properties we are considering—the higher-level properties—are only aspects of the lower-level properties.

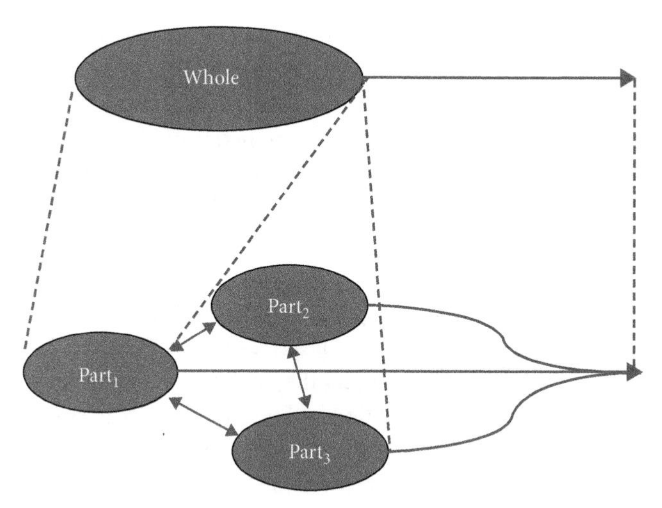

Figure 1.3 Wholes are invariants over certain changes in parts, properties of wholes are aspects of their lower-level realizers. Ovals represent objects, arrows represent properties. Dash lines between a whole and its parts indicate composition. Dash lines between higher-level and lower-level arrows indicate realization. The higher-level property is an aspect of its realizer. This scheme iterates downward with respect to the decomposition of parts into subparts and their properties, as well as upwards with respect to the composition of wholes into larger wholes and their properties.

Consider the cuboid in Figure 1.2.A. It has six sides that form a cuboid structure, which interact with the environment in characteristic ways (e.g., by causing sensations of a cuboid). Now consider the cuboid's four cubic parts and their properties, taken collectively (Figure 1.2.C). Since they form the cuboid, collectively they have all the properties of the cuboid. In addition, each part also has its own sides and relations to one another. Thus, the properties of the cuboid taken in abstraction from its parts (1.2.A) are some and only some of the properties of the cuboid's parts (1.2.C). At this point this may sound obvious and almost trivial but, as we'll see in subsequent chapters, it's enormously useful and important in constructing multilevel mechanistic explanations.

In addition, when we let time flow, the parts that make up a whole can change in many ways. Some parts—with all their properties—are lost, others acquired, and even those parts that remain within the whole can change their specific properties. And yet, through all these changes, many of the global properties of the whole remain the same, or they may change in a way that is invariant under changes in the specific values of lower-level properties.

For example, the temperature of a gas may remain the same through myriad changes in the specific kinetic energies of the particular molecules that make up that gas. Many scientific techniques, such as coarse-graining and the renormalization group—aim to identify and describe higher-level properties in abstraction from changes in their lower-level realizers. That is, they aim to identify a specific and stable aspect of many—often very, very many—lower-level realizers (cf. Batterman 2002). Again, once you see this point clearly, it may seem obvious and conceptually trivial. But understanding it clearly took some nontrivial conceptual work, which sheds light on the metaphysics of both science and commonsense.

Being an aspect of lower-level properties allows higher-level properties to be invariant under changes to their lower-level realizers. They can be aspects of different realizers. Like the whole objects they belong to, higher-level properties are also invariant under (some) addition, subtraction, and rearrangement of the parts of a whole object. For example, my car's ability to carry passengers is preserved when I add a navigator, throw away a mat, replace the oil, or rotate the tires.

There may appear to be a problem in the way I combine the *aspect of* relation between different levels and an egalitarian ontology. Given that, on the aspect view, lower-level properties can do everything that higher-level properties can do, but not vice versa, it seems that they aren't on equal footing after all. The lower-level properties are more powerful, as it were, than the higher-level ones. Hence, the lower-level properties appear to be more fundamental. But being able to do more or fewer things does not entail being ontologically more or less fundamental. Aspects are neither more nor less fundamental than what they are an aspect of.

What matters is what something can do relative to what needs to be done. If what needs to be done is X, it doesn't matter whether something can also do Y and Z. All that matters is whether it can do X. Therefore, when we are interested in higher-level phenomena and their explanation, we home in on wholes and higher-level properties.

1.13 An Egalitarian Ontology of Levels

Much remains to be said to flesh out this egalitarian ontology, but its contours should already be clear enough for us to proceed to the next chapter. Higher levels are just invariant aspects of lower levels. Neither wholes nor their parts are ontologically prior to one another. Neither higher-level properties nor lower-level properties are prior to one another. Neither is more fundamental; neither grounds the other.

Instead, whole objects are portions of reality considered in one of two ways. If they are considered with all of their structure at a given time, they are identical to their parts, and their higher-level properties are identical to their lower-level properties. For most purposes, we consider wholes in abstraction from most of their parts and most of their parts' properties. When we do this, whole objects are subtractions of being from their parts—they are invariants under addition, subtraction, and rearrangement of some parts.

The limits to what lower level changes are acceptable are established by the preservation of properties that individuate a given whole. When a change in parts preserves the properties that individuate a whole, the whole survives; when individuative properties are lost by a change in parts or their properties, the whole is destroyed. By the same token, higher-level properties are subtractions of being from lower-level properties—they are aspects of their realizers and are also invariant under some changes in their lower level realizers.

This account solves the puzzle of causal exclusion without making any property redundant. Higher-level properties produce effects, though not as many as their realizers. Lower-level properties also produce effects, though more than the properties they realize. For higher-level properties are aspects of their realizers. There is no conflict and no redundancy between them causing the same effect.

As long as we focus on the right sorts of effects—effects for which higher-level properties are sufficient causes—to explain effects in terms of higher-level causes is more informative than in terms of lower-level ones. The reason is that adding lower-level details adds nonexplanatory information. In addition, tracking myriad lower-level parts and their properties is often practically unfeasible. In many cases, we don't even know what the relevant parts and their properties are. That's why special sciences are both necessary and useful: to find the sorts of abstractions that provide the best explanation of higher-level phenomena, whereas tracking the lower-level details may be both unfeasible and less informative.

Given this egalitarian ontology, reductionism fails because, for most scientific and everyday purposes, there is no identity between higher levels and lower levels. Traditional anti-reductionism also fails because higher levels are not entirely distinct from lower levels. Ontological hierarchy is rejected wholesale. Yet each scientific discipline and subdiscipline has a job to do—finding explanations of phenomena at any given level—and no explanatory job is more important than any other because they are all getting at some objective aspect of reality.

This egalitarian ontology helps elucidate a number of basic issues in the metaphysics of mind and science. I will continue in the next chapter by clarifying three foundational concepts for the science of cognition: multilevel mechanisms, multiple realizability, and medium independence.

2
Mechanisms, Multiple Realizability, and Medium Independence

2.1 Multiple Realizability: Different Realizers Share an Aspect

In Chapter 1, I introduced an egalitarian ontology of levels. Neither wholes and their properties nor parts and their properties are more fundamental. Instead, wholes are invariants under certain transformations of their parts, and their properties are aspects of their realizers. This egalitarian ontology explains how multiple levels are causally efficacious without being redundant.[1]

In the same chapter, we encountered multiple realizability (MR). To a first approximation, a property P is multiply realizable if and only if there are multiple properties $P_1, \ldots P_n$, each one of which can realize P, with P, P_1, $P_2, \ldots P_n$ not identical to one another. Some philosophers prefer to discuss multiple realiz*ation* rather than multiple realiz*ability*. A property is multiply realiz*ed* if and only if it is multiply realiz*able* and at least two of its possible realizations actually exist. Thus, whether a multiply realizable property is multiply realized depends on contingent facts about the world that I'd rather not worry about. Notice that putative multiply realiz*able* properties are typically individuated by their effects. Thus, if they are multiply realiz*able*, not much stands in the way of their being multiply realiz*ed*. Either way, the same conceptual issues arise for both multiple realization and multiple realizability. I focus primarily on multiple realizability.

The idea that mental properties are multiply realizable was introduced in the philosophy of mind in the early functionalist writings of Putnam and Fodor (Fodor 1965, 1968a; Putnam 1960, 1967a). They argued that since mental states are multiply realizable, mental states are not reducible to brain states. Since then, MR has been an important consideration in favor of anti-reductionism in psychology and other special sciences (e.g., Fodor 1974, 1997). Some higher-level properties appear to be realizable by many distinct lower-level properties. For example, there are different ways to build a mousetrap, a corkscrew, or a computer. So, there appear to be different ways of realizing the property of catching mice, lifting corks out of bottlenecks, or computing. If these higher-level

[1] This chapter includes a substantially revised and expanded descendant of Piccinini and Maley 2014, so Corey Maley deserves partial credit for much of what is correct here. Sections 2.1, 2.3, and 2.5 are new.

Neurocognitive Mechanisms: Explaining Biological Cognition. Gualtiero Piccinini, Oxford University Press (2020).

DOI: 10.1093/oso/9780198866282.001.0001

properties are multiply realizable, then they are not identical—and therefore not reducible—to lower-level properties. This is the most common argument against reductionism.

Reductionists have responded with a dilemma.[2] Either a putative higher-level property is merely a disjunction of lower-level properties, or it is a genuine property with genuine causal powers. If the former, then all the causal work is done by the lower-level realizers. If the latter, then the higher-level property must be identical to some realizer after all. Either way, reductionism holds (Kim 1992; Sober 1999; Shapiro 2000). This dilemma has been so effective that in recent years the tide has turned against MR. Critics point out that MR had neither been clearly analyzed nor cogently defended. It has become common to deny MR, either about mental properties or about any properties at all (Bechtel and Mundale 1999; Bickle 2003; Couch 2009a; Keeley 2000; Klein 2008, 2013; Shagrir 1998; Shapiro 2004; Polger 2009; Zangwill 1992; Polger and Shapiro 2016). But this recoil into skepticism about MR is an overreaction.

We can now see what the existing dialectic misses: the possibility that higher-level properties are an aspect of their realizers. Different lower-level properties can share an aspect.[3] I will argue that MR occurs when the same type of higher-level property is an aspect of different types of lower-level properties that constitute different mechanisms. Both the higher-level property and its realizers retain their causal efficacy without taking anything away from each other and without being redundant (Figure 2.1).

In this egalitarian picture of MR, neither traditional reductionism nor traditional anti-reductionism holds. Reductionism fails because higher levels are not identical to lower levels. Anti-reductionism fails because higher levels are not wholly distinct

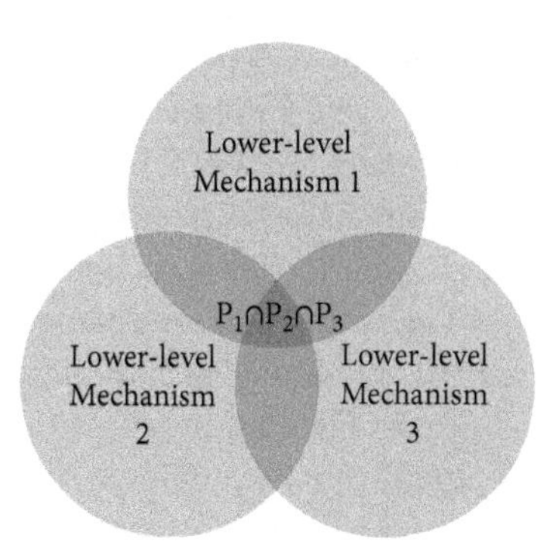

Figure 2.1 The overlap between lower-level property 1 and lower-level property 2 is a higher-level property realized by both properties 1 and 2, and so forth. The overlap between the three lower-level properties is a higher-level property realized by properties 1, 2, and 3; thus, the higher-level property is an aspect of each of its lower-level realizers. The symbol '∩' represents overlap.

[2] The dilemma assumes no overdetermination and no strong emergence. See Chapter 1, Sections 1.3 and 1.5.

[3] Reminder: I use the term "aspect" to mean (proper) part. A precursor of this point—that different types of property can share the same aspect—is the view that properties are causal powers and higher-level properties are proper subsets of the causal powers of their realizers. This is the subset view of realization, which I discussed in Chapter 1, Section 1.8.

from lower levels—they are *an aspect of* their realizers. One job of the special sciences is to find out which aspect of lower-level properties higher-level properties are. Needless to say, this requires studying both higher and lower levels. Thus, disciplines studying different levels are inter-dependent—not autonomous from one another.

There is a special kind of MR that will help us understand neurocognitive mechanisms. Following Garson (2003), I call it *medium independence.* To see what's special about it, let's take a step back.

Objects are often individuated by their higher-level properties, and many higher-level properties are individuated by the specific effects they have under certain conditions. For example, mousetraps are individuated by their catching mice, corkscrews by their lifting corks out of bottlenecks, and computers by their performing computations. There is an important difference between the first two examples and the third.

Mice, corks, and bottles are specific physical media with specific physical properties of their own. Mice have a size, mass, cognitive abilities, locomotive powers, etc., which fall within certain ranges. Mutatis mutandis for corkscrews and bottles. These specific higher-level properties are what mousetraps and corkscrews interact with. These properties of mice, corks, and bottles constrain what mousetraps and corkscrews can be made of and how they can work.

By contrast, computers need not manipulate any specific physical medium. Their physical effects are individuated purely in terms of the relation between vehicles with certain degrees of freedom. For example, digital computers are individuated in terms of strings of digits—namely, vehicles of finitely many types that can be concatenated into strings—and the specific relations between strings of digits that must obtain for a certain function to be computed (more on this in Chapter 6). Any physical medium can count as a string of digits so long as it comes in the right number of types, there is a way to concatenate the digits into strings, and there is a way for the mechanism to manipulate the strings and yield the right relations between strings.

Here is a more specific example. Ordinary computers are complex arrangements of logic gates. One type of logic gate is a NOT gate, which is a binary device that returns an output signal of opposite type to its input. As long as a device (i) takes two possible input types, (ii) yields two possible output types, and (iii) its output type is the opposite of its input type, it counts as a NOT gate. It doesn't matter whether the signals are electrical, mechanical, acoustic, or anything else, and it doesn't matter by what lower-level mechanism they are physically manipulated.

This feature of computers and their computing components—medium independence—is a stronger condition than MR. Medium independence entails MR, because any medium independent property—such as performing a computation—can be realized by different kinds of medium. But MR does not entail medium independence, because typical multiply realizable properties—such as catching mice and lifting corks out of bottlenecks—can only be realized by mechanisms that manipulate the right kind of media—i.e., mice, corks, and bottles.

Given how central MR is to debates about explanation in the special sciences—especially the science of cognition—any adequate account of neurocognitive mechanisms must address MR. And given that I will define physical computation in terms of medium independence (Chapter 6), I need clarity about medium independence. In this chapter, I provide a sound ontological basis for MR. I explain why MR needs to be scrutinized with care, introduce mechanisms, distinguish MR from mere variable realizability, analyze MR and medium independence in terms of different mechanisms for the same function, explore two different sources of MR and how they can be combined, and reiterate that both traditional reductionism and traditional anti-reductionism should be abandoned in favor of an egalitarian, integrationist framework.

2.2 Troubles with Multiple Realizability

Discussions of MR used to revolve around intuitions about certain cases. The prototypical example is the same software running on different types of hardware (Putnam 1960). Another prominent class of examples are mental states such as pain, which are supposed to be realizable in different types of creature, such as nonhuman animals, Martians, and robots (Putnam 1967a, 1967b).[4] A third class of examples includes artifacts such as engines and clocks, which are supposed to be realizable by different mechanisms (Fodor 1968a). A fourth class of examples includes biological traits, which may be thought to be realized in different ways in different species (Block and Fodor 1972). A final class of examples includes certain physical properties, such as being a metal, which are allegedly realized by different physical substances (Lycan 1981).

On a first pass it may seem that, in all these examples, the same higher-level property (being in a certain computational state, being in pain, etc.) is realizable by different lower-level properties (being in different types of hardware states, being in different physiological states, etc.). For these intuitions to prove correct, at least two conditions must be satisfied. First, the property that is putatively realizable by the realizers must be the same property, or else there wouldn't be any *one* property that is multiply realizable. Second, the putative realizers must be

[4] Putnam mentions even angels as possible realizers of pain. I assume everything is physical and leave angels aside.

relevantly different from one another, or else they wouldn't constitute *multiple* realizations of the same property.

Unfortunately, there is little consensus on what counts as realizing the same property or what counts as being relevantly different realizers of that same property (cf. Sullivan 2008 and Weiskopf 2011a, among others). To make matters worse, supporters of MR often talk about MR of functional properties without clarifying which notion of function is in play and whether the many disparate examples of putative MR are instances of the same phenomenon or different phenomena. Thus, the notion of function is often vague, and there is a tacit assumption that there is just one variety of MR. As a consequence, critics of MR have put pressure on the canonical examples of MR and the intuitions behind them.

Three observations will help motivate the present account. First, properties can be described with higher or lower resolution—in other words, properties can be described in more specific or more general ways (cf. Bechtel and Mundale 1999). If a description is sufficiently specific, the property picked out by the description might have only one possible lower-level realizer, so it may not be multiply realizable. If a description is sufficiently general, the property picked out by that description might have many lower-level realizers, but perhaps only trivially so; its multiple realizability might be an artifact of such a broad higher-level description that the property plays no useful role in scientific taxonomy or explanation.

Consider keeping time. What counts as a clock? By what mechanism does it keep time? How precise does it have to be? If we answer these questions liberally enough, almost anything counts as a clock. The "property" of keeping time might be multiply realizable, but trivially so. If we answer these questions more restrictively, however, the property we pick out might be realizable by only a specific kind of clock, or perhaps only one particular clock. Then, the property of keeping time would not be multiply realizable. Can properties be specified so that they turn out to be multiply realizable in a nontrivial way?

A second important observation is that things are similar and different in many ways, not all of which are relevant to MR. For two things to realize the same property in different ways, it is not enough that they are different in just some respect or other. The way they are different might be irrelevant. For example, the differences between the lower level realizers might be too small to amount to multiple realizations of the same higher-level property. Or the differences between the lower-level realizers might contribute nothing to the higher-level property (cf. Shapiro 2000).

The latter case is especially evident in the case of properties realized by entities made of different kinds of material. Two otherwise identical chairs may be made of different metals, woods, or plastics. Yet these two chairs may be instances of the same realizer of the chair type, because the different materials may contribute the same *relevant* properties (such as rigidity) to realization. The same point applies to the property of being a metal. There are many kinds of metal, and this has suggested to some that the property of being a metal is multiply realizable. But this is so only if the property of being a metal is realized in relevantly different ways. If it turns out that all metals are such in virtue of the same realizing properties, then, despite appearances, the property of being a metal is not multiply realizable after all. An easier case along the same lines is differently colored objects: two cups made of the same material in the same organization but differing only in their color do not count as different realizations of a cup. More generally, given two things A and B, which are different in some respects but realize the same property P, it does *not* follow that A and B are *different* realizations of P, and that P is therefore multiply realizable.

A third observation is that the realization relation itself may be construed more strictly or loosely. A heavily discussed example is that of computer programs. If all it takes to realize a computer program is some mapping from the states of the program to the microphysical states of a putative realizer, then most programs are realized by most physical systems (Putnam 1988). This unlimited realizability result is stronger than MR: it entails MR at the cost of trivializing it, because MR is now a consequence of an intuitively unattractive result. Is there a way of constraining the realization relation so that MR comes out true and nontrivial?[5]

The easy way out of this conundrum is to deny MR. According to MR skeptics, a higher-level property can be related to their realizers in one of two ways. On one hand, a higher-level property can be realized by entities with the same relevant lower-level property, as in chairs made of different materials all of which have the right degree of rigidity. These do not count as multiple realizations of the higher-level property because the differences between the realizers are irrelevant. On the other hand, a higher-level property can be realized by entities with different relevant lower-level properties, as in corkscrews that operate by different causal mechanisms. But then, the different causal mechanisms are the genuine kinds, whereas the putative property that they realize is eliminated in favor of those different kinds. For example, one should eliminate the general kind *corkscrew* in favor of, say, the more specific kinds *winged corkscrew* and *waiter's corkscrew* (cf. Kim 1992, 1998, 2005; Shapiro 2000).

Rejecting MR is tempting but premature. To obtain an adequate account, we must specify three things: the properties to be realized, the properties to serve as realizers, and the realization relation. Once we have that, we can see whether any properties are multiply realized. Before we can have that, we need to say more about mechanisms.

[5] I refute unlimited pancomputationalism—the thesis that most programs are realized by most physical systems—in Piccinini 2015, chap. 4.

2.3 Systems and Mechanisms

There are many ways to carve a portion of reality. We can divide it into top half and bottom half, left half and right half, front half and back half, and so forth. Ditto for thirds, quarters, fifths, etc. Ditto for equal parts of 1 cm^3, 1 mm^3, etc. Most of these partitions are of no practical or scientific interest (cf. Haugeland 1998).

Among the many ways that reality can be carved into objects and properties, and objects can be carved into parts, some partitions carve reality at its joints, which makes such partitions explanatory. The notion of MR makes the most sense when we consider not just *any* partition of reality, but partitions identified by the special sciences for their explanatory value. Special scientists observe natural phenomena. They use phenomena to classify portions of reality into systems that partake in such phenomena. Within those systems, they identify mechanisms that explain those phenomena (cf. Glennan 2017). Thus, the special sciences partition reality into systems and mechanisms.

Systems and mechanisms have properties, whose manifestations are the mechanisms' activities (Piccinini 2017b). These are the natural properties I introduced in Chapter 1, Section 1.8. A natural property is a way that an object is: a quality, a disposition, or a combination of both. Either way, a natural property is associated with some causal powers. If a property is a disposition, it *just is* a plurality of causal powers. If a property is a quality, it is the categorical basis for some causal powers.

From now on, I'll focus on systems, mechanisms, and their properties. Systems and mechanisms in the present sense are not arbitrary partitions of reality; they are partitions that explain phenomena because of their natural properties.

A system can partake in many phenomena. For example, a human being can partake in cognition, locomotion, digestion, respiration, reproduction, and so forth. Each phenomenon is explained by its own mechanism (Glennan 2002), which is a part of the whole system. Systems are classified based on which phenomena they partake in as well as their intrinsic similarities. For example, systems may be classified as multicellular organisms, cells, molecules, atoms, and so forth.

Systems that cover a portion of reality without gaps and without overlaps constitute what in Chapter 1 I called a level of being. But systems are not arbitrary partitions; they are partitions based on objective similarities and differences between the parts. And systems are *organized* in the sense that they are made of subsystems that stand in specific relations to one another and have, in turn, objective similarities and differences between them. Because of this, partitions of reality in terms of systems are *levels of organization*—ways that reality is objectively organized into systems and subsystems.[6]

[6] By defining levels of organization in terms of similarities and differences between systems, and mechanisms as parts of systems that explain properties of systems, I am bypassing the debate about

Whole systems and mechanisms have higher-level properties, which give rise to their causal powers. A mechanism's higher-level properties, and the causal powers they give rise to, can be constitutively explained by its components and their properties, which include the way the components are organized. This is another way to see that mechanistic partitions are not arbitrary: they involve active components with specific properties that give rise to the properties of the whole mechanism. It's also another way of seeing that the properties of the whole mechanism are *aspects* of the components' properties. For one thing, the components are many, each with its own properties, while the whole mechanism is one. For another thing, some components of a mechanism can be added or lost, and their properties can change to a degree, while the whole and some of its properties remain the same. The whole and its properties are invariants under such transformations in the components and their properties.

This point iterates through as many levels of organization as there are. Going downward from a mechanism to its components, mechanistic components can be further decomposed and their properties constitutively explained in terms of *their* components. Going upward from a mechanism to its containing system, a mechanism's properties contribute to constitutively explaining the properties of the larger mechanisms and systems the mechanism is part of.

This iterative structure of mechanisms within mechanisms and systems within systems is often described as a *hierarchy*; this label may be ok so long as it's merely a hierarchy of descriptions, not of levels of being. In other words, it may be ok to call it a hierarchy as long as we don't imply that some mechanistic levels are ontologically more fundamental than others. But the ontological implication is probably hard to avoid. Since I argued in Chapter 1 that there is no ontological hierarchy, I will avoid hierarchy talk. That's not to say that mechanisms are mere descriptions—it's to say that all levels of mechanistic organization are equally real.

A mechanism's causal powers may be specified more finely or more coarsely. At one extreme, where mechanisms are described in a maximally specific way, no two mechanisms possess the exact same causal powers—no two particular mechanisms are functionally equivalent. At the other extreme, where mechanisms are described in a maximally general way, everything is functionally equivalent to everything else. Special sciences find some useful middle ground, where causal powers are specified in such a way that some mechanisms (not all) are functionally equivalent—that is, they have causal powers of the same type. The grain that is

when two components of a mechanism are at the same or different mechanistic levels (Craver 2007, sect. 3.3). An explicit exploration of the relation between levels of organization in the present sense and levels of mechanisms in Craver's sense goes beyond the scope of this chapter and will have to wait for another occasion. At any rate, Fazeka and Kertesz (2019) helpfully argue that higher mechanistic levels possess *no* novel causal powers (contra strong synchronic emergence). Their conclusion is in line with the framework defended here. In what follows, I will use "levels of organization" and "levels of mechanism" interchangeably.

most relevant to a certain level of organization is the type of causal power that explains the phenomena in question—e.g., how much blood-pumping is enough to keep an organism alive under normal conditions, as opposed to the exact amount of pumping that is performed by a given heart (cf. Couch 2009b).

In summary, a system's properties are explained mechanistically in terms of some of the system's components and their properties, including their organization. The explaining mechanism is the set of components and properties that produce the properties in question. The same explanatory strategy iterates for the properties of the components. The relation between a mechanism's (higher-level) properties and its components' (lower-level) properties is realization.

2.4 Realization

Claims of realization, whether singular or multiple, presuppose an appropriate specification of the property to be realized and a property (or plurality of properties) doing the realizing. If the kind *corkscrew* is defined as a device with a part that screws into corks and pulls them out, then winged corkscrews and waiter's corkscrews count as different realizations—because they employ different lifting mechanisms. If it is defined more generally as something with a part that only *pulls* corks out of bottlenecks, then two-pronged "corkscrews"—which have no screw at all, but instead have two blades that slide on opposite sides of the cork—also count as a realizer of the kind. If it is defined even more generally as something that simply *takes* corks out of bottlenecks, then air pump "corkscrews" count as yet another realizer of the kind. Whether something counts as a realization of a property depends in part on whether the property is defined more generally or more specifically.

Someone might object that this seems to introduce a bothersome element of observer-relativity to functional descriptions. By contrast, the objector says, the identification "water = H_2O" does not seem to have the same kind of observer-relativity. But this objection is confused. Functional descriptions are neither more nor less observer-dependent than descriptions of substances. Whether more fine- or coarse-grained, if true they are objectively true. It is both true of (some) corkscrews that they pull corks out of bottlenecks and that they pull corks out of bottlenecks by having one of their parts screwed into the cork: there is no observer-relativity to either of these facts.

In fact, nonfunctional descriptions can also be specified with higher or lower resolution. You can give more or fewer decimals in a measurement, require particular isotopes when referring to a chemical element, or get more or less specific about the impurities present in a substance such as water. None of this impugns the observer-independence of the descriptions.

MR depends on mechanisms. If the same capacity of two systems is explained by two relevantly different mechanisms, the two systems count as different realizations. As we've seen, critics of MR object that if there are different causal mechanisms, then the different properties of those causal mechanisms, and not the putatively realized property, are the only real properties at work. According to them, there is nothing more to a system possessing a higher-level property beyond possessing its lower-level realizer (Kim 1992; Shapiro 2000).

My answer is that there is something *less*, not more, to a system's possessing a higher-level property. The worry that higher-level properties are redundant disappears when we realize that higher-level properties are *subtractions* of being from, as opposed to *additions* of being to, lower-level properties. MR is simply a relation that can obtain when there are relevantly different kinds of lower-level properties that realize the same higher-level property.

The objection that the realizing properties, rather than the realized properties, are doing all the work depends on a hierarchical ontology in which parts ground wholes and, therefore, the properties of parts ground the properties of wholes. This hierarchy may be reversed in favor of the view that wholes ground parts (e.g., Schaffer 2010) and, therefore, the properties of wholes ground the properties of parts. In Chapter 1, I rejected hierarchical ontologies in favor of a neglected third option—an egalitarian ontology. According to ontological egalitarianism, neither parts and their properties nor wholes and their properties are prior to one another.

This egalitarian framework allows me to cut through the debate about realization. According to the flat view (Polger 2007; Polger and Shapiro 2008; Shoemaker 2007, 12), realization is a relation between two different properties of a whole in which a proper subset of the realizing properties of that whole constitute the causal powers individuative of the realized property. By contrast, according to the dimensioned view (Gillett 2003, 594), realization is a relation between a property of a whole and a distinct set of properties possessed (either by the whole or) by its parts, such that the causal powers individuative of the realized property are possessed by the whole "in virtue of" (either it or) its parts possessing the causal powers individuative of the realizing properties.

There is something appealing about both the flat and the dimensioned views of realization. The flat view makes MR nontrivial and provides a clear explanation of the relation between the realized and realizer properties: the former's powers are a proper subset of the latter's powers. This accounts for the intuitive idea that realizations of, say, corkscrews are themselves corkscrews, even though corkscrews have properties irrelevant to their being corkscrews (e.g., their color). At the same time, the dimensioned view connects different mechanistic levels and thus fits well with multilevel mechanistic explanation, the prevailing form of explanation about the phenomena that motivates talk of realization and MR in the first place. The firing of a single neuron is realized (in part) by ions flowing through ion channels; the properties of the ions and the channels they move through are not properties of the neuron as a whole, although they are constitutive of a property of the whole neuron.

There is also something unattractive about both the flat and the dimensioned views. The flat view makes it sound like realized properties are superfluous, since it assumes that realizers are more fundamental than realized properties and yet realizers are enough to do all the causal work. We might as well eliminate the realized property. On the other hand, the dimensioned view lacks a clear account of the relation between realized properties and realizers. Realized properties are said to have their causal powers "in virtue of" the causal powers of their realizers, even though such causal powers are allegedly distinct from those of the realizers. What does "in virtue of" means, and where do the distinct causal powers come from? This is the kind of mysterious anti-reductionism that I rejected in Chapter 1.

Another problem is that the theory of MR that accompanies the dimensioned view suggests that (almost) *any* difference in a property's realizers generates MR (Gillett 2003; Aizawa and Gillett 2009; Aizawa 2017). For example, as Gillett (2003, 600) argues, the dimensioned view entails that two corkscrews whose only difference is being made of aluminum versus steel count as different realizations, even though, as Shapiro (2000) notes, the aluminum and the steel make the same causal contribution to lifting corks out of bottlenecks. This makes MR too easy to come by. For MR was supposed to refute reductionism. Being made of different materials that make the same causal contribution is consistent with reductionism—according to reductionism, the causal power shared by the materials is what the realized property reduces to.

In other words, if two realizers realize the same higher-level property in the same way—by making the same causal contribution—then this does not refute reductionism. If MR is going to refute reductionism, the multiple realizers must realize the same higher-level property *in different ways*—that is, by making different causal contributions. I will soon argue that these different ways can be cashed out in terms of different mechanisms. If there are different causal mechanisms for a higher-level property, then this is genuine MR—the kind of MR that has a chance of refuting reductionism.

On both the flat and dimensioned views, there are worries about causal or explanatory exclusion, whereby causes or explanations at a lower level render causes or explanations at a higher level superfluous (cf. Chapter 1). On the flat view, what is left to cause or explain if realizing properties provide causal explanations of interest? And on the dimensioned view, what could the properties of wholes cause or explain if the realizing properties of their parts constitute the mechanism of interest?

Egalitarian ontology accommodates what is appealing about both the flat and dimensioned views without inheriting their unattractive features:

> Property P of system S is *realized* by properties **Q** if and only if **Q** belongs to S's components and P is an aspect of **Q**.

According to this egalitarian account of realization, realization is a relation between a property of a whole and the properties of its component parts (per dimensioned view). The relational properties of the components are what give them organization; hence, realization is a relation between a property of a whole and the properties of its component parts *in an organization*. The realized property is nothing but an aspect of the properties possessed by the organized parts (per flat view, modulo the appeal to parts and their organization).

The relation between realized property and realizing properties is clear: it's the *aspect of* relation (cf. the flat view). The account fits like a glove the kind of mechanistic explanation that motivated this dialectic in the first place (as per the dimensioned view). Whether there is MR remains a nontrivial matter, because it depends on whether different mechanisms generate the higher-level properties in relevantly different ways (more on this below). And yet realized properties are not superfluous, because they are not posterior to (nor are they prior to) the properties of their parts. They are simply aspects of them (Figure 2.2).

This egalitarian account departs from Craver's (2007, 212ff.). Craver distinguishes between levels of realization and levels of mechanistic organization. Craver holds that realization is a relation between two properties of one and the same whole system, not to be confused with the relation that holds between levels of mechanistic organization. (According to Craver, as according to me, levels of mechanistic organization are systems of components, their capacities, and their organizational relations, and they are related compositionally to other levels of mechanistic organization.) I reject Craver's account of realization; I hold that each

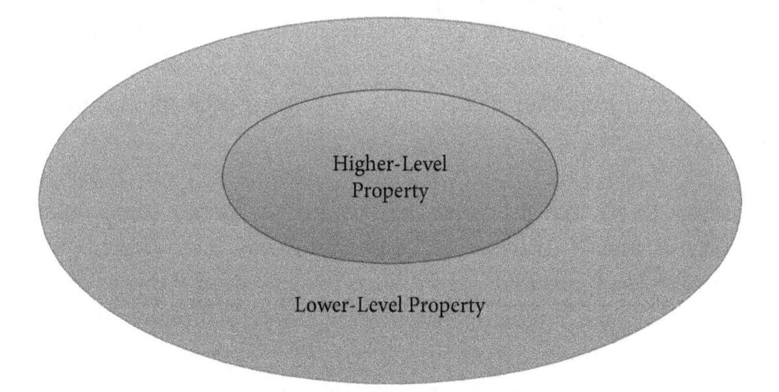

Figure 2.2 Higher-level properties are aspects of their lower-level realizers.

level of mechanistic organization realizes the mechanistic level above it and is realized by the mechanistic level below it. Realization, in its most useful sense, is precisely the relation that obtains between two adjacent mechanistic levels in a multi-level mechanism and is thus a compositional relation.

Clarification: since a whole can also be considered together with its structural properties, which include its components, then a formulation analogous to the flat view can also be subsumed under the present framework. That is, we can say that property P belonging to system S *considered in abstraction from its parts* is realized by properties Q if and only if Q belongs to system S considered in virtue of its structural properties and P is an aspect of Q. But the point of looking at lower-level properties is to find mechanisms, which means decomposing the system. So, the flat view is recoverable only insofar as it incorporates a mechanistic framework to the point of becoming a notational variant of the present view.

A proponent of the dimensioned view of realization might object to my construing realization in terms of aspects. According to this objection, entities at different levels of organization are qualitatively *distinct*—they have distinct kinds of properties and relations that contribute different powers to them (Gillett 2002, 2010). Carbon atoms do not scratch glass, though diamonds do; corkscrew handles do not lift corks, though whole corkscrews do; etc. If the higher-level mechanisms are distinct from the lower-level ones, then their higher-level properties are not just an aspect of the lower-level ones—rather, they are distinct properties.

This objection is a non sequitur. Sure, an individual carbon atom taken in isolation cannot scratch. Nor can a bunch of individual carbon atoms taken separately from one another. But large enough arrays of carbon atoms held together by appropriate covalent bonds into an appropriate crystalline structure do scratch: on the aspect view, under appropriate conditions, one aspect of the properties of such an organized structure of carbon atoms *just is* the property that enables a diamond to scratch. This organized plurality of atoms does many other things besides scratching, including, say, maintaining the bonds between the individual atoms, reflecting and refracting electromagnetic radiation, and so on. That's why, when we consider the whole in abstraction from the parts (as we normally do), the properties of the whole are one aspect among others of the properties of the parts.

By the same token, a single corkscrew lever, taken in isolation, cannot lift corks. But corkscrew levers that are attached to other corkscrew components in an appropriately organized structure, in cooperation with those other components, do lift corks: under appropriate conditions, one aspect of the properties of that organized structure *just is* the property that enables corkscrews to lift corks out of bottles. A whole's parts and their properties, when appropriately organized, do what the whole and its properties do, and they do much more besides. Hence, what the whole and its properties do is an aspect of what the components and their properties do (when appropriately organized and taken collectively).

Here a proponent of the dimensioned view might reply that I make it sound like there are (i) parts with their properties, (ii) wholes with their properties, and then (iii) a further object/property hybrid, parts in an organization. But why believe in (iii)? An ontology that includes (iii) is clearly profligate and therefore should be rejected (cf. Gillett 2010).

This objection leads us back to the metaphysics of composition. As I pointed out in Chapter 1, a whole can be considered in two ways: as consisting of all its parts organized together, or in abstraction from its parts. The parts of a whole may change over time. Nevertheless, when a whole is considered as consisting of all of its parts organized together, at any time instant the whole is identical to its organized parts. Therefore, at any time instant, the organized parts are nothing over and above the whole, and the whole is nothing over and above the organized parts. The organized parts are no addition of being over the whole, and vice versa. Since the whole and its organized parts are the same thing, an ontology that includes organized parts is no more profligate than an ontology that includes wholes.

But a whole can also be considered in abstraction from its organized parts, such that the whole remains "the same" through some additions, subtractions, and rearrangement of its parts. When a whole is considered in abstraction from its organized parts, the whole is an invariant under limited addition, subtraction, or rearrangement of its (organized) parts. That is, a whole is that aspect of its organized parts that remains constant when a part is lost, added, etc., within limits. Thus, even when a whole is considered in abstraction from its parts, a whole is nothing over and above its organized parts. Rather, a whole is *less* than its organized parts—a *subtraction of being* from its organized parts. Since wholes considered in abstraction from their organized parts are less than their organized parts, there is nothing profligate about positing wholes as well as organized parts.

Take the example of a winged corkscrew. The property of *lifting corks out of bottlenecks* is a property of the whole object, and it is realized by the parts (the worm, the lever arms, rack, pinions, etc.) in a particular organization (the rack connected to the worm, the pinions connected to the lever arms, etc.). Those parts in that organization *lift corks out of bottlenecks*, and because *lifting corks out of bottlenecks* is not, in any sense, a property over and above what those parts in that organization do, we can also say that those parts in that organization are a realization *of* a corkscrew. In this ontologically egalitarian account, the existence of higher-level properties does not entail that they are over and above their realizers. They are aspects of their realizers—that is, parts of their realizers—that are worth singling out and focusing on. This notion of higher-level property, and the related notion of MR, is useful for several reasons.

First, higher-level properties allow us to *individuate a phenomenon of interest*, such as removing corks from bottles, which might be difficult or impossible to individuate on the basis of lower-level properties of corkscrews. Shapiro points out that in the case of corkscrews made of different metals, rigidity screens off composition (Shapiro 2000). True enough. And for something to be a functional corkscrew, it's not sufficient that it be rigid. It must be hard enough to penetrate cork but not so brittle that it will break when the cork is lifted. Many different substances can be used to build corkscrews, but their only insightful, predictive, explanatory, non-wildly-disjunctive specification is that they must lift corks out of bottlenecks (or do so in such-and-such a way).

Second, this notion of a higher-level property allows for the *explanation of the phenomenon in question in terms of a relevant property*, e.g., we explain the removal of corks from bottlenecks in terms of corkscrews' power to remove corks from bottlenecks, rather than any of the other properties of corkscrews. And they explain what systems with that property can do when organized with other systems (i.e., in a higher-level context). Higher-level properties have nontrivial consequences, e.g., about what a system can and cannot do. To take just one example, results from computability theory specify precise limits to what can and cannot be computed by those things that realize various kinds of automata (and Turing's original universality and uncomputability results themselves were the foundations of the field). Two things that realize the same finite state automaton—even if they realize that automaton in very different ways—will have the same computational limits in virtue of their sharing that higher-level property.

Third, this notion of a higher-level property supports an *informative taxonomy* of systems that differ in their lower-level properties. What these systems have in common, which is not immediately revealed by listing their lower-level properties, is a higher-level property. In other words, although the lower-level properties that realize the higher-level property in these different systems are different, they also have something in common, and what they have in common is the aspect of the lower-level properties that we call the higher-level property. In other words, different lower-level properties realize the same higher-level property when they share aspects of the same type. The same type of aspect (higher-level property) may be embedded in different wholes (lower-level properties), as illustrated in Figure 2.1.

Finally, a higher-level property calls for its *explanation to be provided in terms of appropriate combinations of lower-level properties* (i.e., by mechanistic explanation). The dual role of higher-level properties as explanantia of higher-level phenomena as well as explananda in terms of lower-level properties adds to our understanding of a system.

2.5 Variable Realizability

Lower-level properties may realize a lot of higher-level ones, none of which are identical to the lower-level properties that realize them. And different lower-level properties may realize the same higher-level property without being identical to it. For instance, storing a '1' (as opposed to a '0') within a computer circuit is a higher-level property of a memory cell that may be realized by a large number of voltages (all of which must fall within a narrow range; e.g., 4 ± 1 volts). Each particular voltage, in turn, may be realized by an enormous variety of charge distributions within a capacitor. Each particular distribution of charges that corresponds to a '1' is a very determinate property: we obtain a particular voltage by abstracting away from the details of the charge distribution, and we obtain a '1' by abstracting away from the particular value of the voltage. Thus, a higher-level property is an aspect of a lower-level property.

That's not yet to say that different charge distributions amount to *multiple realizations* of a given voltage (in the sense that refutes reductionism), or that different voltages within the relevant range amount to *multiple realizations* of a '1'. Invariance of a higher-level property under transformations in lower level realizers is more general than multiple realizability, because—as we will see presently—it does not depend on different causal mechanisms. To distinguish it from MR, I call it *variable realizability* (Figure 2.3).[7]

Variable realizability is the ubiquitous relation that obtains when (the value of) a higher-level property is indifferent to certain lower level details. A straightforward case is aggregates of lower-level properties (cf. Wimsatt

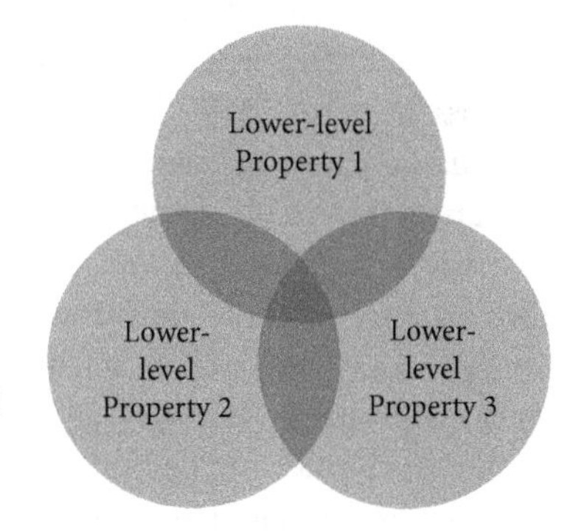

Figure 2.3 The intersections of lower-level properties are variably realizable higher-level properties.

[7] I borrow 'variable realizability' from Zangwill (1992) but I depart from his use; Zangwill uses it as synonymous with 'MR'. As Aizawa (2017) points out, the terminology in this area is somewhat unfortunate, because the term 'multiple realizability' does not sound any more restrictive than 'variable

1997). For example, consider an object of mass 1 kg. It may be composed of two parts ½ kg each, three parts 1/3 kg each, one part of 1/3 kg and one part of 2/3 kg, and so on. There are indefinitely many combinations of masses that sum up to 1 kg. The higher-level property—1 kg—is invariant under any transformation in the masses of the parts as long as their sum remains the same. This is not MR but merely variable realizability.

Aggregating properties by operations more complex than addition gives rise to more complex cases. For example, in entities similar enough to ideal gases, temperature (a higher-level property) equals average molecular kinetic energy, which is a kind of aggregate of lower-level properties. Specifically, such an aggregate is formed by averaging the kinetic energy of the molecules making up the gas. The specific values of the lower-level properties—the specific kinetic energies of the specific molecules—don't matter as long as their aggregate remains the same; the higher-level property remains the same under myriad combinations of specific lower-level properties. This is not yet MR but merely variable realizability.

An even more complex case involves universality in a sense used in physics—not to be confused with computational universality in Alan Turing's sense. As Batterman (2000) explains, universality occurs when a few parameters characterize the behavior of a wide range of systems under certain conditions, regardless of the details of the interactions among the systems' parts. For example, the same few parameters characterize the thermodynamics of both fluids and magnets near phase transitions, regardless of what they are made of and how their parts interact. Universality in this sense can be explained by techniques such as the renormalization group, by which many of a system's degree of freedom are systematically shown to be irrelevant to the behavior in question.

Batterman (2000) suggests that universality is a case of MR. On the contrary, given how the term "MR" is historically used, "MR" should be reserved for something else (to be discussed shortly). Instead, universality is a species of the much more pervasive phenomenon of *variable* realizability. Although universality is not mere aggregation—that's why explaining universality requires specialized mathematical techniques—universality is like aggregation of lower-level properties in that many lower-level details are simply irrelevant to the value of the higher-level property. For any given higher-level property, there are myriad lower-level properties of the system whose changes make no difference to the higher-level property, and any system with certain characteristics will exhibit the same higher-level property. Once again, this illustrates that higher-level properties are an *aspect* of their realizers, an aspect that remains invariant under many changes in realizers.

realizability'. As I explain in the main text, I use MR more restrictively because MR is supposed to refute reductionism, but what I call variable realizability is consistent with reductionism.

MR is more than mere lower-level differences—more than mere variable realizability. As I will soon argue, multiple realizability of a property occurs when there are differences *in the types of mechanism* for that property. In other words, MR is the special case of variable realizability that requires different types of causal mechanism for the same higher-level property.

Comparing MR to variable realizability removes much of the mystery that surrounds it. Some argue that MR is a "metaphysical mystery" (Fodor 1997, 159). Perhaps this putative mystery is why so many others argue against MR. But once we see that variable realizability is the ubiquitous embedding of an aspect (a higher-level property) within different realizers (many lower-level properties) and MR is the special case in which there are different types of mechanism for the same higher-level property, there is no longer any mystery about it or reason to deny it.

2.6 Sources of Multiple Realizability

MR has two sources. Each source is sufficient to give rise to MR, but the two sources can also combine to form a composite form of MR.

2.6.1 Multiple Realizability$_1$: Multiple Organizations

The first kind of MR results when the same component types exhibit the same capacities, but the organization of those components differs. Suppose that a certain plurality of components, S_1, when organized in a certain way, O_1, form a whole that exhibits a certain capacity. It may be that those same components (S_1) can be organized in different ways (O_2, O_3, . . .) and still form wholes that exhibit the same capacity. If so, then the property is multiply realizable.

A simple example is what you can do with a round tabletop and three straight bars of equal length. If the three bars are arranged so as to support the tabletop in three different spots far enough from the tabletop's center, the result is a table. Alternatively, two of the legs may be arranged to form a cross, and the remaining leg may be used to connect the center of the cross to the center of the tabletop. The result of this different organization is still a table.

For this proposal to have bite, I need to say something about when two organizations of the same components are different. Relevant differences include spatial, temporal, operational, and causal differences. Spatial differences are differences in the way the components are spatially arranged. Temporal differences are differences in the way the components' operations are sequenced. Operational differences are differences in the operations needed to exhibit a capacity. Finally, causal differences are differences in the components' causal powers that contribute to the capacity and the way such causal powers affect one another. Two organizations are relevantly different just in case they include some combination of the following: components spatially arranged in different ways, performing different operations or the same operations in different orders, such that the causal powers that contribute to the capacity or the way the causal powers affect one another are different.

MR_1 is ubiquitous in computer science. Consider two programs (running on the same computer) for multiplying very large integers, stored as arrays of bits. The first program uses a simple algorithm such as what children learn in school, and the second uses a more sophisticated (and faster) algorithm, such as the Fast Fourier Transform. These programs compute the same function (i.e., they multiply two integers) using the same hardware components (memory registers, processor, etc.). But the temporal organization of the components mandated by the two programs differs considerably: many children could understand the first, but understanding the second requires nontrivial mathematical training. Thus, the processes generated by the two programs counts as two different realizations of the operation of multiplication.

The notion of MR_1 allows us to make one of Shapiro's conclusions more precise. Shapiro (2000) is right that components made of different materials (e.g., aluminum versus steel) need not count as different realizations of the same property (e.g., lifting corks out of bottlenecks) because they contribute the same property (e.g., rigidity) that effectively screens off the difference in materials. But this is only true if the different groups of components are organized in the same way. If two groups of components that are made of different materials (or even the same material) give rise to the same functional property by contributing the *same* properties through *different* functional organizations, those are *multiple* realizations of the same property.

2.6.2 Multiple Realizability$_2$: Multiple Component Types

The second kind of MR results when components of different types are organized in the same way to exhibit the same capacity. By components of different types, I mean components with different capacities or powers. Again, suppose that a certain plurality of components, S_1, when organized in a certain way, O_1, form a whole that exhibits a certain property. It may be that a plurality of different components, S_2, can be organized in the same way, O_1, and yet still form a whole that exhibits the same property.

As in the case of different organizations mentioned above, I need to say something about when two components belong to different types. Two components are different in type just in case they contribute different causal powers to the performance of the whole's capacity. Here is a simple test for when two components contribute the same or different causal powers: to a first approximation, if two components of similar size can be substituted for one another in their respective systems without losing the capacity in either system, then these components contribute the same causal powers. If not, then they contribute different causal powers.[8]

For example, consider two pendulum clocks with the same organization, where one has metal gears while the other has wooden gears. We could take a gear from the wooden clock and its counterpart from the metal one and switch them. Assuming a few insignificant details (their sizes are the same, the number of teeth in each gear is the same, etc.), both clocks would work as before, at least for a while, and thus these gears contribute the same causal powers. But of course this would not work if we were to swap a quartz crystal from a digital clock with the pendulum of a cuckoo clock: these components contribute different causal powers. Neither of these is a case of MR_2. The pair of pendulum clocks is not a case of MR at all, whereas the pendulum and quartz pair is a case of MR_3 (see below).

An example of MR_2 is a system of signs for communicating messages, such as Morse code. A longer and a shorter sound may be used as Morse signaling units, and so may a louder and a quieter sound, or sounds of different pitches. Other physical media can be used as well, such as various kinds of electromagnetic radiation, or a sequence of rigid rods of two distinct lengths. In each of these cases, the components that realize the system are such that the recombination technique mentioned above would not work.

Computer science once again provides examples of MR_2. Consider computer organization, which is concerned with designing processors (and processor components) to implement a given instruction-set architecture (the low-level, basic instructions a processor is capable of performing). This is all done at the level of digital logic design, in which the most basic (or atomic) components of the design are individual logic gates. Logic gates can be realized in different ways, meaning that a particular computer organization is not specific to how the components are realized. For example, a logic gate might be realized by a silicon circuit in one computer, a gallium-arsenide circuit in another, relays in a third, mechanical gears in a fourth, and so forth. In each of these cases, our test for differences in causal powers is passed (or failed, as it were): replacing a mechanical gear with a silicon circuit will not result in a working computer.

A biological case of MR_2 is the phenomenon of circadian rhythms. A diverse set of organisms exhibit circadian rhythms, and the organization of the systems

[8] What about components of different size, such as two levers from two winged corkscrews, one of which is twice the size of the other? In this case, we may have to adjust the scale of the components before recombining them. See Wimsatt (2002) for a detailed discussion of functional equivalence, isomorphism, and similarity.

responsible for generating the oscillations constitutive of circadian rhythms is the same. For a large class of organisms, circadian rhythms are driven by a transcriptional/translational feedback loop (TTFL). As described by Dunlap (1999, 273), "circadian oscillators use loops that close within cells. . . . and that rely on positive and negative elements in oscillators in which transcription of clock genes yields clock proteins (negative elements) which act in some way to block the action of positive element(s) whose role is to activate the clock gene(s)." Some organisms violate this generalization, but the generalization remains true of many kinds of organisms, including plants, animals, and fungi (see Buhr and Takahashi 2013 for a review). Thus, we have the same organization, even though the clock proteins and clock genes are different in different species. As before, our causal-difference test yields the correct result: replacing a gene in the circadian rhythm mechanism from one species with an analogous gene from a different species will typically result in an organism without a properly functioning circadian rhythm, even though the two genes play the same role in their respective organisms.

2.6.3 Multiple Realizability$_3$: Multiple Component Types in Multiple Organizations

A third kind of MR combines MR_1 and MR_2: different component types with different organizations. As before, suppose that a certain plurality of components, S_1, when organized in a certain way, O_1, form a whole that exhibits a certain property. It may be that a plurality of different components, S_2, can be organized in another way, O_2, different from O_1, and yet still form a whole that exhibits the same property.

MR_3 is probably the typical case of MR, and the one that applies to most of the standard examples described by philosophers. One much-discussed example that we've already encountered is the corkscrew. Proponents of both the flat and dimensioned views usually agree that a waiter's corkscrew and a winged corkscrew are multiple realizations of the kind *corkscrew*. The components of the two corkscrews are different, as is the organization of those components: a waiter's corkscrew has a folding piece of metal that serves as a fulcrum, which is sometimes hinged, and often doubles as a bottle opener; a winged corkscrew has a rack and pinion connecting its levers to the shaft of the screw (or worm). So, both the components differ and their organizations differ.

Eyes—another much-discussed case—follow a similar pattern. Several authors have noted that there are many different ways in which components can be organized to form eyes (e.g., Shapiro 2000), and many different types of components that can be so organized (e.g., Aizawa and Gillett 2011). The same pattern can be found in many other examples, such as engines, mousetraps, rifles, and so on. I'll mention just one more.

Computer science provides examples of systems that exhibit MR_3. In some computers, the logic circuits are all realized using only NAND gates, which, because they are functionally complete, can implement all other gates. In other computers, the logic circuits are realized using only NOR gates. In yet other computers, AND, OR and NOT gates might realize the logic circuits. The same computer design can be realized by different technologies, which in turn can be organized to perform the same computations in different ways. Computers are anomalous because their satisfaction of both MR_1 and MR_2 are independent of one another (more on this below), whereas in most cases, the two are mixed together inextricably because the different pluralities of components that form two different realizations can exhibit the same capacity only by being organized in different ways.

2.6.4 An Egalitarian Account of Multiple Realizability

What I've said so far is enough to explain when two realizations of a higher-level property are different enough to constitute MR. Let $\mathbf{P_1}$ be the properties and $\mathbf{O_1}$ the organizational relations that realize property P in system S_1 and let $\mathbf{P_2}$ be the properties and $\mathbf{O_2}$ the organizational relations that realize property P in system S_2. That is to say, $\mathbf{P_1}(S_1)\&\mathbf{O_1}(S_1)$ mechanistically explains P in S_1, whereas $\mathbf{P_2}(S_2)\&\mathbf{O_2}$ (S_2) mechanistically explains P in S_2. Then, **P1+O1** and **P2+O2** are relevantly different realizations of P when and only when:

1. $\mathbf{P_1}+\mathbf{O_1}$ and $\mathbf{P_2}+\mathbf{O_2}$ are at the mechanistic level immediately below S
2. One of the following is satisfied:
   a. $[MR_1]$ $\mathbf{P_1} = \mathbf{P_2}$ but $\mathbf{O_1} \neq \mathbf{O_2}$
   b. $[MR_2]$ $\mathbf{P_1} \neq \mathbf{P_2}$ but $\mathbf{O_1} = \mathbf{O_2}$
   c. $[MR_3]$ $\mathbf{P_1} \neq \mathbf{P_2}$ and $\mathbf{O_1} \neq \mathbf{O_2}$.

In other words, a property is multiply realizable just in case there are at least two different *types* of causal mechanism that can realize it at the immediately lower mechanistic level (Figure 2.1).

The restriction to the immediately lower mechanistic level is needed to avoid trivialization. Consider a property P whose only possible immediately lower-level (−1) mechanism is Q. Suppose that Q is multiply realizable by different lower-level (−2) mechanisms $R_1, \ldots R_n$. If we don't restrict MR to the immediately lower mechanistic level (-1), by transitivity of realization P will be multiply realizable by $R_1, \ldots R_n$. This is not what we want. In other words, MR is a relation between a property and the *kind* of realizers it has at the immediately lower mechanistic level. Realizers at levels below that shouldn't count.

To find out whether a property is multiply realizable, proceed as follows: fix the higher-level property in question by identifying a relevant set of causal powers. Find the immediately lower-level mechanisms realizing different instances of the property. Figure out whether the lower-level mechanisms have either components or organizations of different types (or both). If, and only if, they do, you have a case of MR.

This account of MR bears similarities to Aizawa and Gillett's (2009, 2011) account but with two crucial differences. First, I rely on the ontologically egalitarian account of realization I defended above rather than Aizawa and Gillett's dimensioned view. Second, not all lower-level differences count as cases of MR, even when such differences occur between realizer properties at the same level; the lower-level differences must amount to differences either in the component types or in the way the components are organized (or both) at the mechanistic level immediately below T. Thus, I rule out many overly easy cases of putative MR that Aizawa and Gillett accept.

For example, corkscrews of different colors are not a case of MR because color is mechanistically irrelevant to lifting corks. Here Aizawa and Gillett agree. But contra Aizawa and Gillett, on the present account a steel and an aluminum corkscrew are not multiple realizations of the kind corkscrew because steel versus aluminum composition is not the mechanistic level immediately below lifting corks; i.e., being made of one metal or another is not a relevant aspect of the level that explains how the corks are lifted. But rigidity is a relevant aspect of that level, and many different kinds of material are rigid. Here Shapiro is right: rigidity screens off composition.

Equally important is that changes in the number of components, their rate of functioning, or their arrangement (while preserving component and organization type) may not amount to MR. Consider a three-legged stool versus a four-legged stool versus a five-legged stool... versus an n-legged stool. Are these cases of MR? There are differences between stools with a different number of legs. But they do not require significantly different mechanistic explanations, because they all rely on the same types of component organized in (roughly) the same way: an n-legged stool has approximately $1/n$th of its weight supported by each leg, supposing the legs are distributed equidistantly on the outside edge of the seat. Thus, this is not a case of MR, and the present account of MR accommodates this fact.

By the same token, so-called MR by compensatory differences (Aizawa 2013) is not MR properly so called. In cases of compensatory differences, the same higher-level property is realized by different combinations of lower-level properties whereby quantitative changes in one lower-level property can be compensated by corresponding quantitative changes in another lower-level property. For example, suppose that applying a certain amount of force to a lever of a certain length is enough to lift a certain weight. You can lift the same weight by applying less force if you use a longer lever. The mechanism is the same. Therefore, this is a case of mere variable realizability.

While some lower-level differences are clear cases of MR, other lower-level differences are cases of mere variable realizability, and yet other lower-level differences are simply irrelevant to realizing a given higher-level property. As it often happens in real life, between the clear cases there is a grey area. The present account works well for the clear cases in which MR occurs or fails to occur, and that's all that I hoped to accomplish.

2.7 Multiple Realizability and Levels of Organization

MR can iterate through a system's levels of organization. To illustrate, start with a pump, a prototypical example of a functionally defined artifact. One way to realize a pump is to have a device with chambers and moving parts to fill and empty the chambers. A chamber is, among other things, something that won't leak too much; to make one, you need sufficiently impermeable materials. The materials need other properties, such as sufficient elasticity. Many materials have these properties, but that is beside the point. The point is that there are certain properties of materials that are relevant to their being impermeable to certain fluids and elastic enough to move properly without breaking. Presumably this requires molecules with certain properties, organized together in appropriate ways. Perhaps there are many relevant properties of molecules that can be exploited for this purpose (using the same organizing principle); this is MR_2. And perhaps there are many kinds of organizations that can be exploited for this purpose (organizing the same property of molecules); this is MR_1. And perhaps the two kinds of MR can be combined.

The process can iterate as follows. The properties of molecules may be realized by different kinds of atoms organized in the same way (MR_2) or by the same kinds of atoms in different ways (MR_1), or both (MR_3). The properties of atoms may be realized by different kinds of subatomic particles organized in the same way (MR_2) or by the same kinds of particles organized in different ways (MR_1), or both (MR_3).

Another example is a neuron's capacity to fire—to generate an action potential. Is this capacity multiply realizable? No, because neural firing is always explained by the movement of ions into and out of the axon through ion channels, although there are many different kinds of ion channels. So, this is mere variable realizability. Are ion channels multiply realized? Yes, at least in some cases. For instance, there are two kinds of potassium ion channels, the voltage-gated channels and the calcium-activated channels, and the capacity to selectively allow ions to pass through the channel is explained by different mechanisms in each case. Because they have different components (one has a voltage sensor, the other has a calcium sensor) organized in different ways, this is a case of MR_3. Now take the voltage-gated potassium ion channels. Are they multiply realized? It seems not, although there are variations: there are differences in the molecules that constitute the structure allowing these channels to inactivate, but they seem to operate in the same way, so this appears to be mere variable realizability (although research into these structures continues; e.g., Jensen et al. 2012).

Where does MR stop? Either at a level where there is no MR, because there is only one type of component and one type of organization that realizes a certain property, or at the level of the smallest physical components (if there is one).

2.8 Medium Independence

As we have seen, the typical kind of property that is multiply realizable is a functional property. That is, it's a property individuated by the specific effects it produces under certain conditions. For example, corkscrews have the property of lifting corks out of bottlenecks, mousetraps have the property of catching mice, and computers have the property of yielding outputs that stand in certain mathematical relations to certain inputs. These are all multiply realizable because there are many different mechanisms that can exhibit the relevant functional property. Yet the third example is different from the first two in an important respect.

Lifting corks out of bottles and catching mice are defined, in turn, in terms of specific physical objects with specific physical properties: bottles, corks, and mice. I call them physical media. Corks, bottles, and mice have specific sizes, masses, rigidity, elasticity, etc., which must fall within certain ranges for them to exhibit their capacities. All corkscrews operate on the same physical medium: corks and bottles. All mousetraps operate on the same physical medium: mice. Ditto for other ordinary functional properties.

The specific physical properties of a physical medium put specific constraints on anything that operates on it. Any corkscrew must have enough rigidity to penetrate corks, must exert enough force to lift corks out of bottlenecks, etc. Any mousetrap must be large enough to catch a mouse, must exert enough force to keep the mouse from escaping, etc. The physical properties of the medium constrain what mousetraps and corkscrews can be made of and how they can work.

By contrast, there are functions that are defined without reference to any particular physical medium. Computing functions are an example, as are communication functions. Their physical effects are individuated purely in terms of the relation between their inputs, internal states, and outputs. Given a certain input and internal state, a (correct) computation yields a certain output. What the inputs, internal states, and outputs are physically made of doesn't matter, so long as they can be differentiated from one another and manipulated accordingly by the computing mechanism. For example, a digital adder need not have any specific physical effect. All it needs to do is possess components with enough degrees of freedom to exhibit the relevant computational states and be organized in one of the many ways that yields sums out of addends. Because computers, communication systems, and the like are not limited to a specific physical medium, I call them *medium independent*.

One way to see the difference between MR and medium independence is to consider the inputs and outputs of different systems. Consider the following examples:

Free mouse → trapped mouse

Cork in bottleneck → cork out of bottleneck

$x, y \rightarrow x^y$

In the first two cases, the property of yielding the right output from the right input is multiply realizable, but the inputs and outputs are not multiply realizable. By contrast, in the third case, not only is the property of yielding the right outputs from the right inputs multiply realizable—the inputs and outputs themselves are multiply realizable$_2$. This is why medium independence is a stronger condition than multiple realizability. Medium independence entails MR, but not vice versa (Figure 2.4).

Another way to see the difference between MR and medium independence is to consider that medium independent systems exhibit both MR_1 and MR_2 independently of one another. Consider computers again. They exhibit MR of computed function by algorithm, of algorithm by program (using different programming languages), of program by memory locations, of memory locations (and processing of programs) by architecture, of architecture by technology.

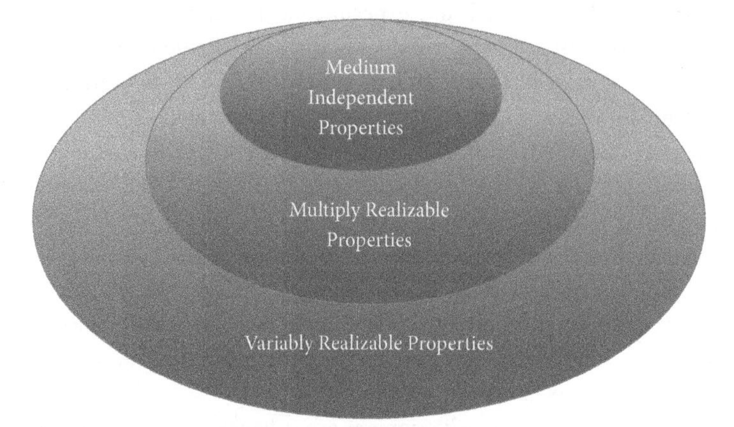

Figure 2.4 Different types of higher-level properties.

As I pointed out above, computing systems exhibit all forms of MR. You can realize the same computation type using the same component types arranged in different ways (MR_1), different component types arranged in the same way (MR_2), as well as different component types arranged in different ways (MR_3).

When typical multiply realizable systems exhibit MR_3, it's because different components can only exhibit the same capacity by being organized in different ways. By contrast, computing systems are such that the same components can be organized in different ways to exhibit the same computational capacity, different components can be organized in the same way to exhibit the same computational capacity and, as a consequence, different components can be organized in different ways to exhibit the same computational capacity. In other words, in medium independent systems the two aspects of MR_3 are independent of one another.

2.9 Multiple Realizability, Reductionism, and Autonomy

One original motivation for focusing on MR was to rule out reductionism, thereby ensuring the autonomy of the special sciences (Fodor 1974). Reductionists retorted that either MR fails to occur or, if it does occur, it fails to undermine reductionism. While I have argued that MR is a genuine phenomenon and articulated its sources, I will now circle back to a conclusion drawn in Chapter 1, which may be the most surprising moral so far: neither reductionism nor autonomy (in most of their traditional guises) holds.

To be sure, some forms of reduction and autonomy are sufficiently weak that they are compatible with one another and relatively uncontroversial. On the reductionist side, every concrete object is made out of physical components and the organized activities of a system's components explain the activities of the whole. On the autonomist side, special scientists choose (i) which phenomena to study, (ii) which observational and experimental techniques to use, (iii) which vocabulary to adopt, and (iv) some of the ways in which evidence from other fields constrain their explanations (Aizawa and Gillett 2011). The genuine controversy between reductionism and anti-reductionism lies elsewhere.

The form of reductionism most commonly discussed in this context is ontological: the identity of higher-level properties to their lower-level realizers. In general, we have seen that there is no such identity. Rather, higher-level properties are an aspect of their lower-level realizers, and MR is neither the source of this fact nor does it alter it. MR is just a natural byproduct, insofar as when it comes to functional properties—properties defined by a specific effect under specific conditions—the same higher-level property can often be produced via different types of lower-level mechanisms.

Other traditional forms of reductionism are epistemological: the derivation of higher-level theories from lower-level theories and of higher-level laws from lower-level laws. The special sciences rarely, if ever, yield the kinds of theory or law that lend themselves to derivation of a higher-level one from a lower-level one (cf. Cummins 2000). MR may well be a reason for that. Contrary to what both reductionists and many anti-reductionists maintain, laws play at best a minor role in the special sciences, and their being reducible or irreducible is of little importance. As I said above, discovering and articulating mechanisms, rather than discovering and articulating laws, is the most widespread and valuable form of explanation in the biological and psychological sciences, and perhaps the special sciences in general (Glennan 2017; Glennan and Illari 2017).

But the failure of traditional reductionism is no solace to the traditional anti-reductionist. For traditional anti-reductionism is predicated on the autonomy of higher-level sciences and their explanations from lower-level ones. MR, properly understood, provides no support for autonomy.

A common form of autonomy discussed in this literature is ontological: the distinctness of higher-level properties from lower-level ones. As we have seen, higher-level properties are not identical to their lower-level realizers (contra reductionism). It doesn't follow that higher-level properties are distinct from their lower-level realizers in the sense of being additions of being to them, as distinctness is usually understood. Rather, higher-level properties are *subtractions* of being from their lower-level realizers. Higher-level properties are aspects of their lower-level realizers. What remains after subtracting appropriate aspects from lower-level (organized pluralities of) properties is the higher-level properties they realize (cf. also Ney 2010). This is not to say that lower-level properties are ontologically more fundamental than higher-level ones; they are ontologically on a par.

Another popular form of autonomy is epistemological: the underivability of higher-level theories or laws from lower level ones. While in general there is no such derivability, the main reason is simply that the relevant kind of law and theory play little or no role in the special sciences. Thus, this putative kind of autonomy is irrelevant to the special sciences.

Yet another form of epistemological autonomy is the lack of direct constraints between higher-level explanations and lower level ones (Cummins 1983). In Chapter 7 I will argue that this kind of autonomy fails too.

2.10 The Role of Multiple Realizability and Medium Independence

Instead of either reductionism or autonomy, we should endorse ontological egalitarianism cum multi-level explanatory integration. Understanding multiple realizability helps us understand the structure of mechanistic explanation within the special sciences and define the stronger relation of medium independence, which in turn will help us understand computational explanation. In this chapter I have built an account of realization, multiple realizability, and medium independence on the egalitarian ontology I proposed in Chapter 1. Parts are neither prior to nor posterior to wholes; and properties of parts are neither prior to nor posterior to properties of wholes. Realization is a relation between a property of a whole and the properties of its parts; the realized property is nothing but an aspect of the properties possessed by the organized parts.

A higher-level property can typically be embedded into many lower-level ones; this is not yet multiple realizability but mere variable realizability. Multiple realizability occurs when the same higher-level property is realized by the properties of different lower-level mechanisms. A higher-level property can be multiply realized in three ways: the components responsible for that property can differ in type while the organization of the components remains the same, the components can remain the same while the organization differs, or both the components and the organization can differ.

Finally, medium independence occurs when even the inputs and outputs of a higher-level property are multiply realizable, so that the higher-level property can be realized by any lower-level structure with the right degrees of freedom organized in the right way. Thus, medium independence entails multiple realizability but not vice versa.

This account captures the kinds of multiple realizability of interest to the special sciences without metaphysically prioritizing either parts or wholes. It is ontologically serious without being mysterious, and it fits well with multilevel mechanistic explanation in the special sciences. Multilevel mechanistic explanation is explanation in terms of multilevel mechanisms. The type of multilevel mechanism that is most relevant to cognition is mechanisms that perform *teleological functions*. What are teleological functions? That's what the next chapter is about.

3
Functional Mechanisms

3.1 Mechanisms with Teleological Functions

In Chapters 1 and 2, I introduced an egalitarian ontology of systems and mechanisms. Systems do things, and what they do is a manifestation of their properties. Some of a system's properties are invariant over time; they individuate systems as belonging to kinds. A system's capacities are constitutively explained by the properties and activities of the mechanisms contained within the system. The properties of a whole mechanism are aspects of the properties of its components. Such properties are typically variably realizable and, if they can be produced by different mechanisms, they are multiply realizable too. Some properties are even medium independent.

By building on this ontological foundation, I will now improve on existing accounts of multilevel mechanisms that have *teleological* functions.[1] This is crucial to the present project because neurocognitive mechanisms have functions, so any adequate account of neurocognitive mechanisms must be grounded in an adequate account of functions. As I will argue presently, the functions of neurocognitive mechanisms are teleological.

Most mechanists have been either hostile or at least cautious about whether mechanisms have teleological functions (e.g., Craver 2013). Yet there is an important distinction between organisms and their artifacts, on one hand, and other mechanisms, on the other. Organismic traits and artifacts have teleological functions; e.g., firing action potentials (neurons) and cooling enclosed living spaces (air conditioners). Such mechanisms can malfunction; e.g., neurons can fail to fire action potentials and air conditioners can fail to cool enclosed spaces in appropriate circumstances—when that happens, something is going wrong. Other mechanisms, such as the mechanisms of volcanic eruptions or hurricanes, lack teleology altogether. They are not *aimed* at anything, so they can't go wrong in any way no matter what they do.

Teleological functions play a role in scientific practices and they can do philosophical work, so it's important to have an adequate account of them. Specifically, neurocognitive systems have teleological functions like processing information and

[1] This section presents a revised and improved version of the goal-contribution account of functions defended in Maley and Piccinini 2017. Even though the present account improves on the account presented in that paper, I omit many details that are adequately covered there.

Neurocognitive Mechanisms: Explaining Biological Cognition. Gualtiero Piccinini, Oxford University Press (2020).

DOI: 10.1093/oso/9780198866282.001.0001

controlling organisms. I will appeal to such teleological functions later in the book. To prepare for that, I will now provide an account of teleological functions. In this section I will sketch a goal-contribution account of teleological functions. In subsequent sections I will argue that this goal-contribution account is more adequate than other accounts.

Let's begin by focusing on a special class of physical systems—living organisms. For present purposes, I assume that there is an objective difference between living organisms and nonliving systems, which has to do with extracting energy from the environment and using it to perform physical work that has certain specific effects. In virtue of their specific organization, populations of living organisms have special capacities that typical nonliving systems lack. The following are especially salient. First, living organisms *stay* alive—they perform specific actions that preserve their specific organization—for a while. Second, some of them reproduce. Third, some of them help others—especially offspring, relatives, and symbionts (cf. Piccinini and Schultz 2018). Fourth, they go through a life cycle that allows them to survive and sometimes reproduce, help others, or both.

These special capacities of living organisms are mutually dependent. Surviving is necessary for completing one's life cycle, reproducing, and helping others. Going through one's life cycle—that is, growing and developing into a mature organism—is necessary for becoming more self-sufficient, reproducing, and helping others. (From now on, I will use "development" in a broad sense that includes growth.) Finally, reproducing and helping others (the latter only for some organisms) is necessary for the continued existence and survival of similar organisms.[2]

These capacities—survival, development, reproduction, and helping—are invariants that individuate living organisms as a kind. As long as organisms survive, develop, reproduce, and possibly help others, subsequent generations of organisms will continue to exist. As soon as organisms stop surviving, they go out of existence qua individual living organisms. As soon as juvenile organisms stop developing, they impair their survival and reproduction. And as soon as all similar organisms—all members of a population—stop reproducing, receiving help, or both, that population is on its way to extinction.

These special invariants—survival, development, reproduction, and help—do not occur at random or due to some basic physical law. Rather, organisms possess them to the extent that their special functional organization allows them to

[2] Some biologists and philosophers of biology have independently posited a similar kind of mutual dependence between processes *internal* to organisms. Briefly, organisms have the following three closure properties: they contain autocatalytic sets, namely sets of entities each of which can be produced catalytically from other entities within the set (Kauffman 1993), they exert work to maintain internal constraints that in turn are necessary to produce the work (Kauffman 2002), and their processes are mutually constrained in such a way that each constraint is generated by at least one other constraint (Montévil and Mossio 2015). This mutual dependence of internal processes is complementary to the mutual dependence of survival, development, reproduction, and helping that I describe in the main text.

successfully *pursue* them. Specifically, organisms must extract energy from the environment and expend it to survive, develop, reproduce, and help, on pain of extinction of their population. This is not an ethical imperative but a biological one. It means different things for each capacity. Pursuing survival means expending energy to maintain internal structure in the face of environmental perturbations. Pursuing development means expending energy to create a specific internal structure or transform it in a specific way, reaching a certain outcome in spite of environmental variation. Pursuing reproduction means expending energy to create new organisms similar to oneself, possibly in cooperation with a reproductive partner. Pursuing care of others means expending energy in ways that benefit other organisms.

Because these special invariants must be pursued, I call them biological *goals*. Goals in this sense need not be represented by the organism and need not be reached. In addition, I can't stress enough that there is *no ethical obligation* to pursue biological goals. It's just that if biological goals cease to be pursued, organisms eventually cease to exist. In this limited sense, organisms considered collectively must pursue their biological goals.

To pursue their biological goals, organisms possess a number of interdependent mechanisms. Each mechanism provides a *regular contribution* towards one or more biological goals—that is, a contribution that occurs reliably enough at appropriate rates in appropriate situations. For instance, the circulatory system circulates blood through the body, whereby blood transports many of the things the body needs where it needs them. The locomotive system moves the organism in its environment, which in turn is necessary to forage, find shelter, escape dangers, pursue mates, and take care of others. The nervous system processes information and uses it to control the locomotive system as well as other systems. Et cetera.

These regular contributions, which each biological mechanism provides towards one or more biological goals, are *biological teleological functions*. There are several notions of function; this is the core biological one. It's a teleological notion because functions in this sense are regular contributions *to goals*. It's a biological notion because the goals are *biological*: their lack of fulfillment leads to reduced chances of survival, development, reproduction, and helping others. Thus, if under normal background conditions a biological mechanism does not fulfill its teleological functions or does not fulfill them at the appropriate rate in an appropriate situation, it *malfunctions*. To count as a function, a contribution must be regular enough—that is, it must be produced reliably at appropriate rates in appropriate situations. In other words, functions are not contributions to a goal that happen by random coincidence. Lucky accidents happen at random times by coincidence; functions happen regularly at appropriate rates in appropriate situations.

Teleological functions are a special subset of all the *causal roles* performed by mechanisms. Most of a mechanism's effects, even if they are regular enough to occur reliably under certain conditions, do not contribute to any goals of organisms. For instance, stomachs make gurgling sounds under specific circumstances. Making gurgling sounds is one of stomachs' causal roles. But such a causal role need not make any contribution to the goals of organisms. If it doesn't, then making gurgling noises is not a teleological function of stomachs. Occasionally, mechanisms and their traits make accidental contributions to goals by random coincidence, but those are not teleological functions either. It's only when an effect is both regular enough and contributes to a goal that it counts as a teleological function.

Teleological functions are contributions made by the mechanisms of successful traits—traits that reach their biological goals well enough that they could contribute to generating subsequent generations of organisms. Similar traits in similar organisms are said to *have* the same functions even though they may not perform them or perform them to a lesser extent.

Teleological functions, including biological ones, exist at the same levels at which mechanisms exist. Mechanisms perform functions in virtue of their components fulfilling subfunctions, which in turn perform their subfunctions in virtue of their subcomponents performing sub-subfunctions, and so forth. Thus, functions can be decomposed into subfunctions, which in turn can be decomposed into sub-subfunctions, and so forth. Each function, subfunction, etc. is performed by a mechanism, sub-mechanism, etc. Mechanisms with teleological functions may be called *functional* mechanisms (cf. Garson 2013). Functional mechanisms explain the capacities of the systems they are part of in virtue of the organized performance of teleological subfunctions by their components.

Teleological functions can be described more widely (distally) or more narrowly (proximally) depending on how much of a mechanism and its context is included in describing the function. A heart's narrowest teleological function is to expand and contract. That is its function in the strictest sense. The inability to do that given normal background conditions is a malfunction of the heart, which we can fix only by fixing the heart (cf. Garson 2019a, chap. 7). When filled with blood, though, the heart has the slightly wider function of pulling blood from certain holes and pushing it into other holes. When attached to veins and arteries, the heart has the wider function of sucking blood from the veins and pumping it into the arteries. When the veins and arteries are attached to one another via capillaries, the heart has the even wider function of circulating the blood through the circulatory system. When the circulatory system is part of the body, the function of the heart is to circulate the blood through the body. This, of course, makes many contributions to the survival of the organism, which is the ultimate function of the heart.

Organisms modify their environment, sometimes in very specific and complex ways. For example, they dig burrows to protect themselves and their offspring from the weather and predators. When such modifications are relatively manipulable by the organism, they are called tools. Examples include sticks and stones used by chimps to hunt ants and crack nuts. When such modifications involve complex manipulations that create novel structures, they are called artifacts. Examples include spider webs and beaver dams. Many tools, artifacts, and other environmental modifications provide regular contributions to the organisms' goals. Such regular contributions are also teleological functions. Thus, we obtain a unified account of teleological functions in biological organisms and artifacts (and other stable environmental modifications). Biological teleological functions are regular contributions to the biological goals of organisms.

Some organisms have enough sapience and sentience to develop nonbiological goals. Examples include pleasure, the absence of pain, knowledge, truth, beauty, justice, wealth, and power. Nonbiological goals are states that satisfy two conditions. First, some organisms *pursue* nonbiological goals—organisms expend energy towards such states in the face of environmental perturbations. Second, nonbiological goals are *distinct* from biological goals—life goes on whether or not organisms pursue nonbiological goals and pursuing nonbiological goals does not always contribute to biological goals, and sometimes it may even be detrimental. Any regular contributions to these goals, by either organisms themselves or their artifacts and other environmental modifications, may be considered teleological functions. Thus, we obtain a unified account of teleological functions, both biological and nonbiological, in both organisms and artifacts. Teleological functions are regular contributions to the goals of organisms.

One last point is that functions are relative to the goals they contribute to, and such goals don't always align with one another. A function of guns is to kill, which may be a goal of an organism. One possible target of a gun is oneself. But shooting oneself is antithetical to survival. Thus, something may perform one function, which contributes to one goal of an organism, while defeating or being detrimental to another goal.

Let's see where this account fits within the broader debate about functions. This will put us in a position to see its virtues.

3.2 Three Accounts of Function: Causal Roles, Selected Effects, and Goal Contributions

Historically, the most influential view about functions within the philosophy of mind is that functions are causal roles—*any* causal roles, without qualifications. A causal role is the production of certain effects under certain conditions by a state or subsystem within a containing system.

Causal role account: X has function F if and only if F is a causal role and X performs F.[3]

The causal role account is clear, simple, and based on the ubiquitous and uncontroversial notion of causal role. Just about anything can be analyzed in terms of the causal role it plays within a containing system (cf. Amundson and Lauder 1994). Another great advantage is that causal roles can be investigated here and now, by observing what a system does under various conditions. Thus, functions as causal roles provide a solid foundation for the special sciences.

This liberality of the causal role account—the fact that just about anything has plenty of causal roles—is also its Achilles heel. Biologists and other life scientists distinguish between causal roles in general and causal roles that have biological significance. Among biologically significant roles, some are harmful to organisms while others are beneficial. The latter are often the reason why organisms possess their characteristic biological traits (Wright 1973).

For example, neuronal activity sends electrical signals, increases blood flow, and warms up the brain, among other effects. These three causal roles are not on a par. Neurons are in the brain *in order to* produce electrical signals, not in order to increase blood flow let alone produce heat. Blood flow increases *in order to* nourish the neurons; heat is merely a physical byproduct to be dispersed. So, sending electrical signals is a function of neurons in a sense in which increased blood flow and heat production are not. In addition, increased blood flow has its own function in a sense in which heat production does not. What sense is that? Perhaps it has something to do with what traits are selected for.

Some causal roles are the outcome of a selection process, while others are not. Living organisms are the product of natural selection, which selects traits based on their differential contribution to the ability of organisms to survive, develop, reproduce, and help others. Immune systems select certain antibodies because they can fight antigens. Neurodevelopmental processes select some neurons and synapses because they play specific causal roles better than others. Many organisms learn by trial-and-error, which may be seen as a selection process. Finally, human beings themselves select artifacts and continue to produce them because those artifacts perform a certain job better than other artifacts. These are all selection processes, and selection partially explains why certain objects and their

[3] The causal role account of functions goes back at least to the beginning of functionalism in the philosophy of mind (more on functionalism in Chapter 4). Putnam (1967a) argued that the essence of mental states is a function, by which he meant a causal role. Fodor (1968a) argued that psychological explanation proceeds by giving functional analyses of psychological capacities, that is, analyses in terms of functions, by which he meant causal roles (more on functional analysis in Chapter 7). Shortly thereafter, Cummins (1975, 1983) gave functional analysis and functions qua causal roles their canonical account. More recently, Craver (2001) merged the causal role account with mechanisms by arguing that mechanisms and their components perform functions qua causal roles.

states are where they are—because they play *specific* causal roles, not just *any* causal roles.[4]

The role of selection in the origin of functional traits and artifacts gives rise to the selectionist account of function. According to the selectionist account, a function is a *selected* causal role that belongs to traits in virtue of their type.

> *Selectionism*: Tokens of type X within a population have function F if and only if F is a causal role and tokens of X were recently selected because they performed F.[5]

That tokens of X were selected because they performed F means one of two things. Either systems that had an X that performed F within previous generations left more descendants than systems that lacked an X that performed F, and this is at least in part because the Xs performed F, or the Xs themselves were differentially retained by their containing system because they performed F (Garson 2019a).

Biological adaptations are the most common example of the first kind of selection: adaptations have the functions they have because traits of the same type were selected for in their recent ancestors thanks to the causal role they played. Neuronal selection is an example of the second kind of selection: nervous systems differentially retain certain synapses over others because of the causal role they play.

The selectionist account is more powerful than the causal role account in three respects. First, it explains why only some causal roles are functions in a special, teleological sense—only causal roles that were selected for are teleological functions. Second, it accounts for the normativity of teleological functions: traits have teleological functions because they were selected for; therefore, all tokens of those types have teleological functions whether or not they perform them; if they don't perform their function, they malfunction. Third, it partially explains why functional attributions are sometimes used as etiological explanations of traits. Traits are where they are in part because they or their ancestors played that causal role, and that caused either those traits' own persistence (differential retention) or the persistence of the same trait in subsequent generations (differential reproduction) via a selection process.

[4] I'm indebted to Garson 2019a for this list of selection processes. He describes them in more detail.

[5] For classic selectionist accounts of biological functions, see Millikan 1984; Neander 1983. Precursors include Ayala 1970; Wimsatt 1972; and Wright 1973. The selection process posited by the selectionist account must continue until the present (Griffiths 1993; Godfrey-Smith 1994), otherwise the selectionist account would entail that vestigial traits such as the human appendix—which are now functionless—still have the function their ancestors did. Garson 2019a provides a generalized selectionist account of biological functions similar to the formulation I gave in the main text. Preston 2013 offers a selectionist account of human artifact functions. Garson 2016 provides a thorough review of the literature on biological functions.

This ability to partially explain the etiology of traits comes with a drawback. Given the selectionist account, strictly speaking, discovering whether something has a teleological function requires investigating whether it was selected and what it was selected for. That is, it requires investigating its selection history and providing an etiological explanation of its persistence. Investigating selection history is wonderful but it's not everyone's favorite, and sometimes it's very hard (Gould and Lewontin 1979). It's especially hard for biological traits like brains, which leave no fossil records.

Discovering teleological functions is often easier than discovering whether a causal role was selected or what it was selected for. For instance, when William Harvey (1628) discovered that hearts are pumps that circulate blood through the body, he used a combination of experiments and calculations. He did not investigate the selection history of hearts. Evolutionary biology didn't even exist yet. Countless other teleological functions have been discovered by analogous means, without having the slightest clue about their selection history. This suggests that teleological functions are not only something more specific than a mere causal role (contra the functional role account); they are also something more general than a *selected* causal role (contra the selectionist account).

Selectionists might respond that the above is just an epistemic problem. Functions are selected effects, whether or not we can find out they were selected. Whether or not making functions dependent on selection history undermines selectionism, a more serious problem is that being selected is not necessary for having a teleological function. Consider mutations. Most mutations are either harmful or make no difference to the fitness of organisms. Once in a while, though, adaptive mutations arise. Such mutations make regular contributions to the fitness of organisms. These organisms are lucky, but they are not lucky in the sense in which organisms that narrowly escape death due to a random coincidence are lucky (contra Garson 2019a). Lucky accidents are too irregular and they are not inheritable, so they cannot be selected for. By contrast, adaptive mutations are heritable and produce reliable adaptive effects, which is why they are subject to selection. In fact, they are a crucial driver of evolution by natural selection. Therefore, they perform valuable biological functions. But they haven't been selected for yet, so the selectionist account has no means to assign them biological functions.

Another example is novelty in artifacts. Some artifacts are invented from scratch; they may have teleological functions even though they were not selected for. Even artifacts that are modifications of previous artifacts typically contain useful novel features that previous generations of similar artifacts lacked. This is how corporations motivate us to buy the latest generation of gadgets—by adding novel features. Some novel features are useless, but some are useful. The useful novel features are analogous to adaptive mutations. They haven't been selected for yet, so they have no functions by the light of selectionism. But, patently, they do have functions—in the teleological sense.[6]

These examples show that selectionism gets the relationship between teleological functions and selection backwards. It's not that traits have teleological functions because they were selected for. Rather, traits are selected for because they perform teleological functions. Teleological functions must be performed *before* selection can possibly take place, precisely so that selection can operate on the traits that perform them. But these teleological functions cannot be just *any* causal roles, as the causal role account would have it. For most causal roles are either useless or harmful and, therefore, should not and usually are not selected for. Generally, only causal roles that contribute to a goal are selected for.[7]

Enter the goal-contribution account. According to it, a function is a *teleological* causal role—a causal role that contributes to a goal.

> *Classical goal-contribution account*: X has function F if and only if F is a causal role and F contributes to a goal of the systems that contain X.[8]

The virtue of the goal-contribution account is that it steers a middle ground between the causal role account and selectionism. The goal-contribution account provides a stronger notion of function than the causal role account, without appealing to etiology like selectionism does. Of course, the goal-contribution account needs to explain what a goal is and what it means to contribute to a goal.

There are a number of accounts of goals and goal-directedness. According to one such account, a goal is a state towards which a system's behavior is directed, and being directed towards a goal is to correct one's behavior in the face of environmental perturbations until either the goal is achieved or the pursuit of the goal is defeated. Accounts of this sort may be used to define (teleological) functions as contributions to the goals of a goal-directed system (e.g., McShea 2012).

As stated, the classic goal-contribution account does not account for malfunction. Nothing in the account says that a goal-directed system should reach its

[6] Garson (2019b) has a novel reply to objections of this sort against selectionism. The objections are that, contra selectionism, (i) functions can be discovered without knowing a trait's selection history and (ii) traits can have functions without being selected for. Garson replies that there is a sense in which functions depend on history according to all viable accounts of function. True, but the history required by other accounts is not *selection* history. The history required by other accounts is just enough history to show that some organisms are successful within a type of environment. This is relatively easy to investigate and is compatible with novel functions arising without selection. Therefore, the fact that all accounts of functions rely on history does not undermine objections to selectionism.

[7] One possible exception is harmful drugs, which an addict may select even though they are harmful. Selection by human beings involves layers of cognitive processes that complicate the story of how some things get selected. At any rate, any exceptions to the principle that only beneficial causal roles are selected for reinforces the argument in the main text: contra selectionism, being selected for is not the same as having a teleological function.

[8] Goal-contribution accounts include Nagel 1953, 1977; Wimsatt 1972; Adams 1979; Schaffner 1993; Boorse 1977, 2002.

goals, or that something goes wrong if it doesn't. Either a goal-directed system reaches its goals or it doesn't, and that's that. Later we'll see that a prominent solution to this problem is to identify functions with behaviors that are statistically typical in a population, and call any departure from what is statistically typical a malfunction. We'll also see that that is an inadequate account of malfunction. For now, let's look at an even more serious problem with the classic goal-contribution account.

The problem is that goal-directedness in the present sense is neither necessary nor sufficient for having teleological functions. Against necessity, many biological traits and artifacts have teleological functions without being goal-directed. For example, coasters have the function to protect flat surfaces from drink spills and melanin has the function to protect skin from sunburns. Yet neither coasters nor melanin correct their behavior in the face of environmental perturbations in any serious sense. They barely behave at all. Ditto for countless other biological traits and artifacts. Against sufficiency, many systems are goal-directed in the above sense even though they lack teleological functions. A classic example is a dynamical system with an attractor, such as a marble rolling to the bottom of a vase. If the marble encounters a small obstacle along its path, it goes around the obstacle until it reaches the bottom of the vase, regardless of where it starts—unless it finds too large an obstacle, in which case it stops before reaching the bottom. So the marble satisfies the above account of goal-directedness, even though it has no function (for a more detailed argument to this effect, see Garson 2016, chap. 2).

This is where the improved goal-contribution account I presented in Section 3.1 comes to the rescue. I began with living systems as such and defined goals in terms of living systems. Specifically, I pointed out that living systems have special properties—surviving, developing, reproducing, and helping—that require pursuit. That is, living organisms must expend energy so as to reach those biological goals at least some of the time, on pain of population extinction. Other biological goals are means towards those overarching goals. Some organisms develop other, nonbiological goals. Nonbiological goals *are* defined in terms of goal-directed behavior on the part of organisms, but traits and artifacts may still contribute to nonbiological goals without being goal-directed. Teleological functions are causal roles that provide regular contributions to goals of organisms.

> *Improved goal-contribution account*: Tokens of type X have function F if and only if F is a causal role and performing F by tokens of X provides a regular contribution to a goal of organisms.

This improved goal-contribution account identifies just those causal roles that constitute (teleological) functions among all the causal roles that traits perform. It distinguishes between teleological functions and lucky accidents, which are too irregular to count as functions in the relevant sense. It also accounts for malfunctions: tokens of type X that do not perform F or do not perform it at the appropriate rate in appropriate situations are malfunctioning tokens. Thus, the improved goal-contribution account does justice to the notion of teleological function without getting bogged down in selection histories and without facing the objections encountered by classical goal-contribution accounts. As we've seen, classical goal-contribution accounts define all goals in terms of goal-directedness, and all functions as contributions to the goals of a goal-directed system containing a given trait. On the contrary, both biological traits and artifacts can contribute to goals of organisms whether or not the systems containing them are goal-directed, and biological goals are states that organisms must pursue on pain of extinction.

And yet we are not done with teleological functions because there is one remaining problem to solve—a big one. From now on, by "function" I mean teleological function unless otherwise indicated.

3.3 Functions Must Be Performed at Appropriate Rates in Appropriate Situations by Appropriate Members of a Population

Most functions cannot and should not be performed at a constant rate at all times by all members of a population. Rather, most functions must be performed at appropriate rates in appropriate situations. Many functions are performed only at appropriate developmental stages, and some functions are performed only by members of one class of organisms or another (e.g., male vs. female, or worker bee vs. queen vs. drone). Let's set development and biological specializations aside for now and focus on appropriate situations and rates. What's an appropriate situation for the performance of a function, and what's an appropriate rate of performance? These important questions have received inadequate attention.

Causal role theorists have not discussed these questions at all, perhaps because they have no resources with which to address them. Causal roles simpliciter are either performed or not—there is nothing appropriate or inappropriate about their performance. In other words, if functions are just causal roles simpliciter, without anything normative about them, then the question of whether they are performed at appropriate rates in appropriate situations does not arise.

Selectionists have not done justice to these questions either. They focus primarily on what it is for a trait to have a function, how functions are distinct from accidents, and why the inability to perform a function is a malfunction. They explain these features by the selection of traits for specific causal roles. Selectionists say little about how selection processes could operate not simply on the performance of causal roles, but on the performance of causal roles at appropriate rates in appropriate situations (by appropriate members of a population), especially since most traits are the product of complex interactions between innate and environmental influences. There is a complex story to tell about how all of this plays into selection processes. To my knowledge, the full story has yet to be told.

Goal-contribution theorists are the only ones who have tried to explicate what count as appropriate rates of functioning in appropriate situations. Christopher Boorse (1977, 2002) takes one step in this direction. He defines biological functions as statistically typical contributions by a trait to survival and reproduction within a reference class. A reference class is a class of organisms of a certain age and sex.[9] According to Boorse, whatever contribution traits (of members of a reference class) typically make to survival and reproduction is their function. Since such contributions are typically made at certain rates in certain situations, an implication of Boorse's biostatistical account is that those are the appropriate rates and situations for the performance of biological functions (Schwartz 2007).

Unfortunately, Boorse's account is hard to reconcile with our ordinary and medical judgments about what is an appropriate rate of functioning in an appropriate situation (Kingma 2010). Specifically, there are situations in which we judge that a function should be performed at a certain rate even though it is statistically typical for traits (within a reference class) to function at a different—intuitively inadequate—rate.

To illustrate, consider a modified version of the digestion example discussed by Kingma (2010, see Table 3.1; whether the example is physiologically accurate is inconsequential). Suppose that, in situation s_1, there is food in the stomach but the stomach is not digesting due to extremely strenuous exercise; in situation s_2, there is food in the stomach but the stomach is digesting at a very low rate due to strenuous, though not extremely strenuous, exercise; in situation s_3, there is food in the stomach but the stomach is not digesting due to poison-induced paralysis; in situation s_4, there is food in the stomach but the stomach is functioning at a very low rate due to poison-induced impairment.

Table 3.1 Four situations in which there is food in the stomach but digestion is either not occurring or slower than usual

	Exercise	Poison
No digestion	s_1	s_3
Slow digestion	s_2	s_4

[9] Boorse's account could be generalized to include other biological goals, such as development and helping others, and other specializations besides sex. Kingma (2007) argues persuasively that Boorse's reference classes are chosen in an ad hoc way to suit the account's needs, without any principled way of deriving them. I'll address this issue in the next section.

Intuitively, s_1 is not an appropriate situation for digestion, meaning that strenuous exercise makes it appropriate for the organism to divert blood to the locomotive system, which results in digestion being prevented. By contrast, s_2–s_4 are appropriate situations for digestion. In addition, the stomach is functioning at an appropriate rate in s_2 but not in s_3 and s_4. In other words, the stomach is functioning correctly in s_2, incorrectly in s_3 and s_4. And yet, in all cases, the stomach is performing in a statistically typical way. It is entirely typical for stomachs to be either paralyzed or impaired during strenuous exercise or after the ingestion of poison. Thus, Boorse's biostatistical account deems such performances appropriate, even though two of them are not. What Boorse's account counts as a properly performed function of a trait, our ordinary and medical judgments deem a malfunction.

Justin Garson and I (Garson and Piccinini 2014) attempted to address this problem by formulating an explicit account of what it is to perform a biological function at an appropriate rate in an appropriate situation (cf. also Hausman 2011, 2012; Kraemer 2013). As I will now show, our solution is inadequate.

There are two interrelated constraints on a good account of teleological functions: first, it should rule that s_3 (poison) but not s_1 (strenuous exercise) is an appropriate situation for digestion even though in both cases the function is not performed at all; second, it should rule that the stomach is functioning at an appropriate rate in s_2 (exercise) but not s_4 (poison) even though the rate of performance is reduced in both cases.

With respect to the first issue, here is how Garson and I define the set of appropriate situations for the performance of a function F by a trait X:

(i) Inclusivity: P(X is in S|X's doing F contributes to a biological goal) $\approx$ 1.
(ii) Specificity: there is no S' which is a proper subset of S such that (i) is true of S'. (Garson and Piccinini 2014, 8; slightly amended here)

We wrote $\approx$ instead of = to make room for accidental contributions to biological goals; that detail is irrelevant here so we may ignore it. To see that our account is inadequate, consider the set of situations S in which there is food in the stomach. The probability that the stomach is in a member of S (appropriate situations) given that it's contributing to survival or inclusive fitness by digesting is pretty close to 1, so having food in the stomach arguably satisfies clause (i), which is what we want. However, there is a proper subset of S such that (i) is true of it, namely the set S' of situations in which there is food in the stomach but the stomach is neither paralyzed by poison nor by extremely strenuous exercise. Therefore, clause (ii) mandates that S' should be taken as the set of appropriate situations. Now we have ruled that both s_3 (poison-induced paralysis) and s_1 (exercise-induced paralysis) are outside the set of appropriate situations for digestion, which means that the stomach is not malfunctioning in either of those. That is not what we wanted: we wanted s_3 to be ruled appropriate for digestion but s_1 inappropriate, so that the stomach is malfunctioning in s_3 but not in s_1. Interim conclusion: Garson and Piccinini (2014) have no principled way to rule s_1 but not s_3 out of the class of appropriate situations.

Now set the first constraint aside for a moment and focus on the second constraint. Here is Garson and Piccinini's account of appropriate rate of functioning:

> Trait X performs function F at a rate of functioning that is appropriate in a situation s $=_{def}$
>
> (i) If an organism in the reference class possessing X is in s and $s \notin S$, then X performs function F at a rate of zero (or close to zero).
>
> (ii) If an organism in the reference class possessing X is in s and $s \in S$, then X's rate of functioning provides an adequate contribution to a biological goal in s, relative to other rates that are physiologically possible for X in s. (Garson and Piccinini 2014, 10, slightly amended here)

Given this account, the follow-up question is whether a rate of functioning provides an adequate contribution to a biological goal in s. Garson and Piccinini say that to determine whether a given rate of functioning is adequate, we should compare it to the rates of functioning that are physiologically possible for that trait in the relevant situation. But what is the relevant type of situation? Is it the type of situation the trait is currently in (as Garson and Piccinini's official definition suggests), any situation in which performing the function is appropriate, or some other situation type? The answer makes a big difference, but none of the options is adequate.

If the appropriate rate of functioning is defined based on what is physiologically possible in the current situation, then the stomach's rate of functioning in both s_2 (low rate of digestion due to exercise) and s_4 (low rate of digestion due to poison) turns out to be appropriate, which is not the desired outcome. If, instead, the appropriate rate of functioning is defined based on what is physiologically possible in a broader range of situations, including when the organism is not exercising and its stomach is not poisoned, then the stomach's rate of functioning in both s_2 and s_4 turns out to be inappropriate, which is also not the desired outcome.

The source of our difficulty is that there are two types of factor that push the rate of functioning of a trait outside the range that is adequate to the pursuit of biological goals: *healthy* defeaters of a trait's function, such as the metabolic cost of digesting while exercising, and *pathological* defeaters, such as poison. The former do not give rise to inappropriate rates of functioning, the latter do. Garson and Piccinini's account cannot distinguish between these two cases. Other accounts in the literature have similar limitations (Kingma 2016; Casini 2017).

To improve on Garson and Piccinini (2014), we need to distinguish between healthy defeaters such as exercise, sleep, pregnancy, and giving birth, which diminish the performance of certain functions without thereby causing malfunctions, and pathological defeaters, which cause genuine malfunctions. Healthy defeaters affect whether a situation is appropriate for the performance of a function and what rate of functioning is appropriate in that situation. Pathological defeaters do not. To draw the right distinction, we need to introduce functional tradeoffs.

3.4 Functional Tradeoffs

If performing multiple functions requires shared resources, performing one of them above a certain threshold may interfere with performing the others. This is what I call *functional tradeoffs*. Functional tradeoffs require systems to prioritize some functions over others. This prioritization is not a malfunction. On the contrary, it is an unavoidable consequence of operating with limited resources, which every organism must do. Thus, we need to incorporate functional tradeoffs within an adequate account of teleological functions.

To do that, let's go back to the guiding idea I introduced in Section 3.1. Biological functions are causal roles that explain the peculiar capacities of living organisms: survival, development (including growth), reproduction, and help. The manifestations of these peculiar capacities are biological goals—pursuing them is necessary to the continued existence of living organisms. We've also seen that some organisms pursue other, nonbiological goals. So, teleological functions in general are causal roles that make a regular contribution to some goal—either biological or nonbiological—of organisms. I will now focus on biological goals of organisms; mutatis mutandis, the resulting account can be applied to artifacts and nonbiological goals.

Functions and subfunctions are organized in a means-ends hierarchy that mirrors the functional organization of mechanisms. Each mechanistic subsystem of a living organism has one or more wide functions, which contribute to the organism's goals. Each component of the mechanism has subfunctions that contribute to the mechanism's functions, and so forth. Thus, explaining how organisms pursue their goals—how they survive, develop, reproduce, and help others—requires identifying the mechanisms that contribute to such goals and the ways in which they contribute.

Goals can be reached to varying degrees, which means that functions qua contributions to goals can be performed to varying degrees. Organisms develop fully, partially, or little. They live lives in which their organs function either optimally or sub-optimally to varying degrees. They help many others, just a few, or no one. They have many offspring, few, or none. In all this variation, there are minimal thresholds that must be met. In order for organisms to function reasonably well, hearts must pump at certain rates (within certain ranges), lungs must exchange oxygen and carbon dioxide at certain rates, stomachs must digest food at certain rates, kidneys and livers must filter blood at certain rates, and so forth. In order to at least *be able* to reproduce, reproductive organs must develop in certain ways and operate in certain ways. Below certain thresholds of functioning, the organism struggles or dies, does not develop fully, cannot reproduce, or cannot help others.

As a consequence, explaining the success of organisms that reach their goals to a certain degree requires identifying both situations that call for mechanisms to perform their functions and the relevant thresholds—the ranges within which such functions must be performed to provide an adequate contribution to the goals of organisms. What counts as an adequate contribution depends on what, exactly, we aim to explain. If we aim to explain bare survival at any cost, certain rates of functioning are adequate. If we aim to explain survival with fully functional organs, other rates are adequate. Similarly for other goals. What matters is that once we fix the explanandum, the explanans—the range of situations that call for performing a function and the rates of adequate performance—are objective matters.[10]

Organisms must perform many functions at once. Performing them at adequate rates requires expending energy and other resources. Since organisms possess bounded resources, performing multiple functions above certain thresholds involves tradeoffs. That is, performing one function above a certain threshold may interfere with performing other functions above certain thresholds. This must be taken into account when identifying both appropriate situations for the performance of a function and appropriate rates of functioning.

Other things being equal, a situation calls for the performance of a function just in case performing that function in that situation would contribute to a goal of the organism. But things are often unequal: performing some functions takes priority over performing others. Taking functional tradeoffs into account, a situation calls for the performance of function F just in case performing F in that situation would contribute to a goal of the organism *and* performing more urgent functions does not prevent F from being performed.

Other things being equal, a rate of performance is adequate (in a situation that is appropriate for performing a function) just in case performing at that rate in

[10] This should put to rest persistent worries that the notion of teleological function contains a pernicious circularity, perhaps because finding a mechanism's (teleological) function(s) requires knowing its structure, yet identifying mechanistic structures requires knowing a mechanism's (teleological) function(s) (Dewhurst 2018a; cf. Nanay 2010 and our response in Maley and Piccinini 2017). On the contrary, functions and structures can be identified partially independently of one another and mutually constrain one another; in natural systems such as organisms, we must study both functions and structures until we discover which structures perform which functions. More on the mutual constrains between functions and structures in Chapters 7 and 8.

that situation would contribute adequately to a goal of the organism—meaning, it would explain how the goal is reached to the degree that requires explaining. Again, things are often unequal. Some functions take priority over others even in situations that call for their collective performance. Taking functional tradeoffs into account, a rate of performance is adequate in a situation that is appropriate for performing function F just in case performing F at that rate in that situation would contribute adequately to a goal of the organism and performing more urgent functions does not prevent F from being performed at that rate—if it does, F's rate of adequate performance must be adjusted to what a successful system can muster while prioritizing more urgent functions.

For example, the following functions share common resources: inhaling, talking, swallowing, and vomiting. If you do one, you can't do the others, or you can't do the others at the same rate. Breathing regularly is necessary for survival. Talking piggybacks on exhaling, and sometimes we delay the next inhalation so that we can finish expressing our thought—without any negative consequences. If we need to swallow, we can't breathe at the same time. Luckily, swallowing is usually quick: we can delay breathing a little without overly negative consequences. If we are out of breath, however, we might delay swallowing and prioritize breathing instead. Finally, if we need to vomit, other activities just have to wait. The point is this: scarce resources need to be allocated to different functions under different circumstances, and this affects which situations are appropriate for performing a function and which rates of performance are appropriate.

Here is how the point applies to our digestion example. Other things being equal, ingesting food is a situation in which digestion should take place, because that's how the organism provides energy for itself, which in turn contributes to survival. Other things being equal, strenuous physical exertion is a situation in which directing blood to the muscles should take place, because that's what the organism requires, and strenuous exertion contributes to survival too. If there's food in the stomach while the organism is exerting itself strenuously, a functional tradeoff occurs. The tradeoff must be resolved by determining which function takes priority. By contrast, if poison is ingested, the result is not a functional tradeoff but a situation in which (by hypothesis) the very function of digestion is prevented from being performed with no independent functional priority being satisfied.[11]

There is yet another way that things are unequal. Different biological goals impose different constraints on different classes of organisms according to the role they play in the life cycle. Survival is the most basic goal; it applies to all organisms.

[11] In fact, depending on the poison, stomach paralysis may well have a function—e.g., the function of preventing damage to the organism that would occur if the stomach kept operating with poison in it. This just reinforces the point that functional tradeoffs must be taken into account when assessing which rates are appropriate for performing functions in various situations.

Nevertheless, different contributions to survival are appropriate at different stages of development. The other biological goals—development, reproduction, and helping others—require different contributions by different types of organisms. Specifically, development requires different contributions at different stages of development, reproduction requires different contributions by (fully developed) male and female reproductive traits, and helping others requires a sufficient degree of development and possibly biological polymorphism (as in the case of workers and soldiers in certain insect species). Because of this, explaining development, reproduction, and helping requires different contributions by different classes of organisms, which in turn requires the performance of different functions at different rates by traits belonging to different classes of organisms.[12]

To summarize, we identify relevant classes of organisms and their traits' functions by considering the whole developmental trajectory of *successful* members of a population, including how different developmental stages, different sexes, and (if applicable) other polymorphisms contribute to the life cycle. As we do that, we determine the ways traits of members of each developmental stage, sex, or (if applicable) other polymorphisms contribute to survival, development, reproduction, or helping. Those are their functions. As we identify such functions, we also determine when some functions must be prioritized over others, so that other functions must be delayed or performed at a lower rate. Taking functional tradeoffs into account, the circumstances in which traits of successful organisms perform their functions are the appropriate circumstances for performing such functions, and the rates at which they perform them are the appropriate rates.

When anything other than a functional tradeoff prevents a function from being performed at an adequate rate, or from being performed at all, we finally enter the realm of malfunction. Malfunction occurs when, in situations that are appropriate for the performance of a function, either the function is not performed or it's performed at an inadequate rate. As before, what counts as adequate depends on the explanandum. Fix the explanandum, identify the functional tradeoffs, and you establish adequate performance rates. Anything else that prevents functions from being performed adequately is a source of malfunction.

Here is a more general way to articulate this account, for both organism and artifacts, for both biological and nonbiological functions. Begin by providing a mechanistic explanation of the whole population of organisms, looking at all the functions performed, by either biological traits of organisms or their artifacts, in the service of the organisms' goals. Consider how different classes of organisms have different goals or different roles in pursuing biological goals, and consider

[12] These are the reference classes that Boorse (1977, 1997, 2002, 2014) talks about, here derived directly from the improved goal-contribution account of functions rather than conveniently assumed to exist in a way that suits the account's needs.

the organisms that are successful in pursuing their goals. Juveniles have to develop biologically in a way that adults do not. For reproduction to be possible, female reproductive organs must function in different ways than male organs. And so forth. At any given time, there are also tradeoffs between the functions performed. Being active prevents resting and restoring, resting and restoring prevents being active. Giving birth prevents other functions from being performed.

We can identify functions without considering functional tradeoffs, simply based on what contributions are necessary to reach organisms' goals. Identifying appropriate situations for the performance of functions and appropriate rates of performance, however, requires taking functional tradeoffs into account.

To determine the class of relevant situations, take all the possible situations in which a trait or artifact can be found (besides the accidentally beneficial effect situations), subtract those in which the function need not be performed (e.g., no food in stomach), then subtract those in which the trait cannot exercise its function at all because it's prevented from doing so by the performance of some other functions that takes priority at that time (e.g., food in stomach but extremely strenuous exercise). You are left with the appropriate situations for that function's performance.

(At this point, if you were to subtract the situations in which a trait or artifact is prevented from performing its function by some other factor, you would get the set of situations in which the trait or artifact actually performs its function.)

Now, to determine the rates of functioning that are appropriate in any situation, eliminate the last part of clause (ii) in Garson and Piccinini 2014's account of appropriate rate of functioning. (That is, eliminate "relative to other rates that are physiologically possible for X in s"). The result is the following generalized account of appropriate rate of functioning:

> Trait or artifact X possessed by organism O performs function F at a rate of functioning that is appropriate in a situation s $=_{\text{def}}$
>
> (i) If X is in s and s $\notin$ S, then X performs function F at a rate of zero (or close to zero).
>
> (ii) If X is in s and s $\in$ S, then X's rate of functioning provides an adequate contribution to one of O's goals.

To identify what counts as adequate contribution to one of O's goals, proceed as follows. First, look at X's that can perform function F well enough to pursue O's relevant goal to the degree that needs explaining. Next, look at what rates of functioning are possible in various situations (which is, roughly, the comparison suggested by Garson and Piccinini 2014 but performed over a wide range of situations) to make sure that in such organisms F provides a relevant contribution to reaching the relevant goal. While doing so, consider the way such rates of functioning are affected by the performance of other functions in various situations. Whatever rates provide appropriate contributions to reaching the relevant goals, when you consider only the ways such rates are affected by other functional priorities, are the appropriate rates. In other words, you must consider the functional trade-offs that are in play within such organisms and artifacts and how they affect the appropriate rate of performance of F in various situations including s. If the situation is s_2 (slowed digestion during exercise), the slowed rate of functioning is due to a functional trade-off, hence it's appropriate. If the situation is s_4 (slowed digestion during poison-induced impairment), it should not be included in the analysis of appropriate rates because by hypothesis we are only looking at organisms that can perform F well. This fixes the appropriate rates of functioning of F in various situations.

Now we can look at other situations, in which rates of functioning are affected by other factors, and see whether they are appropriate. In s_4, the low rate of functioning is not due to a functional trade-off, hence it is inappropriate. More generally, any rate of functioning that departs from the range of rates identified by taking functional tradeoffs into account is inappropriate and constitutes a malfunction.

3.5 Functions as Goal Contributions

To round out our discussion, let's look again at three ways in which attributions of functions to traits are used both in ordinary language and in science, which give rise to three desiderata for accounts of functions (Garson 2019a). First, functional attributions have normative import: traits that do not perform their function when they should exhibit malfunction. Second, functional attributions identify special causal roles that are not accidental: accidental contributions to a goal are not functions. Third, functional attributions may contribute to explaining the origin of traits: they may be there because that is their function. How do the three accounts address these three roles of functional attributions?

The causal role account either rejects the desiderata or deals with them in terms of explanatory interests. Insofar as we are interested in some effects not others, we may feel that, when those effects are lacking, something went wrong. That's all a malfunction is. Ditto for the distinction between functions and accidents. There is nothing objective about malfunctions or the distinction between functions and accidents. And functions have nothing to do with the origin of traits. In sum, the causal role account deflates the three desiderata. Naturally, anyone who takes ordinary function talk seriously will look for an alternative account.

The selectionist account fulfills the three desiderata in terms of selection processes. Accidents are not selected for, functions are. Being selected for is also what gives functions normativity: when traits don't do what they were selected for, they malfunction. As to the third desideratum, that they were selected for is (part of) the explanation for the origin of traits.

Finally, the goal-contribution account fulfills the first desideratum in terms of contribution to a goal. Functions are causal roles that make regular contributions to some goal of organisms; when those contributions are lacking, the goals are not contributed to, and that's what a malfunction is. The second desideratum is accounted for in terms of the regularity and reliability of causal roles that constitute functions. An accidental contribution to a goal is not regular and reliable enough to be a function, and therefore it cannot be selected for. Finally, the goal-contribution account does not come with an immediate explanation of the origin of traits. Far from being a weakness of the account, however, here is where the goal-contribution account has an additional advantage over selectionism.

Many traits originate via selection processes—the selectionist account has a perfectly adequate partial explanation for those. As we've seen, however, many useful traits originate in other ways. Some originate by random mutation. Some originate by deliberate planning. Some originate by genetic drift. Some originate via bacterial conjugation. Some strands of an organism's DNA are absorbed from viral invaders; how that happens is not well understood, but the important point is that DNA of viral origin can benefit the host organism—it can perform a function. By staying neutral on the explanation for a trait's origin, the goal-contribution account avoids the false implication that all functional traits are selected for.

Even when traits are selected for, selection is only one part of the explanation. Development is another big part, and development involves complex interactions between innate and environmental influences (Northcott and Piccinini 2018). So, explaining why traits are there, even when selection is part of the story, involves a lot more than selection. By staying neutral on the explanation for a trait's origin, the goal-contribution account avoids the false implication that selection is a sufficient etiological explanation for traits that are selected for.

This being said, there are plenty of traits for which selection is an important part of the explanation—specifically, selection alone explains why certain traits prevail over others within certain populations. Thus, identifying what a trait was selected for is illuminating and worth pursuing. The goal-contribution account is perfectly consistent with explaining why selected traits are where they are via selection. Yet the goal-contribution account avoids the mistake of attributing functions when and only when a trait was selected for. Instead, it's precisely the performance of a function that explains why a trait was selected for. Selectionism points at one useful (part of) an explanation of some traits' origin, but it mistakes a sound etiological explanation of the origin of current traits for the ground of their function.

In conclusion, the three accounts of function are not so much in competition with one another as they are complementary. The core notion of teleological function is that of a causal role that contributes to a goal. In some contexts, especially outside the life sciences, we may use a more austere, nonteleological notion of function as mere causal role. In other contexts, especially when selectionist considerations are pertinent, we may use a stronger notion of function, such that functions are selected causal roles (cf. Garson 2019a, chap. 9). Nevertheless, we should remember that those causal roles were selected for in the first place *because* they made a regular contribution to a goal.

Now that we have a viable notion functional mechanism—a mechanism with teleological functions—we are in a position to articulate an adequate functionalist metaphysics that will be the foundation for the rest of this book. That's the next chapter's job.

4
Mechanistic Functionalism

4.1.1 Functional Natures within Mechanisms

Functionalism about X is the view that the nature of X is functional. Mechanistic functionalism embeds this claim in the functions of mechanisms and their components. We can now leverage the framework introduced so far to make sense of functionalism.

In the first three chapters, I introduced mechanisms, including functional mechanisms. Functional mechanisms perform teleological functions, which in turn are regular contributions to the goals of organisms. Teleological functions can be decomposed into subfunctions, which are performed by components of mechanisms. For now, though, let's focus on mechanisms in general; we'll get back to functional mechanisms in a bit. Recall that neither whole mechanisms and their properties nor mechanistic components and their properties are more fundamental; instead, whole mechanisms and their properties are invariants under certain transformations of components and their properties. Higher-level properties are aspects of their lower-level realizers.

When a higher-level property can be an aspect of different lower-level realizers that constitute different mechanisms, we call it multiply realizable. When a multiply realizable property is defined solely in terms of the manipulation of certain degrees of freedom, without reference to any other properties of the medium in which it can be realized, we call it medium independent.

Each level of organization within a mechanism consists of structures possessing causal powers, standing in certain organizational relations to one another. The whole mechanism possesses its own specific causal powers in virtue of the causal powers of its components and how they are organized. Specifically, the causal powers of the whole mechanism are aspects of the causal powers of the organized components.

If we consider a system as a whole, we can focus on one of its specific causal powers and use it to define a kind of system. For example, we can single out the causal power to break very suddenly and violently into pieces, regardless of the specific mechanism by which the causal power is achieved. Such a causal power may be called a *functional property*. Functional properties are to be distinguished from qualities, such as having a certain mass or shape, and structural properties, such as having components of a certain kind. This notion of functional property

Neurocognitive Mechanisms: Explaining Biological Cognition. Gualtiero Piccinini, Oxford University Press (2020).

DOI: 10.1093/oso/9780198866282.001.0001

does not require that there be any *teleological* function involved; it's the more encompassing notion of function as causal role (Chapter 3). If we define a kind of system by a functional property in this broad sense, we thereby define a functional kind—that is, a kind that has a *functional nature*. In our example, anything that possesses the power to break sufficiently suddenly and violently into pieces is called an *explosive*. Exploding is a functional property; therefore, by definition, explosives as such have a functional nature.

Since functional kinds are defined solely in terms of specific causal powers, they are typically multiply realizable. That is, there are different mechanisms by which the causal power can be obtained and the causal role can be performed. In our example, the three main explosion mechanisms are deflagration, detonation, and certain nuclear reactions. In special cases, there may be only one mechanism that provides the required causal power. For example, if we define a functional kind as the explosive that releases the greatest amount of energy in the shortest amount of time, there is likely to be only one type of structure that fits this description. In this sort of exotic case, a functional kind is not multiply realizable. In the ordinary case of interest to most laypeople and scientists, multiple realizability is the rule.

There is a special type of causal power that is defined without reference to any specific concrete cause or effect, except for the possession of certain degrees of freedom and relations between them. For example, we may write a system of differential equations and define a functional kind as anything that satisfies that system of equations. Anything that possesses enough variables standing in the right dynamical relations counts as a member of that kind. Except for exotic cases analogous to those discussed in the previous paragraph, functional kinds that are defined like that—without reference to any specific concrete effect besides satisfying a system of equations—are not only multiply realizable but also medium independent. That is, whereas ordinary functional kinds must all share the concrete causal power that defines them, mechanisms that possess a causal power defined without reference to any concrete cause or effect need not share any specific physical properties besides the possession of certain degrees of freedom and relations between them. They form a medium-independent kind.

We can now go back to systems and mechanisms that have teleological functions. The same distinctions apply. We can use a teleological function to define a functional kind, which is typically going to be multiply realizable. Alternatively, if the kind is defined with no reference to specific causes or effects beyond certain degrees of freedom, it is medium-independent. For example, we can single out the teleological function to lift corks out of bottlenecks and thereby define the functional kind *corkscrew*. We can do it more generally, so that any way of lifting corks out of bottlenecks suffices, or more specifically, e.g., by requiring that the lifting of corks out of bottlenecks relies on screwing something into the cork. Either way, using a specific causal power to define a kind gives rise to a functional kind, and that kind is typically multiply realizable. Another example:

we can single out the teleological function to compute prime factorization and thereby define the functional kind *prime factorization computing system*. Unlike lifting corks out of bottlenecks, which requires manipulating corks and bottlenecks, computing prime factorization does not require any particular physical property, except for possessing enough degrees of freedom to encode numbers and manipulating the encodings in the right way. Thus, the kind *prime factorization computing system* is medium independent.

As I will use the term, a mechanism's *functional organization* includes the states and activities of components, the spatial relations between components, the temporal relations between the components' states and activities, and the specific ways the components' states and activities affect one another. For example, the heart *sucks* blood *from* the veins and *pumps* it *into* the arteries. This simple mechanistic description can begin to be unpacked as follows: (i) the mechanism includes a heart (component), veins and arteries (components), and blood (medium), (ii) the heart sucks and pumps the blood (activities of the heart), (iii) the heart is attached to the veins and arteries in a certain way (spatial relation), (iv) blood enters the heart after it leaves the veins and enters the arteries after it leaves the heart (temporal relations), (v) the heart's sucking and pumping causes the blood to exit the veins and enter the arteries (causal relations). The relevant spatial relations between components may continue to hold even when the mechanism is not functioning. Not so for most temporal and causal relations. Much more could be said about functional organization. The important point is that the functional organization of a mechanism is a necessary condition for the mechanism to do what it does. This notion of functional organization is what we need in order to explicate functionalism.

4.1.2 Mechanistic Functionalism

Functionalism in the present sense originates in the 1960s as a solution to the mind–body problem—that is, the problem of understanding the relation between mind and body. Functionalism says that the mind has a functional nature. Another way to put this is that the mind is the functional organization of the organism (Putnam 1967a; Fodor 1968a). If we generalize, functionalism about X says that X has a functional nature, or that X is the functional organization of its realizers. We can now clarify what this means, or at least what it *should* mean, and what does and does not follow from it.

Roughly, functionalism says that mental states are functional states or, equivalently, that mental properties are functional properties. That is, mental properties are defined by their causal powers within certain systems at certain levels of organization. The relevant systems are presumably human beings and systems that are sufficiently similar, such as cognitively advanced animals and robots. The relevant levels are the levels at which we can identify mental states—whatever, exactly, those are (more on that in later chapters).

Functional states occur within mechanisms, which are the subject matter of most special sciences—including neuroscience, psychology, and computer science. Investigators in these disciplines analyze systems (e.g., trees) by breaking them down into component parts (e.g., roots) and discovering (and, in engineering, designing) the functions of those parts (e.g., supporting the tree and absorbing water from the soil). Neuroscientists and psychologists elaborate their theories in the same way: they partition the brain or mind into components (e.g., the suprachiasmatic nuclei or episodic memory) and they ascribe them functions (respectively, regulating circadian rhythms and storing records of events). Mutatis mutandis, computer scientists do the same thing: they partition a computer into components (e.g., the memory and the processor) and ascribe them functions (respectively, storing data as well as instructions and executing instructions on the data).

Since mechanisms give us the notion of functional organization that is relevant to understanding theories in psychology, neuroscience, and computer science, we should adopt this notion of functional organization in our formulation of functionalism. We can now give a clear formulation of functionalism, which does justice to its original motivations.

> Functionalism about *X*: *X* is the functional organization of the mechanism that possesses *X*.

We can now explicate the claim that the mind is the functional organization of the organism. In biological organisms, the main organ of the mind is the brain: the brain contains the most relevant mechanisms, and the mind is its functional organization. If there are other minded systems, which are not biological or at any rate have a mind but lack a brain, functionalism says that the mind is a type of functional organization that can be shared between brains and the mental mechanisms possessed by such systems. This *mechanistic functionalism* preserves functionalism's insight while doing justice to the relevant scientific practices.[1]

Under this mechanistic version of functionalism, a mechanism is individuated by its component parts, their functions, and their relevant causal and spatiotemporal relations. The functional states of the system are individuated by their role in fulfilling the system's functions, as specified by a mechanistic explanation of the system. The states of the system are not only individuated by their relevant causal relations to other states, inputs, and outputs, but also by the component to which they belong and the function performed by that component when it is in that state.

[1] Carl Gillett (2013) has independently developed a proposal similar to what I'm calling "mechanistic functionalism," with which he addresses other aspects of functionalism. See also Bartlett 2017.

This applies to all mechanisms, including computing mechanisms. For example, pace Putnam (1967a), ordinary Turing machine states are individuated not only as having the function of generating certain outputs and other internal states on the basis of certain inputs and states, but also as being states *of* the active device (as opposed to the tape), which is a component of the Turing machine and has the functions of moving along the tape, reading the tape, and writing on it.

Functionalism is especially plausible for so-called cognitive states and functions. These are states and functions defined in terms of the kind of internal states and outputs they produce in response to certain inputs and internal states. In particular, cognitive states and functions are generally defined in a way that does not presuppose they are phenomenally conscious. Since their inputs-internal states-outputs relations define them without reference to any qualitative properties, it is plausible that cognitive states and functions have a functional nature. Functionalism plausibly holds about them.

Functionalism is less plausible for so-called phenomenally conscious states. These are states defined in terms of how they feel, or what it is like to be in them. Phenomenally conscious states appear to be qualitative, so many authors find it hard to believe that their nature could be wholly functional. There seems to be something utterly qualitative about them, which cannot be reduced to functional organization alone. I will say a bit more about this later in this chapter (see Section 4.3.24: Functionalism 6.5) and in Chapter 14. From now on, except when otherwise noted, I will focus on functionalism about cognitive states and functions, although many of the following points apply to functionalism about any X.

We can immediately establish a number of points about functionalism.

First, unless cognition is an exotic case in the same league as *most powerful explosive*, *hardest material*, and the like, cognition is multiply realizable. It is very unlikely that cognition is such an exotic case. There are many cognitive systems with varying degrees of sophistication, so cognition is not any kind of superlative concept with a unique value. In addition, cognition evolved independently in different taxa, giving rise to rather different kinds of nervous system. Furthermore, many artifacts exhibit some cognitive capacities, and the machinery behind their cognitive functions is quite different from the neural machinery biological cognizers employ. Therefore, contra Polger and Shapiro (2016) and other skeptics about the multiple realizability of cognition, we should expect that cognition be multiply realizable—that different kinds of mechanism give rise to the same or similar cognitive states and functions.

Second, cognition is likely to be not only multiply realizable but also medium independent to a large extent. This is because the inputs and outputs over which cognitive states and functions operate can often be defined rather abstractly, with no reference to specific physical realizations. Of course, inputs and outputs can also be defined in terms of a specific realization—e.g., in terms of specific patterns of light waves coming into the system and specific movements of specific bodily parts coming out. But they don't have to. We can define at least some cognitive functions in terms of relations between abstractly specified inputs and outputs—e.g., chess positions as inputs and chess moves as outputs—without any reference to any specific way to encode such positions and moves. As a consequence, cognition is likely to be medium independent to a large extent.

And yet, third, functionalism—even cum multiple realizability—does not have many of the corollaries that it is often taken to have. It does not entail traditional anti-reductionism, the autonomy of higher-level sciences, or computationalism.

Functionalism does not entail traditional anti-reductionism for reasons I canvassed in Chapter 1. Traditional anti-reductionism asserts that higher-level properties are distinct from their lower-level realizers. On the contrary, I argued that higher-level properties are aspects of their realizers. Needless to say, aspects are not wholly distinct from the realizers of which they are aspects. This aspect view applies to functional properties as much as any properties. More precisely, functional properties are aspects of the organized plurality of properties of the parts of the system that possesses them. In yet other words, suppose that system S possesses functional property F. And suppose that S is composed by subsystems $S_1, \ldots S_n$, which have properties $F_1, \ldots. F_m$. For simplicity, I assume that $F_1, \ldots F_m$ include the organizational relations between $S_1, \ldots S_n$. According to the aspect view, F is an aspect of $F_1, \ldots F_n$ taken collectively. All of this is consistent with functionalism. Therefore, functionalism does not entail traditional anti-reductionism. Instead, functionalism fits ontological egalitarianism perfectly well.

Functionalism does not entail the autonomy of higher-level sciences for a closely related reason. The autonomy of higher-level sciences is best defended via the kind of traditional anti-reductionism that functionalism and multiple realizability are mistakenly thought to entail. For if higher-level properties are distinct from their lower-level realizers, then studying their realizers is neither necessary nor sufficient for understanding higher-level properties. By the same token, if higher-level properties are distinct from their lower-level realizers, there seem to be no direct constraints from such realizers to the properties they realize. At any rate, this is what many think. If, on the contrary, higher-level properties are aspects of their realizers, understanding them fully requires understanding what they are aspects of, which in turn requires understanding their realizers. Realizers constrain what can be an aspect of them, so lower-level properties directly constrain higher-level properties. Therefore, higher-level sciences are not autonomous from lower-level sciences. I will discuss the autonomy of higher-level sciences in more detail in Chapter 7. For now, let's just register that functionalism does not entail the autonomy of higher-level sciences.

Finally, functionalism does not entail computationalism—the view that cognitive states are computational—for the simple reason that most functional properties are not computational. Or, at least, most functional properties are not computational in the interesting sense that motivates those who conflate functionalism and computationalism. Nevertheless, some functional properties are medium independent. I also mentioned that cognition looks medium independent to a large extent. If this is correct, and if medium independence has something to do with computation in the interesting sense, there is an argument that cognition is computational. Before we can appreciate all of this, we need to wait until Chapter 6, where I argue that a process is a computation if and only if it is the processing of medium independent vehicles by a functional mechanism in accordance with a rule and that neurocognitive processes are medium independent and therefore computational.

To recap, multilevel mechanisms have causal powers, some of which perform teleological functions. Causal powers, including those that perform teleological functions, can be used to define functional kinds. According to functionalism about property X, X is a functional kind. According to functionalism about cognition, cognition is a functional kind. Functionalism makes it likely that cognition is multiply realizable and perhaps even medium independent. It does not entail traditional anti-reductionism, computationalism, or the autonomy of higher-level sciences.

In the rest of this chapter, I will situate mechanistic functionalism within the broader debate about the mind–body problem. Functionalism comes in many versions. I will list the main versions of functionalism and their relations as if they were subsequent releases of a software package, describing some of their important features as well as some of the bugs that plague them. Many have been pessimistic about functionalism's prospects, but most criticisms have missed the latest upgrades. I end by suggesting a version of functionalism that, in combination with the egalitarian ontology defended in previous chapters, potentially provides a complete account of the mind.

4.2 The Mind–Body Problem

The mind–body problem is how the mental relates to the physical. We know they are intimately involved: the physical affects the mental when our body is injured and we feel pain, or when we drink alcohol and feel intoxicated; the mental affects the physical when we decide to retrieve a book and grab it, or when we perceive the signal to cross the street and begin walking.[2]

The first solution that comes to mind is dualism: the view that the mental and physical are two different kinds of thing. While dualism is intuitively appealing, it faces such serious difficulties (how to account for the intimate connections

[2] The rest of this chapter includes a substantially revised and expanded descendant of Maley and Piccinini 2013, so Corey Maley deserves partial credit for much of what is correct here.

between the mental and the physical, how two very different kinds of thing could possibly interact, etc.) that it's been mostly abandoned.

The other option is monism, according to which the mental and the physical are the same kind of thing. What might seem like two kinds of thing are actually only one. The unpopular version of monism is idealism, which holds that the mental is basic and the physical is to be explained in mental terms. The popular version of monism is physicalism, which holds that the physical is basic and the mental is to be explained in physical terms.

Bare physicalism does not completely solve the mind–body problem: we must still account for the precise nature of mental states. Behaviorists hold that mental states are dispositions to behave in certain ways, but behaviorism is another view that has been mostly abandoned because, among other reasons, it has a hard time accounting for occurrent mental states such as conscious feelings. Type identity theorists hold that mental state types are just physical state types; for instance, beliefs are a certain type of physical state, and desires are another. Token identity theorists hold that each individual mental state is just an individual physical state, but mental types need not correspond to any physical type. Type identity and token identity are helpful in grounding the mental in the physical. But they do not explain the interestingly *mental* aspects of those physical states (or state types) as opposed to other physical states (or state types).

Functionalism is a popular option for explaining the uniquely mental aspects of the mind. What makes a state mental is its function, usually understood to be a relationship between inputs, outputs, and other functional states. Functionalism thus allows us to explain certain states as playing a role in the mind, while being consistent with physicalism.

Throughout the history of functionalism, many versions of functionalism have been offered, with new features added, and certain bugs fixed. I provide here an overview of the major releases of functionalism—a sketch of its version history. This "history" is more of a rational reconstruction than an accurate chronology; it does not always respect the order in which things actually happened. I will list different versions of functionalism from its beta version through its first release and subsequent upgrades. In the end I offer a final upgrade: Functionalism 6.5.

4.3 From Functionalism Beta Version to Mechanistic Functionalism

4.3.1 Functionalism Beta Version: Functional States

The beta version of functionalism is simply the view that mental states are functional states (Putnam 1967b). Functional states, according to Functionalism beta, are states of a system individuated by (some of) their causal relations to inputs, outputs, and other functional states. Thus, the picture of the mind painted by Functionalism beta is the picture of a system that receives sensory inputs, manipulates the inputs via a complex network of causally interconnected functional states (i.e., mental states), and then produces behavioral outputs.

Functionalism beta stands in contrast to dualism, according to which mental states are nonphysical states. Thus, Functionalism beta avoids the notorious pitfalls of dualism, including dualism's inability to do justice to mental causation. (But see Functionalism 1.1 below for an important caveat about functionalism vs. dualism.) Functionalism beta also stands in contrast to input–output (or metaphysical) behaviorism, according to which mental states are nothing more than behavioral dispositions. Thus, Functionalism beta avoids the notorious pitfalls of input–output behaviorism, including its inability to make sense of occurrent mental states (such as occurrent sensations and thoughts).

Despite these features, the beta version of functionalism lacks too many features to be a viable software package: What, exactly, are functional states? Are they physically realized? Are functional types identical to physical types? Functional states are individuated by their causal relations, but which causal relations? Presumably not all of them, but then how are the relevant causal relations picked out?

4.3.2 Functionalism 1.0: Folk Psychology

Functionalism 1.0, the first working release, includes the feature that functional states are specified by a folk psychological theory (Lewis 1972). Folk psychology specifies a number of platitudes about the mind. For example, people generally avoid doing things that cause them pain; or people generally try to achieve what they desire. According to Functionalism 1.0, the conjunction of all such folk psychological platitudes constitutes a theory (Lewis 1970). The terms of the theory refer to the functional states that correspond to mental states. The relations specified by the platitudes between the functional states, their inputs, and their outputs are the causal relations that individuate the functional states.

Functionalism 1.0 turned out to have a series of bugs: it is doubtful that there really is a suitable body of folk psychological platitudes about the mind; if there were such a body, it is doubtful that it would constitute a theory in the relevant sense; if it did constitute such a theory, it is doubtful that it would be correct or precise enough to pick out people's (or animals') actual mental states. Because of these bugs, Functionalism 1.0 never ran successfully.

Functionalism 1.0.1 replaces folk psychological theory with analytic truths about the mind (Shoemaker 1981). But Functionalism 1.0.1 never compiled correctly because no one could find a suitable list of analytic truths about the mind, so the project was abandoned.

4.3.3 Functionalism 1.1: Scientific Psychology

Functionalism 1.1 avoids the bugs that plagued Functionalism 1.0 by replacing folk psychological "theory" with scientific psychological theory (Fodor 1968a). The functional states are picked out by psychological laws, or perhaps psychological generalizations, to be discovered empirically by experimental psychologists.

So far so good. But we still don't know whether and how functional states are physically realized. Some early proponents of functionalism went out of their way to point out that functional states *need not* be physically realized. They may be states of a nonphysical substance (Putnam 1967a). If so, then functionalism is consistent with dualism, with all its mysteriousness and notorious problems. Something had to be done.

4.3.4 Functionalism 2.0: Type Identity

Functionalism 2.0 attempts to solve the realization problem that plagued earlier versions by including the feature that functional states are type-identical to physical states. In other words, each type of functional state is identical to a type of physical state (Armstrong 1968; Lewis 1966, 1972). Thus, functionalism 2.0 is consistent with the type-identity theory of mind, which is a version of the standard reductionism I discussed in Chapter 1: mental state types are physical state types (Place 1956; Smart 1959).

A new bug arises with Functionalism 2.0: multiple realizability. Mental states appear to be multiply realizable. For example, pain appears to be realizable in physically different neural (and perhaps even nonneural) systems. Whether mental states are multiply realizable, functional states are multiply realizable by different lower-level types; a smoke detector in a smoke alarm can be made with an optical sensor or one that uses ionizing radiation. Hence, functional types do not appear to be the same as physical types, because functional types are multiply realizable, whereas physical types allegedly are not.[3]

4.3.5 Functionalism 2.1: Token Identity

Functionalism 2.1 fixes the multiple realizability bug by replacing token identity for type identity. According to Functionalism 2.1, functional states are realized by,

[3] This is roughly how anti-reductionist functionalists understand the matter. The debate between reductionists and anti-reductionists was affected by a widespread conflation of the physical and the *micro*physical (cf. Hütteman 2004; Papineau 2008). In Chapter 2, I offered an ontologically egalitarian account of MR that is neither reductionist nor anti-reductionist in the traditional sense. Higher-level physical properties—including functional ones—may be multiply realizable even though their microphysical realizers are not.

but not identical to, physical states. What this comes down to is that each token of a functional state type is identical to a token of a physical state type, but different tokens of the same functional state type may be realized by tokens of different physical state types. Thus Functionalism 2.1 entails token physicalism: each token of a functional state is realized by a token of a physical state (cf. Fodor 1974, building on Davidson's (1970) token identity theory).

4.3.6 Functionalism 3.0: Machine State Functionalism

Previous versions of functionalism asserted that the functional relations between inputs, outputs, and mental states are specified by a theory. After Functionalism 1.1, such a theory is a scientific psychological theory. What kind of scientific theory? How does such a theory specify the functional relations? Functionalism 3.0 adds that the theory is computational. Specifically, the functional relations are computational relations specifiable by a Turing machine table (Putnam 1967a). As a consequence, according to Functionalism 3.0, mental states turn out to be Turing machine states.

One problem with Functionalism 3.0 is that a Turing machine is in one and only one state at a time, whereas minds appear to possess more than one mental state at a time. In addition, cognitive scientists never used Turing machine tables to formulate computational theories of cognition (Sloman 2001). What they did use is computational models of various kinds. So, while Functionalism 3.0 ran, nobody wanted to use it.

4.3.7 Functionalism 3.1: Psycho-functionalism

To address the shortcomings of Functionalism 3.0, Functionalism 3.1 replaces Turing machine tables with a computational psychological theory. The functional relations between inputs, outputs, and mental states are computational relations specified by the computational theory (Block and Fodor 1972). Mental states are thus computational states.

4.3.8 Functionalism 3.2: Functional Analysis

Functionalism 3.1 says that psychological explanation is computational but doesn't articulate how computational explanation is supposed to work. Functionalism 3.2 remedies this bug by adding that psychological explanation is functional analysis (Cummins 1975; Fodor 1968a).

Functional analysis, in turn, comes in three flavors. One flavor is simply the analysis of a system and its capacities in terms of the functional relations between its inputs, outputs, and internal states (Fodor 1968b). Another flavor adds that the internal states are contained in functionally defined components or black boxes, which reside inside the system (Fodor 1983). A third flavor analyzes a system's capacities in terms of that same system's subcapacities; the functions of these subcapacities are nothing more than the contributions they provide to the capacities of the system (Cummins 1983).

Although functional analysis applies to both computational and noncomputational systems, users of Functionalism 3.2 typically also maintain that minds are both functional and computational. According to Functionalism 3.2, then, the mind is a functional system to be functionally analyzed. The analysis is carried out by a computational theory that specifies the computational relations between inputs, mental states, and outputs.

4.3.9 Functionalism 3.3: Teleofunctionalism

Functionalism 3.3 adds the feature that the functions of the subcapacities or components of the mind, in whose terms the functional analysis is carried out, are teleological functions (Lycan 1981, 1987; Wilkes 1982; Millikan 1984, 1989; Sober 1984, 1990; Shapiro 1994; Rupert 2006). Functionalism 3.3 comes in a number of variants. Teleological functions may be characterized as selected effects, but they could also be characterized as regular contributions to a goal of organisms (Chapter 3). Some supporters of Functionalism 3.3 maintain that once teleological functions are added, computational relations should be dropped (Lycan 1981; Millikan 1984; Sober 1990). But this is by no means a necessary aspect of Functionalism 3.3. For present purposes, all that matters is that, according to Functionalism 3.3, the mind is functionally analyzed in terms of the teleological functions of its components.

4.3.10 Functionalism 3.4: Mechanistic Functionalism

Functionalism 3.4 further enriches the account of psychological explanation that grounds functionalism. Instead of functional analysis, Functionalism 3.4 maintains that psychological explanation is mechanistic, that is, it is a decomposition of a system in terms of its components, their activities, and their organization. In Chapter 7, I will argue that functional analysis itself is subsumed under mechanistic explanation, because functional analyses are sketches of mechanisms.

According to Functionalism 3.4, the mind remains a functional system, but it's also a mechanism. The states and capacities of the system still have functions, which may be cashed out either as causal roles or in teleological terms. If teleological, functions may be characterized either as selected causal roles or as causal roles that contribute to goals of organisms. For present purposes, the best account is that teleological functions are causal roles that contribute to goals of organisms (Chapter 3).

4.3.11 Functionalism 4.0: Representationalism

Mental states appear to have intentionality: one's belief that the sun is a star is *about* the sun. So far, functionalism has had nothing to say about this. Functionalism 4.0 combines functionalism with representationalism—the view that mental states represent, and this explains their intentionality. But what kind of representation are mental states? Functionalism 4.0 was never released on its own—it came in different versions, each answering this question in its own way.

4.3.12 Functionalism 4.1: The Language of Thought

One idea for how mental states might represent comes from a close analogy with language. Language is *productive*: starting with some basic elements, an infinite number of sentences can be constructed in structured ways. Similarly, language is *systematic*: if a person can utter the sentence "Alex chased Dana" then she can also utter the sentence "Dana chased Alex." Thought seems to have these same features, suggesting a natural parallel between the structure of language and the structure of thought. This is the language of thought hypothesis (Fodor 1975; Harman 1973; Sellars 1963).

4.3.13 Functionalism 4.2: Connectionism

Connectionism is one putative alternative to the language of thought. Inspired by early work on perceptrons in artificial intelligence, connectionism views the mind as a multi-layered neural network. Mental representations, then, are vectors of activation within the network, established by the particular pattern of connection strengths between the nodes of the network. This pattern of connection strengths is sometimes explicitly provided to the network, but much more often it is the result of learning and training (Rumelhart, McClelland, and PDP Research Group 1986).

While proponents of the language of thought have argued that connectionism lacks the resources to explain the systematicity and productivity of thought (Fodor and Pylyshyn 1988), connectionists have argued that their view of the mind is more "brain-like," and thus a better candidate for how the mind actually works. Additionally, connectionist systems have built-in features that language of thought-based systems lack, such as the ability to learn in supervised and unsupervised ways, and the ability to degrade gracefully in the presence of damage.

4.3.14 Functionalism 4.3: Cognitive Neuroscience

If a system's being "brain-like" is a virtue, one can go further than connectionism. The kind of artificial neural networks employed by connectionist modelers did not resemble neural systems very closely; rather, they were loosely inspired by neural systems, with nodes and connections that resemble highly simplified and idealized neurons. In other words, the artificial neural networks employed by connectionists did not (and were never meant to) model how actual *neural* processes work, but how *cognitive* processes could be implemented in a system of distributed nodes in a network.

Functionalism 4.3 takes off where connectionism leaves off: rather than mental representations as vectors of activation in an artificial neural network, this version takes mental representations to be vectors of activation in *real* neural networks. This can be called neural representation. The view that mental representation is neural is one foundational tenet of cognitive neuroscience (Churchland 1989; O'Reilly and Munakata 2000; Vendler 1972; see Chapter 12).

4.3.15 Functionalism 5.0: Psychosemantics

Explaining intentionality requires more than an account of the format of mental representations. It also requires a naturalistic account of the semantic relation between mental representations and what they represent, i.e., a psychosemantics. Functionalism 5.0 adds to previous versions of functionalism such a naturalistic semantics.

As a first pass at a naturalistic semantics, Functionalism 5.0 is an internalist, or methodologically solipsist (Fodor 1980), theory. It attempts to give a semantics for mental states without leaving the boundaries of the head, that is, solely in terms of the relations between proximal inputs (such as retinal stimulations), proximal outputs (such as motor commands), and internal states. Sometimes this view is expressed by distinguishing between broad and narrow content. Narrow content is semantic content that is contained solely within the head (Segal 2000). Because Functionalism 5.0 accounts for the semantic content of mental representations in terms of functional roles that are confined within the head, it is also called short-arm functionalism.

But some argue that meanings “ain’t in the head” (Putnam 1975a). According to the popular externalist view about semantic content, semantic content depends on relationships between mental states and things outside the head. In fact, it has proven difficult to give an account of mental content that does not include the relationship between mental representations and things external to the system (in the system’s environment). Including such a relationship leads to long-arm functionalism, of which there are several variants depending on how that relationship is construed.

4.3.16 Functionalism 5.1: Functional Role Semantics

The oldest form of psychosemantics is functional role semantics, which accounts for semantic content in terms of functional relationships between environmental stimuli, internal states, and actions (Sellars 1954). It is quite a natural step to combine functional role semantics with functionalism (Block 1986; Field 1978; Harman 1973, 1999; Loar 1981).[4] After all, if mental states just *are* functional states, the thesis that these states get their content by way of their functional role within a cognitive system is explanatorily parsimonious.

4.3.17 Functionalism 5.2: Representational Deflationism

The second oldest form of psychosemantics is deflationism about mental content. One version of deflationism is instrumentalism about mental content, according to which mental content is just a convenient way to predict behavior (Dennett 1969, 1971, 1987; Egan 2014). According to this version, pragmatic concerns might invite, or even require, that we attribute mental contents (and attendant propositional attitudes) to cognitive systems. But this attribution is nothing more than that—an attribution—that is not in need of theoretical explanation.

A related view is eliminativism about mental content, according to which mental content has no role to play within a science of the mind (Stich 1983). Functionalism 5.2.1 sees mental content as a superfluous feature that earlier users only *thought* was important. A sufficiently advanced scientific account of mentality has no use for mental content, because all of the real explanatory work is done by the functional roles of mental stats. The correct account of such functional roles will be specified only in syntactic terms. Therefore, the individuation of mental states need not appeal to semantics.

[4] Some versions of functional role semantics are more internalist, others more externalist. Harman (1999) discusses this in more detail.

4.3.18 Functionalism 5.3: Informational Teleosemantics

A third family of psychosemantic theories accounts for the semantic relation between mental states and what they represent in terms of some combination of information, teleology, and control. Briefly, a mental state has its content in virtue of some combination of the natural information it carries (Dretske 1981, 1988) and its teleological function (Millikan 1984, 1993), which may include the way it is used by the brain to drive behavior (Neander 2017).

4.3.19 Functionalism 6.0: Functionalism and Consciousness

There is still a major hole in functionalism: the lack of an account of phenomenal consciousness. Why should any functional system be conscious? It was pointed out quite early that, prima facie, a system that is functionally equivalent to a human mind need not have any phenomenally conscious mental states (Block 1978; Block and Fodor 1972). Functionalism 6.0 attempts to remedy this by simply reasserting that some combination of functional relations is enough to explain consciousness, perhaps in combination with a deflationary view of what phenomenal consciousness amounts to (Dennett 1988, 1991; Lycan 1987; Rey 2005).

4.3.20 Functionalism 6.1: Representationalism about Consciousness

A related way to account for phenomenal consciousness is representationalism about consciousness, according to which phenomenal consciousness is just a representational feature of mental states (Harman 1990). Rather than an intrinsic feature of experience itself, the character of phenomenally conscious states has to do with their representational content. This is how Functionalism 6.1 attempts to accommodate phenomenal consciousness.

4.3.21 Functionalism 6.2: Property Dualism

A third way to account for phenomenal consciousness is property dualism, according to which phenomenal consciousness is due to nonphysical, nonfunctional properties of mental states. This view may be conjoined with functionalism about nonconscious aspects of mental states (Chalmers 1996b).

4.3.22 Functionalism 6.3: Reductionism about Consciousness

A fourth way to account for consciousness is to conjoin functionalism with a type identity theory of consciousness. Roughly, although many aspects of mental states are multiply realizable and thus may be accounted for in terms of the functional relations between inputs and other mental states, the phenomenally conscious aspect of mental states must be accounted for in a different way, namely, as some lower-level aspect of the mind's realizers (Block 2006; Polger 2004).

4.3.23 Functionalism 6.4: Consciousness and Powerful Qualities

Another view of consciousness is that properties are both qualities and powers, and it is their qualitative aspect—as opposed to their powers—that accounts for phenomenal consciousness (Heil 2003, 2012; Martin 2007). By this view of properties, functional states have a qualitative aspect, and in the case of states with appropriate functional roles this qualitative aspect amounts to their phenomenally conscious status. This combination of functionalism and the powerful qualities view was first proposed by Maley and Piccinini (2013; cf. Carruth 2016; Taylor 2017, 2018). The result is Functionalism 6.4: Functional states have qualities, and it is those qualities that account for the phenomenally conscious aspects of some mental states.

4.3.24 Functionalism 6.5: Consciousness and Ontological Egalitarianism

A final account combines functionalism with an account of consciousness based on ontological egalitarianism (Chapter 1). According to ontological egalitarianism, the properties that give rise to phenomenal consciousness—let's call them phenomenal properties—are one level of being among others. They are neither more fundamental nor less fundamental than their realizers. Rather, they are an aspect of their realizers. Properties come with causal powers, and phenomenal properties are no exception. Some properties have qualitative aspects, and phenomenal properties definitely have qualitative aspects. Insofar as phenomenal properties come with causal powers, they can be accounted for in functional terms. Insofar as they have qualitative aspects, they can be accounted for in qualitative terms. (Ontological egalitarianism can remain neutral about the exact relation between qualitative aspects of properties and causal powers.) I will flesh out this ontologically egalitarian account of consciousness a bit further in Chapter 14.

4.4 Get the Latest Upgrade

Having tested all versions of functionalism, I recommend that you get yourself mechanistic functionalism plus neural representations cum informational teleosemantics (Chapter 12) plus neural computations (Chapter 13) plus ontological egalitarianism about consciousness (Chapter 14). You'll have the outline of a complete account of the mind. Mental states are neural representations within a multilevel neurocomputational mechanism. Some mental states even have a qualitative feel.

It's important to reiterate that functional kinds are very likely to be multiply realizable, but functional kinds need not be medium independent. Therefore, assuming that computational kinds are medium independent (Chapter 6), functionalism about the mind does not entail that the mind is computational. As far as functionalism is concerned, some aspects of the mind, or even all aspects, may or may not be computational. Thus, a weak version of mechanistic functionalism entails that the mind is multiply realizable (or at least makes MR very likely); it takes a stronger version of mechanistic functionalism to entail that the mind is medium independence.

Nevertheless, functionalism was largely inspired by the emergence of the computational theory of cognition and is closely aligned to it. To fully understand both functionalism and the computational theory of cognition, we should take a look at its origin. Accordingly, the next chapter examines the first computational theory of cognition.

5
The First Computational Theory of Cognition
McCulloch and Pitts's "A Logical Calculus of the Ideas Immanent in Nervous Activity"

5.1 The First Computational Theory of Cognition

Previous chapters introduced a theoretical framework for understanding the computational theory of cognition (CTC). Our next step is to investigate how CTC originated. This chapter argues that the creators of CTC, in the modern sense, are Warren McCulloch and Walter Pitts. The view that thinking has something to do with computation may be found in the works of some modern materialists, such as Thomas Hobbes (Boden 2006, 79). But CTC properly so-called could not begin in earnest until a number of logicians (most notably Alonzo Church, Kurt Gödel, Stephen Kleene, Emil Post, and especially Alan Turing) laid the foundations for the mathematical theory of computation.

Modern CTC began when Warren McCulloch and Walter Pitts (1943) connected three things: Turing's work on computation, the explanation of cognitive capacities, and the mathematical study of neural networks. Neural networks are ensembles of connected signal-processing units ("neurons"). Typically, they have units that receive inputs from the environment (input units), units that yield outputs to the environment (output units), and units that communicate only with other units in the system (hidden units). Each unit receives input signals and delivers output signals as a function of its input and current state. As a result of their units' activities and organization, neural networks turn the input received by their input units into the output produced by their output units. A neural network may be either a concrete physical system or an abstract mathematical system. An abstract neural network may be used to model another system (such as a network of actual neurons) to some degree of approximation.

The mathematical study of neural networks using biophysical techniques began around the 1930s (Rashevsky 1938, 1940; Householder and Landahl 1945; see also Aizawa 1996; Abraham 2004, 2016; Schlatter and Aizawa 2007). But before McCulloch and Pitts, no one had suggested that neural networks have something to do with computation. McCulloch and Pitts defined networks that operate on

Neurocognitive Mechanisms: Explaining Biological Cognition. Gualtiero Piccinini, Oxford University Press (2020).

DOI: 10.1093/oso/9780198866282.001.0001

sequences of discrete inputs in discrete time, argued that they are a useful idealization of what is found in the nervous system, and concluded that the activity of their networks explains cognitive phenomena. McCulloch and Pitts also pointed out that their networks can perform computations like those of Turing machines. More precisely, McCulloch–Pitts networks are computationally equivalent to Turing machines without tape or finite state automata (Kleene 1956).

This chapter elucidates McCulloch and Pitts's theory of cognition and shows that their contributions includes (i) a formalism whose refinement and generalization led to the notion of "finite automata," which is an important formalism in computability theory, (ii) a technique that inspired the notion of logic design, which is a fundamental part of modern computer design, (iii) the first use of computation to address the mind–body problem, and (iv) the first modern CTC, which posits that neurons are equivalent to logic gates and neural networks are digital circuits.

McCulloch and Pitts's theory is *modern* in the sense that it employs Turing's mathematical notion of computation. So, for instance, although Kenneth Craik's theory of cognition was published around the same time (Craik 1943), it is not a *modern* computational theory in the present sense because it appeals to computation only in an informal sense. The modern CTC is often credited to Turing himself (e.g., by Fodor 1998). Indeed, Turing talked about the brain first as a "digital computing machine" (Turing 1947, 111, 123) and later as a sort of analog computer (Turing 1948, 5; 1950, 451). But Turing made these statements in passing, without attempting to justify them, and he never articulated a CTC in any detail. More importantly, Turing made these statements well after the publication of McCulloch and Pitts's theory, which Turing knew about.[1] Before McCulloch and Pitts, neither Turing nor anyone else had used the mathematical notion of computation as an ingredient in a theory of cognition. McCulloch and Pitts's theory changed the intellectual landscape, so that many could see neural computations as the most promising way to explain cognition.

McCulloch and Pitts's computational theory rested on two principal moves. First, they simplified and idealized the known properties of networks of neurons so that certain propositional inferences could be mapped onto neural events and

[1] Turing and McCulloch discussed issues pertaining to computation and cognition with some of the same people, such as John von Neumann, Norbert Wiener, and Claude Shannon (Aspray 1985; Hodges 1983; Heims 1991). McCulloch and Pitts's 1943 paper was discussed among those people shortly after it was published, so Turing would likely have heard about it. More specifically, Turing's report on an Automatic Computing Engine (Turing 1945) cites von Neumann's "First Draft of a Report on the EDVAC" (von Neumann 1945) and, according to Andrew Hodges, in 1946 Turing was using the "notation for logical networks" introduced by von Neumann 1945 (Hodges 1983, 343). As we shall see, von Neumann explicitly acknowledged McCulloch and Pitts's work as the source of his notation. At the very least, Turing would have known about McCulloch and Pitts's work from studying von Neumann's paper.

vice versa. Second, they assumed that individual neural pulses had propositional contents that directly explained cognitive processes. Neither of these moves is likely to find supporters today, at least not in the form proposed by McCulloch and Pitts.

In spite of its importance, McCulloch and Pitts's paper is often misrepresented. For instance, a common misconception is that McCulloch and Pitts demonstrated that neural nets could compute anything that Turing machines could:

> McCulloch and Pitts proved that a sufficiently large number of these simple logical devices, wired together in an appropriate manner, are capable of universal computation. That is, a network of such "lineal threshold" units with the appropriate synaptic weights can perform any computation that a digital computer can, though not as rapidly or as conveniently. (Koch and Segev 2000, 1171)

As we shall see, this is incorrect in two respects. First, McCulloch and Pitts did not *prove* any results about what their nets could compute, although they claimed that there were results to prove. Second, McCulloch–Pitts nets—as McCulloch and Pitts explicitly recognized—were computationally less powerful than Turing machines.[2]

5.2 Motivation

McCulloch and Pitts's paper started by rehearsing some established neurophysiological facts: the nervous system is a system of neurons connected through synapses, neurons send excitatory and inhibitory pulses to each other, and each neuron has a threshold determining how many excitatory and inhibitory inputs are necessary and sufficient to excite it at a given time (McCulloch and Pitts 1943, 19–21).

Then, the authors introduced the main premise of their theory: that neuronal signals are "equivalent" to propositions. This was presumably what justified their title, which mentions *ideas* immanent in nervous activity. They introduced this theme in a curious and oblique way, appealing not to some explicit motivation but to "considerations," made by one of the authors, which they did not give:

> Many years ago one of us, by *considerations impertinent to this argument*, was led to conceive of the response of any neuron as *factually equivalent* to a proposition

[2] Some technical aspects of the paper are discussed in Perkel 1988; Arbib 1989; Cowan 1990a, b, and c; and Díaz and Mira 1996.

> which proposed its adequate stimulus. He therefore attempted to record the behavior of complicated nets in the notation of the symbolic logic of propositions. The "all-or-none" law of nervous activity is sufficient to insure that the activity of any neuron may be represented as a proposition. Physiological relations existing among nervous activities correspond, of course, to relations among the propositions; and the utility of the representation depends upon the identity of these relations with those of the logic of propositions. To each reaction of any neuron there is a corresponding assertion of a simple proposition. This, in turn, implies either some other simple proposition or the disjunction or the conjunction, with or without negation, of similar propositions, according to the configuration of the synapses upon and the threshold of the neuron in question.
> (McCulloch and Pitts 1943, 21; emphasis added)

The author of the "considerations" was McCulloch. Before working with Pitts, McCulloch had speculated for years that mental processes were sequences of atomic mental events, which he called "psychons." Psychons were related via logical operations performed by neurons, and the unit of nervous activity was the all-or-none action potential. Thus, McCulloch speculated that psychons were physically realized by action potentials (Piccinini 2004c). Thanks to his collaboration with Pitts, McCulloch's idea was finally coming to fruition. A proposition that "proposes a neuron's adequate stimulus" was a proposition to the effect that the neuron receives a certain input at a certain time. The authors did not explain what they meant by "factual equivalence" between neuronal pulses and propositions, but their language and the conclusions they drew from their theory (see Section 5.6) suggest they meant both that neuronal pulses are represented by propositions and that neuronal pulses have propositional content.

The theory was divided into two parts: one part for nets *without* "circles" (McCulloch and Pitts's term for closed loops of activity); the other for nets *with* circles or "*cyclic* nets." (Today cyclic nets are called "recurrent networks.") The authors pointed out that the nervous system contains many circular, "regenerative" paths (McCulloch and Pitts 1943, 22). The term "circle" may have been borrowed from Turing (1936–7), who had used "machines with circles" for Turing machines whose computations continue forever without producing the desired output, and "machines without circles" for Turing machines that produce the desired output. Like a Turing machine with circles, a net with circles may run forever if left unperturbed.

5.3 Assumptions

In formulating their theory, McCulloch and Pitts made the following five assumptions:

1. The activity of the neuron is an "all-or-none" process.
2. A certain fixed number of synapses must be excited within the period of latent addition in order to excite a neuron at any time, and this number is independent of previous activity and position of the neuron.
3. The only significant delay within the nervous system is synaptic delay.
4. The activity of any inhibitory synapse absolutely prevents excitation of the neuron at that time.
5. The structure of the net does not change with time.

(McCulloch and Pitts 1943, 22)

These assumptions idealize the then known properties of neurons. Assumption (1) is simply the all-or-none law: neurons were believed to either pulse or be at rest. As to (4) and (5), McCulloch and Pitts admitted that they are false of the nervous system. Under the other assumptions, however, they argued that nets that do not satisfy (4) and (5) are functionally equivalent to nets that do (McCulloch and Pitts 1943, 29–30).

As to (2)—that a fixed number of neural stimuli is always necessary and sufficient to generate a neuron's pulse—it is a radical simplification. First, as McCulloch and Pitts discussed in their preliminary review of neurophysiology, the "excitability" of a neuron does vary over time: after a neuron's pulse, there is an absolute refractory period in which the neuron cannot pulse at all, and "[t]hereafter its excitability returns rapidly, in some cases reaching a value above normal from which it sinks again to a subnormal value, whence it returns slowly to normal" (McCulloch and Pitts 1943, 20; neurons also exhibit spontaneous activity, that is, activity in the absence of stimuli). Second, as McCulloch and Pitts also discussed, synapses are plastic, so that which stimuli will activate a neuron is a function of the previous patterns of activity and stimulation of that neuron. Some changes in the responsiveness of neurons to stimuli are temporary, as in "facilitation and extinction," whereas other changes are permanent and constitute "learning" (McCulloch and Pitts 1943, 21).[3] With respect to this second point, however, McCulloch and Pitts argued that a net whose responsiveness to stimuli varies can be replaced by a formally equivalent net whose responsiveness is fixed (McCulloch and Pitts 1943, 29–30). Hence, synaptic plasticity does not affect their conclusion that neural events satisfy the "logic of propositions" (McCulloch and Pitts 1943, 22).

Assumption (3) was crucial. As Pitts had done in his previous theory of neuron networks (Pitts 1942a, b, 1943a, b, c), McCulloch and Pitts assumed that all

[3] Today, many consider synaptic plasticity the neural mechanism for learning and memory. But although speculative explanations of learning and memory in terms of changes in neuronal connections can be traced back to the late nineteenth century (Breidbach 2001), synaptic plasticity did not start to be observed in the laboratory until the 1950s (Craver 2003).

neurons within a network were synchronized so that all the relevant events in the network—conduction of the impulses along nerve fibers, refractory periods, and synaptic delays—occurred within temporal intervals of fixed and uniform length. Moreover, they assumed that all the events within one temporal interval only affected the relevant events within the following temporal interval. Assumption (3) discretized the dynamics of the net: instead of employing differential and integral equations to describe frequencies of impulses, as the members of Rashevsky's group did, McCulloch and Pitts could now describe patterns of individual neural impulses as a discrete function of previous patterns of neural impulses. Logical functions of discrete states could be used to fully describe the transitions between neural events. McCulloch and Pitts could now replace the continuous mathematics of the physicists—the mathematics used by Rashevsky as well as Pitts before he met McCulloch—with the discrete mathematics of the logicians as the appropriate tool for modeling the brain and studying its functions.

McCulloch and Pitts were perfectly aware that the neuron-like units in their theory were distant from real neurons: "Formal neurons were deliberately as impoverished as possible" (McCulloch 1974, 36). Idealization and simplification were, after all, basic aspects of the biophysical methodology advocated by Rashevsky, with whom Pitts had trained. In a letter written to a colleague asking for clarification after a public presentation of the theory, McCulloch further explained that their theory ignored any neural malfunction:

> [W]e in our description restricted ourselves to the regular behavior of the nervous system, knowing full well that irregularities can be and are frequently brought about by physical and chemical alterations of the nervous system. As a psychiatrist, I am perhaps more interested in these than in its regular activity, but they lead rather to a theory of error than a theory of knowledge, and hence were systematically excluded from the description.
>
> (Letter by McCulloch to Ralph Lillie, ca. February 1943. Warren S. McCulloch Papers, Series 1, Box 12, file folder Lillie)

In McCulloch's eyes, the differences between real neurons and the units employed in his theory were inessential. His goal was not to understand neural mechanisms per se, but rather to explain how something close enough to a neural mechanism could—in McCulloch's words—exhibit "knowledge," the kind of "ideational," "rational," "formal," or "logical" aspect that McCulloch associated with the mind. McCulloch's goal was to explain cognition in terms of neural-like mechanisms. Since McCulloch thought the explanation had to involve logic to describe neural activity, what was needed was a set of simplifications and idealizations that allowed logic to be so used.

5.4 Nets Without Circles

McCulloch and Pitts's technical language was cumbersome; here I articulate their theory in a slightly streamlined form that makes it easier to follow. The neurons of a net N are denoted by $c_1, c_2, \ldots c_n$. A primitive expression of the form $N_i(t)$ means that neuron c_i fires at time t. Expressions of the form $N_i(t)$ can be combined by means of logical connectives to form complex expressions that describe the behavior of different neurons at certain times. For example, $N_1(t)\&N_2(t)$ means that neurons c_1 and c_2 fire at time t, $N_1(t-1)\vee N_2(t-2)$ means that either c_1 fires at $t-1$ or c_2 fires at $t-2$ (or both), etc. These complex expressions can in turn be combined by the same logical connectives. As well-formed combinations, McCulloch and Pitts allowed only the use of conjunction ($A\&B$), disjunction ($A\vee B$), conjunction and negation ($A\&{\sim}B$), and a special connective S that shifts the temporal index of an expression backwards in time, so that $S(N_i(t)) = N_i(t-1)$. A complex expression formed from a number of primitive expressions $N_1(t), \ldots N_n(t)$ by means of the above connectives is denoted by $Expression_j(N_1(t), \ldots N_n(t))$. In any net without circles, there are some neurons that receive no inputs from other neurons; these are called *afferent* neurons.

The two main technical problems McCulloch and Pitts formulated and solved were "to calculate the behavior of any net, and to find a net which will behave in a specified way, when such a net exists" (McCulloch and Pitts 1943, 24). These problems were analogous to those formulated by Householder and Pitts within their earlier theories. Alston Householder, another member of Rashevsky's group, had asked which general propositions "express the activity of any nerve-fiber complex in terms of the dynamics of the individual fibers and of the structural constants" (Householder 1941, 64). Pitts had added the inverse network problem of, "given a preassigned pattern of activity over time, to construct when possible a neuron-network having this pattern" (Pitts 1943a, 23). Within McCulloch and Pitts's theory, however, a net's pattern of activity was no longer described as a frequency of impulses, but rather as a precise pattern of neuronal impulses described by a logical formula.

In terms of McCulloch and Pitts's theory, the two problems can be formulated as follows:

> First problem: given a net, find a class of expressions **C** such that for every neuron c_i, in **C** there is a true expression of the form
>
> $N_i(t)$ if and only if $Expression_j((N_{i-g}(t-1), \ldots N_{i-2}(t-1), N_{i-1}(t-1))$,
>
> where neurons $c_{i-g}, \ldots c_{i-2}$, and c_{i-1} have axons inputting c_i.

The significance of this expression is that it describes the behavior of any (nonafferent) neuron in terms of the behavior of the neurons that are afferent to it. If a class *C* of such expressions is found, then propositional logic can describe the behavior of any nonafferent neuron in the net in terms of the behavior of the neurons afferent to it.

> Second problem: given an expression of the form
>
> $N_i(t)$ if and only if $Expression_j((N_{i-g}(t-1), \ldots N_{i-2}(t-1), N_{i-1}(t-1))$,
>
> find a net for which it is true.

McCulloch and Pitts showed that in the case of nets without circles, these problems were easily solved. To solve the first problem, they showed how to write an expression describing the relation between the firing of any neuron in a net and the inputs it received from its afferent neurons. To solve the second problem, they showed how to construct nets that satisfy their four combinatorial schemes (conjunction, disjunction, conjunction-cum-negation, and temporal predecessor), giving diagrams that showed the connections between neurons that satisfy each scheme (Figure 5.1). Then, by induction on the size of the nets, all expressions formed by those combinatorial schemes are satisfiable by McCulloch–Pitts nets.[4]

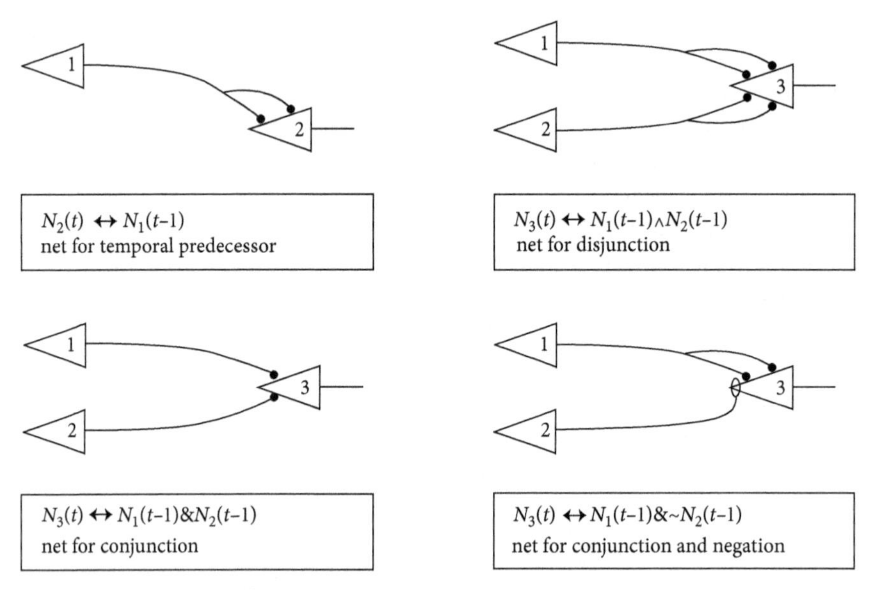

Figure 5.1 Diagrams of McCulloch and Pitts nets. In order to send an output pulse, each neuron must receive two excitatory inputs and no inhibitory inputs. Lines ending in a dot represent excitatory connections; lines ending in a hoop represent inhibitory connections.

[4] Their actual proof was not quite a mathematical induction because they didn't show how to combine nets of arbitrary size, but the technical details are unimportant here.

By giving diagrams of nets that satisfy simple logical relations between propositions and by showing how to combine them to satisfy more complex logical relations, McCulloch and Pitts developed a powerful technique for designing circuits that satisfy given logical functions by using a few primitive building blocks. This is the main aspect of their theory used by von Neumann in describing the design of digital computers. Today, McCulloch and Pitts's technique is part of logic design, an important area of computer design devoted to digital circuits for computers. The building blocks of contemporary logic design are called logic gates (e.g., Patterson and Hennessy 1998). In today's terminology, McCulloch and Pitts's neurons are logic gates and their nets are combinations of logic gates.

The original purpose of McCulloch and Pitts's technique for designing nets was to explain mental phenomena. As an example, they offered a possible explanation of a well-known heat illusion by constructing an appropriate net. Touching a very cold object normally causes a sensation of cold, but touching it very briefly and then removing it can cause a sensation of heat. In designing their net, McCulloch and Pitts reasoned as follows. They started from the known physiological fact that there are different kinds of receptor affected by heat and cold, and they assumed that there are neurons whose activity "implies a sensation" of heat (McCulloch and Pitts 1943, 27). Then, they assigned one neuron to each function: heat reception, cold reception, heat sensation, and cold sensation. Finally, they observed that the heat illusion corresponded to the following relations between three neurons: the heat-sensation neuron fires either in response to the heat receptor or to a brief activity of the cold receptor (Figure 5.2).

McCulloch and Pitts used this example to draw a general conclusion about the relation between perception and the world:

> This illusion makes very clear the dependence of the correspondence between perception and the "external world" upon the specific structural properties of the intervening nervous net. (McCulloch and Pitts 1943, 28)

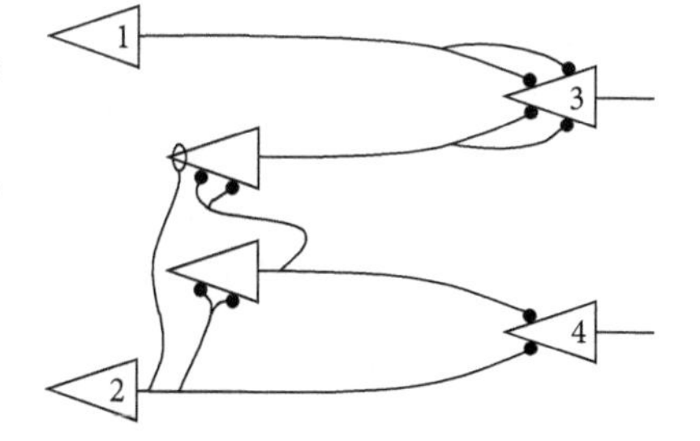

Figure 5.2 Net explaining the heat illusion. Neuron 3 (heat sensation) fires if and only if it receives two inputs, represented by the lines terminating on its body. This happens when either neuron 1 (heat reception) fires or neuron 2 (cold reception) fires once and then immediately stops firing. When neuron 2 fires twice in a row, the intermediate (unnumbered) neurons excite neuron 4 rather than neuron 3, generating a sensation of cold.

Then they pointed out that, under other assumptions about the behavior of the heat and cold receptors, the same illusion could be explained by different nets (McCulloch and Pitts 1943, 28). They were not proposing their network as the actual mechanism behind the heat illusion, but rather as a possible mechanism that explained the illusion.

5.5 Nets with Circles, Computation, and the Church-Turing Thesis

The problems for nets with circles were analogous to those for nets without circles: given the behavior of a neuron's afferents, find a description of the behavior of the neuron; and find the class of expressions and a method of construction such that for any expression in the class, a net could be constructed that satisfies the expression. The authors pointed out that the theory of nets with circles is more difficult than the theory of nets without circles. This is because activity around a circle of neurons may continue for an indefinite amount of time, hence expressions of the form $N_i(t)$ may have to refer to times that are indefinitely remote in the past. For this reason, the expressions describing nets with circles are more complicated, involving quantification over times. McCulloch and Pitts offered solutions to the problems of nets with circles, but their treatment of this part of the theory was obscure, admittedly sketchy (McCulloch and Pitts 1943, 34), and contained some errors. As a consequence, it is almost impenetrable. At any rate, McCulloch and Pitts's mathematical treatment of their nets with circles was superseded a few years later by Kleene's treatment (Kleene 1956).

At the end of this section, McCulloch and Pitts drew the connection between their nets and computation:

> It is easily shown: first, that every net, if furnished with a tape, scanners connected to afferents, and suitable efferents to perform the necessary motor-operations, can compute only such numbers as can a Turing machine; second, that each of the latter numbers can be computed by such a net; and that nets with circles can be computed by such a net; and that nets with circles can compute, without scanners and a tape, some of the numbers the machine can, but no others, and not all of them. This is of interest as affording a psychological justification of the Turing definition of computability and its equivalents, Church's λ-definability and Kleene's primitive recursiveness: If any number can be computed by an organism, it is computable by these definitions, and conversely. (McCulloch and Pitts 1943, 35)

This brief passage is the only one mentioning computation. By stating that McCulloch–Pitts nets can compute, this passage provided the first known published link between the mathematical theory of computation and brain theory. It was a pivotal statement in the history of the computational theory of cognition.

It is often said that McCulloch and Pitts proved that their nets can compute anything that Turing machines can compute (e.g., Koch and Segev 2000). This misconception was propagated by McCulloch himself. For instance, in summarizing the significance of their paper, McCulloch wrote to a colleague:

> [T]he original paper with Pitts entitled "A Logical Calculus of Ideas Immanent in Nervous Activity"... sets up a calculus of propositions subscripted for the time of their appearance for any net handling all-or-none signals, and shows that such nets can compute any computable number or, for that matter, do anything any other net can do by the way of pulling consequences out of premises.
>
> (Letter by McCulloch to Schouten, dated October 18, 1948. Warren S. McCulloch Papers, Series I, Box 17, file folder Schouten)

But in discussing computation in their paper, McCulloch and Pitts did not *prove* any results about the computation power of their nets; they only stated that there were results to prove. And their conjecture was not that their nets can compute anything that can be computed by Turing machines. Rather, they claimed that *if* their nets were provided with a tape, scanners, and "efferents," *then* they would compute what Turing machines could compute; without a tape, McCulloch and Pitts expected even nets with circles to compute a smaller class of functions than the class computable by Turing machines.

Their comment that these conjectures were "easily shown" may suggest that the proofs were easy to come by. On the contrary, the question of which functions are computable by McCulloch–Pitts nets was not even explicitly defined by the authors. Several years later, Stephen Kleene set out to rigorously formulate and solve the problem of what McCulloch–Pitts nets can compute. Kleene proceeded independently of McCulloch and Pitts's treatment of nets with circles because he found it "obscure" and because he found an "apparent counterexample" (Kleene 1956, 4, 17, 22). Kleene defined the notion of "regular events" (today called "regular languages") and proved that McCulloch–Pitts nets can "represent" regular events (in today's terminology, they can accept regular languages). In the same paper, Kleene also defined an alternative formalism, which generalized McCulloch–Pitts nets by allowing the "cells" in the network to take any of a finite number of internal states. Kleene called his new formalism "finite automata," and showed that McCulloch–Pitts nets are computationally equivalent to finite automata.

McCulloch and Pitts did not explain what they meant by saying that nets compute. As far as the first part of the passage is concerned, the sense in which nets compute seems to be a matter of *describing* the behavior of nets by the vocabulary and formalisms of computability theory. Describing McCulloch–Pitts nets in this way turned them into a useful tool for designing circuits for computing mechanisms. This is how von Neumann would later use them in his "First Draft of a Report on the EDVAC" (von Neumann 1945), the first widely-circulated description of the architecture of a general purpose, stored-program digital computer. Von Neumann's paper, which applied McCulloch and Pitts's method for designing digital computing circuits, had a large influence on computer design. Yet if all there was to McCulloch and Pitts's invocation of computation was the use of their formalism for designing computers, which functions are computable by McCulloch–Pitts nets would be an innocent technical question devoid of epistemological significance.

But one intriguing aspect of the above passage about nets and computation is the way it relates McCulloch–Pitts nets to the Church–Turing thesis (CT). Turing and other logicians had justified CT—the thesis that any effectively calculable function is computable by Turing machines—by intuitive mathematical considerations. In their passage, McCulloch and Pitts offered a "psychological" justification for CT, based on the computational limitations of the human brain. Since what their nets can compute (even with the help of "a tape, scanners, and suitable efferents") can be computed by Turing machines, they implicitly suggested that what computing humans can effectively calculate can be computed by Turing machines. By stating that the computational limitations of their nets provide a psychological justification of CT, McCulloch and Pitts presupposed that the computational limitations of their nets capture the computational limitations of brains, and that the computational limitations of brains correspond to the "psychological" limitations of humans engaged in computation. If so, then defining computable functions in terms of Turing machines (or any other computationally equivalent formalism) is justified. McCulloch and Pitts seemed to believe that ordinary human computations using pencil and paper (which is what CT is about), and more generally the "pulling of consequences out of premises," could be explained directly in terms of the computations performed by their nets. Thus, McCulloch and Pitts attributed epistemological significance to the fact that their nets compute.

Indeed, the main purpose of their theory was to account for cognitive functions, such as knowledge and inference, by proposing possible neural-like mechanisms. As McCulloch explained a few years later, he and Pitts were interpreting neural inputs and outputs as if they were symbols written on the tape of a Turing machine:

> What we thought we were doing (and I thought we succeeded fairly well) was treating the brain as a Turing machine... The important thing was, for us, that we had to take a logic and subscript it for the time of occurrence of a signal (which is, if you will, no more than a proposition on the move). This was needed in order to construct theory enough to be able to state how a nervous system

> could do anything. The delightful thing is that the very simplest set of appropriate assumptions is sufficient to show that a nervous system can compute any computable number. It is that kind of a device, if you like – a Turing machine. (von Neumann 1951, 32–3)

In comparing brains to Turing machines, McCulloch thought that by constructing their theory, they showed how brains "could do anything," including performing "the kind of [mental] functions which a brain must perform if it is only to go wrong and have a psychosis" (McCulloch and Pitts 1943). "Treating the brain as a Turing machine" was a crucial part of McCulloch and Pitts's attempt at solving the mind-body problem.

5.6 "Consequences"

McCulloch and Pitts ended their paper by drawing what they called "consequences," in a section that introduced several metaphysical and epistemological themes and related them to McCulloch–Pitts nets. This final section of their paper demonstrates the wide scope of their theory and provides further context. It starts with a general point about the causal structure of nets, which is such that, from a given event in a net, it may be impossible to infer either its cause or the time of its cause's occurrence:

> Causality, which requires description of states and a law of necessary connections relating them, has appeared in several forms in several sciences, but never, except in statistics, has it been as irreciprocal as in this theory. Specification for any one time of afferent stimulation and of the activity of all constituent neurons, each an "all-or-none" affair, determines the state. Specification of the nervous net provides the law of necessary connection whereby one can compute from the description of any state that of the succeeding state, but the inclusion of disjunctive relations prevents complete determination of the one before. Moreover, the regenerative activity of constituent circles renders reference indefinite as to time past. (McCulloch and Pitts 1943, 35)

From this relatively straightforward observation about the causal structure of McCulloch–Pitts nets, they drew striking epistemological conclusions:

> Thus our knowledge of the world, including ourselves, is incomplete as to space and indefinite as to time. This ignorance, implicit in all our brains, is the counterpart of the abstraction which renders our knowledge useful. The role of brains in determining the epistemic relations of our theories to our observations and of these to the facts is all too clear, for *it is apparent that every idea and every*

> *sensation is realized by activity within that net*, and by no such activity are the actual afferents fully determined.
>
> (McCulloch and Pitts 1943, 35–7, emphasis added)

This passage makes clear that McCulloch and Pitts thought of individual neuronal pulses and their relations as realizations of sensations, ideas, and their epistemic relations. This assumption—which had been introduced in the paper by an oblique reference to "considerations impertinent to this argument"—allowed them to draw conclusions about epistemological limitations of the mind directly from the causal structure of their nets (assuming also that brains instantiate the relevant features of nets, namely disjunctive connections and closed loops of activity).

The next passage drew further epistemological conclusions. McCulloch and Pitts noted that changing a network after a stimulus was received would introduce further difficulties in inferring the stimulus from the net's current activity, impoverishing the subject's knowledge and leading to cognitive dysfunctions:

> There is no theory we may hold and no observation we can make that will retain so much as its old defective reference to the facts if the net be altered. Tinnitus, paraesthesias, hallucinations, delusions, confusions and disorientations intervene. Thus empiry [i.e., experience] confirms that if our nets are undefined, our facts are undefined, and to the "real" we can attribute not so much as one quality or "form." (McCulloch and Pitts 1943, 37)

It is worth recalling that McCulloch and Pitts reduced nets that change over time to nets of fixed structure. In their theory, a net's fixed structure determines what a subject can infer about the external world from the net's current activity, so any change in the net's structure diminishes the subject's knowledge and hence is dysfunctional. If the net's structure remains fixed and its past activity is known, however, it is possible to know precisely which patterns of stimulation gave rise to the net's current activity.[5] Perhaps because of this, McCulloch and Pitts thought they had something to say on the Kantian theme that the mind can know only phenomena not things in themselves: "With determination of the net, the unknowable object of knowledge, the 'thing in itself,' ceases to be unknowable" (McCulloch and Pitts 1943, 37). This statement is conspicuous but unclear. It seems to suggest that if a subject can know the structure and past activity of her own net, then she can know things in themselves.

After drawing their epistemological consequences, McCulloch and Pitts went on to offer some morals to the psychologists. They started with two points: first,

[5] This is the case when either there are no disjunctive connections in a net, or one knows the activity of the neurons that send inputs to the disjunctive connections in a net.

they stated a reductionist doctrine according to which their theory had the resources to reduce psychology to neurophysiology; second, they argued that, because of the all-or-none character of neural activity, the most fundamental relations among psychological events are those of two-valued logic. In making their case about psychology, McCulloch and Pitts stated very explicitly that they interpreted nervous activity as having "intentional character":

> To psychology, however defined, specification of the net would contribute all that could be achieved in that field—even if the analysis were pushed to ultimate psychic units or "psychons," for a psychon can be no less than the activity of a single neuron. Since that activity is inherently propositional, all psychic events have an intentional, or "semiotic," character. The "all-or-none" law of these activities, and the conformity of their relations to those of the logic of propositions, insure that the relations of psychons are those of the two-valued logic of propositions. Thus in psychology, introspective, behavioristic or physiological, the fundamental relations are those of two-valued logic.
>
> (McCulloch and Pitts 1943, 37–8)

The long final paragraph begins with a summary of the "consequences" and a restatement that mental phenomena are now derivable from neurophysiology:

> Hence arise constructional solutions of holistic problems involving the differentiated continuum of sense awareness and the normative, perfective and resolvent properties of perception and execution. From the irreciprocity of causality it follows that even if the net be known, though we may predict future from present activities, we can deduce neither afferent from central, nor central from efferent, nor past from present activities—conclusions which are reinforced by the contradictory testimony of eye-witnesses, by the difficulty of diagnosing differentially the organically diseased, the hysteric and the malingerer, and by comparing one's own memories or recollections with his [*sic*] contemporaneous records. Moreover, systems which so respond to the difference between afferents to a regenerative net and certain activity within that net, as to reduce the difference, exhibit purposive behavior; and organisms are known to possess many such systems, subserving homeostasis, appetition and attention. Thus both the formal and the final aspects of that activity which we are want to call *mental* are rigorously deducible from present neurophysiology.
>
> (McCulloch and Pitts 1943, 38)

The same paragraph continues with "consequences" relevant to psychiatry. One is that—contrary to the teachings of psychoanalysis—knowing the history of a patient is unnecessary for treating mental illness. A more general one is that mental diseases reduce to properties of neural nets, and even more generally that the mind–body problem is solved. McCulloch and Pitts were answering Sherrington's statement that the mind–body problem remains a big mystery or, as he put it, "mind goes more ghostly than a ghost" (Sherrington 1940):

> The psychiatrist may take comfort from the obvious conclusion concerning causality—that, for prognosis, history is never necessary. He can take little from the equally valid conclusion that his observables are explicable only in terms of nervous activities which, until recently, have been beyond his ken. The crux of this ignorance is that inference from any sample of overt behavior to nervous nets is not unique, whereas, of imaginable nets, only one in fact exists, and may, at any moment, exhibit some unpredictable activity. Certainly for the psychiatrist it is more to the point that in such systems "Mind" no longer "goes more ghostly than a ghost." Instead, diseased mentality can be understood without loss of scope or rigor, in the scientific terms of neurophysiology.
> (McCulloch and Pitts 1943, 38)

The essay ends with an appeal to neurology and mathematical biophysics:

> For neurology, the theory sharpens the distinction between nets necessary or merely sufficient for given activities, and so clarifies the relations of disturbed structure to disturbed function. In its own domain the difference between equivalent nets and nets equivalent in the narrow sense indicates the appropriate use and importance of temporal studies of nervous activity: and to mathematical biophysics the theory contributes a tool for rigorous symbolic treatment of known nets and an easy method of constructing hypothetical nets of required properties. (McCulloch and Pitts 1943, 38–9)

The last point, the method of construction of "hypothetical nets of required properties," highlights one of the most fruitful legacies of the paper. From then on, McCulloch and Pitts, soon followed by generations of researchers, would use the techniques developed in this paper, and modifications thereof, to design neural networks to explain neural and cognitive phenomena.

5.7 Impact of McCulloch and Pitts's Theory

McCulloch and Pitts's project was not to systematize and explain observations about the nervous system—it was to explain knowledge and other cognitive phenomena in terms of mechanisms that resembled neural ones. To do this, they assumed that mental states can be analyzed in terms of mental atoms endowed with propositional content, the psychons, and that the neural correlates of mental phenomena correspond to precise configurations of neuronal pulses:

individual pulses correspond to individual psychons, and causal relations among pulses correspond to inferential relations among psychons.

McCulloch and Pitts's paper offered a mathematical technique for designing neural nets to implement certain inferential relations among propositions, and suggested that those inferences are mathematically equivalent to certain computations. The paper didn't mention computers, because modern computers didn't exist yet. Nonetheless, their technique for nets without circles could be used in designing circuits for digital computers, because it allowed the design of circuits that compute any desired Boolean function. McCulloch and Pitts's technique was coopted by von Neumann (1945) as part of what is now called logic design, and circuits computing Boolean functions became the building blocks of modern digital computers. As a result, modern digital computers are a kind of McCulloch–Pitts neural network! Digital computers are bunches of logic gates—digital signal-processing units equivalent to McCulloch–Pitts neurons—connected to form a specific architecture.

In the 1950s, the question raised by McCulloch and Pitts about what their nets (with or without circles) could compute led to the development of finite automata, one of the most important formalisms in the theory of (digital) computation.

McCulloch and Pitts's nets were ostensibly "neural" in the sense that the *on* and *off* values of their units were inspired by the all-or-none character of neuronal activity. However, McCulloch–Pitts nets were heavily simplified and idealized relative to the then known properties of neurons and neural nets. The theory did not offer testable predictions or explanations for observable neural phenomena. It was quite removed from what neurophysiologists could do in their labs. This may be why neurophysiologists largely ignored McCulloch and Pitts's theory. Even McCulloch and Pitts, in their later empirical neurophysiological work, did not make direct use of their theory.

But McCulloch and Pitts's theory was well received by people, such as Norbert Wiener and John von Neumann, who were interested in epistemology but trained in mathematics or engineering more than in neurophysiology. For one thing, these scientists liked the claim that the mind had been reduced to the brain; some of their intellectual heirs saw the solution to the mind–body problem as McCulloch and Pitts's great contribution (Arbib 2000, 212 and 213; Lettvin 1989, 514.). For another thing, they liked the operationalist flavor of the theory, whereby the design of nets was seen as all there was to the performance of inferences and more generally to mental phenomena (Cf. Wiener 1948, 147; Shannon and McCarthy 1956). Most of all, they liked the mathematical tools and what they saw as their potential for building intelligent machines. They started exploring the technique offered by McCulloch and Pitts. The mathematical techniques got elaborated, modified, and enriched, but the goal remained to explain knowledge in particular and cognition in general using "computing" mechanisms.

McCulloch and Pitts's theory impressed philosophers with reductionist leanings (Oppenheim and Putnam 1958) and then contributed to inspire functionalism, and the alliance between functionalism and the computational theory of mind, as a relatively novel solution to the mind–body problem (Piccinini 2004b tells that story; I will come back to this theme in Chapter 14).

McCulloch and Pitts's theory was not the only source of modern computational theories of cognition. But McCulloch and Pitts's use of computation to describe neural functions, together with their proposal to explain cognitive phenomena directly in terms of neural computations, contributed to a large shift in the use of computation that occurred around the middle of the twentieth century. Before 1943, computing was thought of as one human activity among others (e.g., cooking, walking, or talking). After 1943, computing could be thought of as, in a sense, all that humans did. Under McCulloch and Pitts's theory, any neural network could be described as performing a computation. In the sense in which McCulloch–Pitts nets compute, and to the extent that McCulloch–Pitts nets are a good model of the brain, every neural activity is a computation. Given that McCulloch and Pitts considered the computations of their nets to be explanations of cognitive processes and human behavior, every cognitive process was turned into a computation, and every behavior into the output of a computation. As a further consequence, the Church–Turing thesis (CT) was turned from a thesis about what functions can be effectively calculated into a thesis about the power and limitations of brains.

If subsequent scientists and philosophers had realized that this shift in the notion of computation and the interpretation of CT was an effect of McCulloch and Pitts's theory, with its idealizations and simplifications of neurons, and its assumptions about the computational and epistemological significance of neural activity, this might be ok. The problem is that after 1943, many took McCulloch and Pitts's reinterpretation of neural activity and CT without question, and thought it based on mathematically proven facts about brains. Invoking CT in support of computational theories of cognition became commonplace. In this changed context, it became natural for many people to read even Turing's own argument for CT (in Turing 1936–7) as a defense of the computational theory of cognition (e.g., by Cleland 1993, 284; Shanker 1995, 55; Fodor 1998; and Webb 1980).

After McCulloch and Pitts's paper, the idea that CT is somehow a psychological thesis about human cognitive faculties, or perhaps a methodological restriction on psychological theories, would stick and would be used time and again to justify computational theories of cognition. For example, von Neumann made statements that resembled this interpretation of CT (von Neumann 1951; for more recent examples, see Chalmers 1996b; Churchland and Churchland 1990; Dennett 1978b; Fodor 1981; Haugeland 1981; McGee 1991; Pylyshyn 1984). A related idea would be that since the brain can only do what is computable, there is a computational theory of cognition (e.g. Webb 1980). I find it ironic that McCulloch and Pitts made many of their simplifying assumptions about networks of neurons in order to solve the mind–body problem by using logic and Turing machines as descriptions of the nervous system but, after their theory was formulated, their theory was used as evidence that the brain is indeed a digital computing mechanism. I will address the relationship between CT and the computational theory of cognition in Chapter 10.

5.8 Three Research Traditions: Classicism, Connectionism, and Computational Neuroscience

McCulloch and Pitts's Computational Theory of Cognition (CTC) has three important aspects: an analogy between neural processes and digital computations, the use of mathematically defined neural networks as models, and an appeal to neurophysiological evidence to support their neural network models. After McCulloch and Pitts, many others linked computation and cognition, though they often abandoned one or more aspects of McCulloch and Pitts's theory. CTC evolved into three main traditions, each emphasizing a different aspect of McCulloch and Pitts's account (cf. Boden 1991).

One tradition, sometimes called classicism, emphasizes the analogy between cognitive systems and digital computers while downplaying the relevance of neuroscience to the theory of cognition (Miller, Galanter, and Pribram 1960; Fodor 1975; Newell and Simon 1976; Pylyshyn 1984; Newell 1990; Pinker 1997; Gallistel and King 2009). When classicists offer computational models of a cognitive capacity, the models take the form of computer programs for producing the capacity in question. One strength of the classicist tradition lies in programming computers to exhibit higher cognitive capacities such as problem solving, language processing, and language-based inference.

A second tradition, most closely associated with the term "connectionism" (although this label can be misleading, see Chapter 9), downplays the analogy between cognitive systems and digital computers in favor of computational explanations of cognition that are "neurally inspired" (Rosenblatt 1958; Feldman and Ballard 1982; Rumelhart and McClelland 1986; Bechtel and Abrahamsen 2002). When connectionists offer computational models of a cognitive capacity, the models take the form of neural networks for producing the capacity in question. Such models are primarily constrained by psychological data, as opposed to neurophysiological and neuroanatomical data. One strength of the connectionist tradition lies in designing artificial neural networks that exhibit cognitive capacities such as perception, motor control, learning, and implicit memory.

A third tradition is most closely associated with the term "computational neuroscience," which in turn is the computational side of theoretical neuroscience. Computational neuroscientists model neurons and neural nets using differential and integral equations, which were the tools pioneered by Rashevsky, and supplement Rashevsky's method with concepts and techniques taken from mathematical physics, control theory, probability and statistics, and information theory. Computational neuroscience downplays the analogy between cognitive systems and digital computers even more than the connectionist tradition. Neurocomputational models aim to describe actual neural systems such as (parts of) the hippocampus, cerebellum, or cerebral cortex, and are constrained by neurophysiological and neuroanatomical data in addition to psychological data (Schwartz 1990; Churchland and Sejnowski 1992; O'Reilly and Munakata 2000; Dayan and Abbott 2001; Eliasmith and Anderson 2003; Ermentrout and Terman 2010).

McCulloch–Pitts networks and many of their "connectionist" descendants are relatively unfaithful to the details of neural activity, whereas other types of neural networks are more biologically realistic. These include the following, in order of decreasing biological detail and increasing computational tractability: conductance based models, which go back to Hodgkin and Huxley's (1952) seminal analysis of the action potential based on conductance changes; networks of integrate-and-fire neurons, which fire simply when the input current reaches a certain threshold (Lapicque 1907/2007; Caianiello 1961; Stein 1965; Knight 1972); and firing rate models, in which there are no individual action potentials—instead, the continuous output of each network unit represents the firing rate of a neuron or neuronal population (Wilson and Cowan 1972). Computational neuroscience offers models of how real neural systems may exhibit cognitive capacities, especially perception, motor control, learning, and implicit memory.

Although the three traditions just outlined are in competition with one another to some extent (more on this in Chapter 9), there is also some fuzziness at their borders. Some cognitive scientists propose hybrid theories, which combine explanatory resources drawn from both the classicist and the connectionist traditions (e.g., Anderson 2007). In addition, biological realism comes in degrees, so there is no sharp divide between connectionist and neurocomputational models.

Classicist explanations and many connectionist explanations appeal to digital computation, but digital computation is not the only kind of computation that theorists of cognition have appealed to. A fully general CTC posits that cognition is a kind of computation in a more general sense that I will introduce in Chapter 6.

I have argued that McCulloch and Pitts's CTC rested on two principal moves, both of which are problematic. On one hand, they simplified and idealized neurons so that propositional inferences could be mapped onto neural events and vice versa. On the other hand, they assumed that neural pulses correspond to atomic mental events endowed with propositional content. McCulloch and Pitts seemed to suggest that their first move justified the second, which is dubious at best. At any rate, the field as a whole quickly abandoned both moves.

Theoretical and computational neuroscientists replaced McCulloch and Pitts's theory of neural nets with more empirically adequate models, which were no longer based on a direct description of neural events in terms of propositional inferences. In spite of the difficulties, both empirical and conceptual, with McCulloch and Pitts's way of ascribing computations to the brain, McCulloch and Pitts's Computational Theory of Cognition—to the effect that neural nets perform computations and that neural computation explains cognition—stuck and became mainstream. In the rest of this book, I will argue that there is something right about this idea. To see why, the next step is to delve into what it takes for a physical system—or, better, a functional mechanism—to perform computations and process information.

6
Computation and Information Processing

6.1 Mechanisms that Compute and Process Information

In previous chapters, I described multilevel mechanisms. Mechanisms are pluralities of components that, being organized in a specific way, possess specific capacities. Mechanisms come in many levels of components nested within components. Smaller components contribute their capacities to larger components. The capacities of higher-level mechanisms are higher-level properties, which are aspects of the lower-level properties that realize them. All levels are ontologically on a par.

Some mechanisms have teleological functions, which are regular contributions to the goals of organisms. I call mechanisms with teleological functions *functional mechanisms*. When functional mechanisms perform their functions, their capacities make regular contributions to the goals of organisms. (In the case of artifacts, their teleological functions are regular contributions to the goals of the organisms that create and use the artifacts.) Such mechanisms can be individuated by their teleological functions. In this sense, they have a functional nature.

Functional natures are typically multiply realizable. That is, different kinds of mechanism can perform the same function. The different kinds of mechanism that can perform the same function may differ in component types, organization type, or both.

Some multiply realizable mechanisms are medium independent. That is, their functions are specified solely in terms of manipulating certain degrees of freedom in certain ways, without specifying the physical medium in which they are realized. Medium-independent functions can be realized in any medium with enough degrees of freedom, provided that such degrees of freedom are organized appropriately.

With this framework as background, I can now explicate physical computation and information processing. This is a confusing topic because there are many different notions of computation and information processing. The notions I'm most interested in are those used in psychology, neuroscience, and computer science.

Let's begin with computation. We are all acquainted with digital computation, which is what digital computers do. It's also what neural systems do according to McCulloch and Pitts's theory. In previous work (Piccinini 2015), I argued that

Neurocognitive Mechanisms: Explaining Biological Cognition. Gualtiero Piccinini, Oxford University Press (2020).

DOI: 10.1093/oso/9780198866282.001.0001

digital computing systems are mechanisms whose teleological function is manipulating strings of bits—binary digits—in accordance with a rule.[1] The rule specifies which strings must come out as output given certain input strings and internal states.

The fact that the digits are binary is dispensable: we could build digital computers that manipulate more than two types of digit. Using more than two types is inconvenient, so we almost never do it. What is indispensable about digital computation is that there be finitely many digit types. If there were infinitely many types of digit, it would become impossible for a physical computing mechanism to keep them apart and manipulate them in the correct way—not to mention, it would be impossible for users to keep the different digits apart. Thus, digital computation requires finitely many types of digit.

Some people think that computation must be digital, more or less by definition. Much literature on the computational theory of cognition is based on this assumption. I grew up reading such literature so I used to assume this too. Yet there are many kinds of process that are said to be computations but are not digital or fully digital: analog computation, quantum computation, and more. There are branches of computer science devoted to studying nondigital computation and building nondigital computing systems. If these are legitimate kinds of computation, what do they have in common? What makes them all kinds of computation?

I have defended the following account (Piccinini 2015). Physical computation is the functional manipulation of a medium-independent vehicle in accordance with a rule. Functional manipulation means manipulation performed by a functional mechanism, which has such manipulation as its teleological function. Medium-independent vehicle means that such vehicles are defined solely in terms of degrees of freedom and their organization, without reference to their composition or other concrete physical properties of the medium that realizes them (Chapter 2). Because computing functions must be specified in terms of medium-independent vehicles, computing functions are medium-independent. In digital computers, the vehicles are the strings of digits; in other kinds of computing mechanisms, there are other kinds of vehicle. A rule is just a mapping from inputs and internal states to outputs.

Some people think that computation is the same as information processing—or even more strongly, that computation *requires* representation. Strictly speaking, this is incorrect. Computation per se may or may not involve information processing. Whether it does depends on whether the computational vehicles carry information. For example, take strings of digits. A digital computer processes information if and only if it manipulates strings of digits that carry

[1] There are other digital structures beside strings—e.g., arrays and graphs—but they can be mathematically reduced to strings. This does not affect our discussion. At any rate, the vehicles physically manipulated by ordinary digital computers are strings of digits.

information. Since it's perfectly possible to manipulate strings of digits or other computational vehicles that carry no information (in the relevant sense), it's perfectly possible to perform computations without processing information.

That being said, information processing is hugely important in computer science as well as psychology and neuroscience. We usually build computers precisely in order to process information. Therefore, it's important to get some clarity on what information is.

The most basic notion of information is reduction of uncertainty. If an event can occur in more than one way, then there is uncertainty about it. For example, a tossed coin can land heads or tail—we don't know which, so we are uncertain about it. Once the event occurs in a certain way, that uncertainty is reduced. That reduction of uncertainty is information in its most basic sense. Thus, when a coin lands, uncertainty about that outcome is eliminated—information is created. In this sense, information is to be found anywhere uncertainty is reduced. Anything that we are uncertain about carries information for us when it occurs. Anything that is inherently probabilistic carries information in an objective sense when it occurs. This is what information theorists call *self-information* or entropy (Shannon and Weaver 1949). In this very limited sense, we may want to say that computation is always information processing—at least for us, because we, or any computing system, may be always somewhat uncertain about the future values of some computational vehicles. Information in this sense—self-information—tells us nothing besides the values of the vehicles of the computing system, nothing about the world outside. It is not an especially interesting or helpful sense in which computation involves information processing.

The second most basic notion of information is correlation. If the values of a signal correlate with the values of a source, the strength of that correlation tells us how much we can learn about the source from the signal. This is what information theorists call *mutual information* between a signal and a source. Mind you: the strength of the correlation alone does not tell us anything specific about the relation between particular states of the signal and particular states of the source. However, it does tell us that given a state of the signal there is a certain probability that the source will be in a certain state. Typically, the vehicles of a computing system correlate with something outside it, so they carry mutual information about it. In this sense, most computation is information processing.

By building on mutual information, we may define a third notion of information: probabilistic *natural semantic information*. This is the informational analog of what Paul Grice (1957) calls natural meaning. Consider a signaling system that carries mutual information about a source. The state of a particular signal carries probabilistic natural semantic information about the state of the source to the extent that its occurrence raises the probability that the source is in a certain state. Typically, signals that enter computing systems and affect the value of computational vehicles do carry natural semantic information about a source. The source may be a variable in the environment, which is being measured by a sensor, or the mental state of a user, which affects what the user types on the keyboard, clicks with a mouse, or taps on the touchscreen.

The mental state of a user, in turn, may carry yet another kind of information—what Andrea Scarantino and I (2010) have called *nonnatural semantic information*. This is the informational analog of what Paul Grice (1957) calls nonnatural meaning. It is the information carried by signaling systems such as natural language, in which there may or may not be a correlation between signals and states of the world. An especially obvious lack of correlation is when we intentionally say something false. The signal occurs, but by design there is no state of the world corresponding to it. Nevertheless, we still say that the signal carries (false) information.

When a vehicle carries semantic information, either natural or nonnatural, often there is an asymmetry between the signal carrying the information and its source: not only does the signal carry semantic information about the source, it also has the *teleological function* of doing so. That is, information processors are typically built and set up so that the vehicles they process have the function of carrying a specific kind of semantic information. Analogously, in subsequent chapters we shall discuss the view that natural computing systems, such as nervous systems, process vehicles whose function is to carry semantic information about the organism and its environment. If the vehicles fail to carry such information, a malfunction occurs. In such cases, we say that what the signal is supposed to carry semantic information about is its *semantic content*, and the state of a signal *represents* its content. Typical computational vehicles are representations in this sense—a fortiori, their vehicles carry semantic information. Therefore, typical computations are forms of information processing. I'll discuss representation in more depth in Chapter 12.

In the other direction, is information processing computation? We've already seen that self-information is everywhere and does not give rise to an interesting sense of information processing. Correlations are everywhere too, so anything that correlates with anything else—that is, virtually everything—could be said to process information in a very minimal sense. This is not very interesting either. Information processing becomes more interesting when we introduce semantic information and, even better, the teleological function of carrying and doing something useful with semantic information. When we have systems with the teleological function of carrying semantic information and doing something useful with it, then we begin to have a sense of information processing relevant to explaining cognition.

For our purposes, the most relevant sense of information processing arises when a system has the teleological function of *integrating* semantic information transmitted by different types of physical signal and using it to control a larger system such as the body of an organism within an environment. When different types of physical signal must be integrated, processing the information in the format in which it comes into the system won't do, because physical signals of different types, such as sound waves and photons, can't be integrated directly with one another. How can you put together the information carried by photons with the information carried by sound waves with the information carried by molecules dispersed in the air? In order to integrate information from different physical signals, a system must first transduce all the different signal types into a common medium. To do that, it needs transducers that transform physically different signals (photons, sound waves, physical pressures, chemical gradients, etc.) into medium-independent vehicles that carry the different types of information. The medium-independent vehicles can then be processed to do something useful with that integrated semantic information. To do something useful with the information, the information must be processed nonrandomly—that is, in accordance with a rule. As we have seen, processing medium-independent vehicles in accordance with a rule is computation. This, in a nutshell, is why processing information, in the most interesting sense, entails computing. In Chapter 9, this will yield an argument for the Computational Theory of Cognition.

In this section, I've briefly reviewed physical computation and information processing and made the following points: computation in the most relevant sense is the processing of medium-independent vehicles by a functional mechanism in accordance with a rule; information in the most relevant sense is the semantic content of a representation; typical computations process information; and typical information processors are computing systems. In the rest of this chapter, I will expand on this picture and explain how it emerges from the broader debate on physical computation.

6.2 The Mathematical Theory of Computation

The notion of computation comes from mathematics. To a first approximation, mathematical computation is the solving of mathematical problems by following an algorithm. A classic example is Euclid's algorithm for finding the greatest common divisor to two numbers. Because algorithms solve problems automatically within finitely many steps, they are sometimes called "mechanical" procedures. This is a hint that computation has something to do with mechanisms.

In the early twentieth century, mathematicians and logicians developed formal logic to answer questions about the foundations of mathematics. One of these questions was the decision problem for first-order logic, namely, the question of whether an algorithm can determine, for any given statement written in a first-order logical system, whether that statement is provable within the system.[2] In 1936, Alonzo Church and Alan Turing proved that the answer is negative—there is no algorithm solving the decision problem for first-order logic (Church 1936; Turing 1936–7).

Turing's proof is especially relevant because it appeals to machines. To show that no algorithm can solve the decision problem for first-order logic, Turing needed to make precise what counts as an algorithm. He did so in terms of simple machines for manipulating symbols on an unbounded tape divided into squares. These *Turing machines*, as they are now called, have a control device that moves relative to the tape and reads, writes, and erases symbols on each square of the tape—one square at a time. The control device can be in one of a finite number of states. It determines what to do based on its state as well as what's on the tape, in accordance with a finite set of instructions.

Turing established three striking conclusions. First, he argued persuasively that any algorithm, suitably encoded, can be followed by one of his machines—this conclusion is now known as the *Church-Turing thesis*. Second, Turing showed how to construct *universal* Turing machines—special Turing machines that can be used to simulate any other Turing machine by executing the relevant instructions, which are written on their tape along with data for the computation. Finally, Turing proved that none of his machines solves the decision problem for first-order logic.

Turing's last result may be generalized as follows. There are denumerably many Turing machines (and therefore algorithms). That is to say, you can list all the Turing machines one after the other and put a natural number next to each. You can do this because you can enumerate all the finite lists of instructions that determine the behavior of each Turing machine. Each Turing machine computes a function from a denumerable domain to a denumerable range of values—for instance, from (numerals representing) natural numbers to (numerals representing) natural numbers. It turns out that there are undenumerably many such functions. Since an undenumerable infinity is much larger than a denumerable one, there are many more functions than Turing machines. Therefore, most functions are not computable by Turing machines. The decision problem for first-order logic is just one of these functions, which cannot be computed by Turing machines. The functions computable by Turing machines are called *Turing computable*; the rest are called *Turing uncomputable*. The existence of

[2] This formulation is due to Church (1936). Hilbert and Ackerman (1928) had asked whether some algorithm could determine whether any given statement written in first-order logic is universally valid—valid in every structure satisfying the axioms. In more modern terms, Hilbert and Ackerman's question is whether any given statement written in first-order logic is logically valid—true under every possible interpretation. Church's formulation is equivalent to Hilbert and Ackerman's due to the completeness of first-order logic (Church 1936, fn. 6).

Turing-uncomputable functions over denumerable domains is very significant. We will run into it again.

6.3 Mechanisms as Turing Machines?

A few years after Turing's paper came out, Warren McCulloch and Walter Pitts argued that brains were a kind of Turing machine (without tapes). As we saw in the previous chapter, McCulloch and Pitts defined circuits of simplified neurons for performing logical operations such as AND, OR, and NOT on digital inputs and outputs (strings of "ones" and "zeros"). They showed how to build neural networks out of such circuits and argued that their networks are equivalent to Turing machines; they also suggested that brains work similarly enough to their artificial neural networks, and therefore to Turing machines (McCulloch and Pitts 1943).

Shortly after that, John von Neumann offered a sweeping interpretation of McCulloch and Pitts's work:

> The McCulloch–Pitts result... proves that anything that can be exhaustively and unambiguously described, anything that can be completely and unambiguously put into words, is ipso facto realizable by a suitable finite neural network.
>
> (von Neumann 1951, 22)

Since McCulloch–Pitts networks are computationally equivalent to Turing machines (with bounded tapes), von Neumann's statement entails that anything that can be exhaustively and unambiguously described is realizable by a Turing machine. It expands the scope of the Church–Turing thesis from covering mathematical algorithms to covering anything that can be described "exhaustively and unambiguously."

Thus was born the idea that Turing machines are not just a special kind of machine for performing computations. According to von Neumann's broader interpretation, Turing machines are a model of anything exhaustively and unambiguously described. Insofar as mechanisms can be exhaustively and unambiguously described, Turing machines are a model of mechanisms. Insofar as physical systems can be exhaustively and unambiguously described, Turing machines are a model of physical systems. (Insofar as physical systems are mechanisms, the previous two statements are equivalent.)

Around the time that von Neumann was making his bold statement, the first digital computers were being designed, built, and put to use. Digital computers are computationally equivalent to *universal* Turing machines (until they run out of memory). The main use of early computers was to run computational simulations of physical systems whose dynamics was too complex to be solved analytically.

This sort of computational model became ubiquitous in many sciences. Computers became a tool for simulating just about any physical system—provided, of course, that their behavior could be described—in von Neumann's words—"exhaustively and unambiguously."

Aided by the spread and popularity of computational models, views along the lines of von Neumann's hype about what McCulloch and Pitts had allegedly shown became influential. One development was the widespread impression that minds must be computable in Turing's sense. After all, McCulloch and Pitts had allegedly shown that brains are basically Turing machines. The theory that minds are computer programs running on neural software was soon to emerge from this milieu (Putnam 1967a; more on this in Chapter 14).

Closely related is the idea that explaining clearly how a system produces a behavior requires providing a computer program for that behavior or, at least, some kind of computational model. Sometimes, this view was explicitly framed in terms of mechanisms—to the effect that explaining a behavior mechanistically requires providing a computer program for that behavior (e.g., Dennett 1978a, 83; more on this in Chapter 10). Conversely, some theorists argued that, if human beings can behave in a way that is Turing-*un*computable, then the mind is not a machine (e.g., Lucas 1961; more on this in Chapter 11).

Other theorists made a similar point using different terminology. They distinguished between the functions being computed and the mechanisms implementing the computation. They argued that the functions performed by a system ought to be explained by functional analysis—a kind of explanation that describes the functions and sub-functions being performed by a system while, supposedly, abstracting away from the implementing mechanisms. For example, multiplying numbers (by the method of partial products) is functionally analyzed into performing single digit multiplications, writing the results so as to form partial products, and then adding the partial products. Since functional analysis supposedly abstracts away from mechanisms, allegedly it is distinct and autonomous from mechanistic explanation. These theorists often maintained that psychology provides functional analyses, while neuroscience studies the implementing mechanisms—therefore, psychological explanations are distinct and autonomous from neuroscientific explanations (more on this in Chapter 7). These same theorists also maintained that functional analysis consists in a computer program or other computational model (Fodor 1968a; Cummins 1983). As a result, again, explaining a behavior requires providing a computer program or other computational model for that behavior.

Even more grandly, von Neumann's statement prefigured pancomputationalism—the view that *every* physical system is computational (e.g., Putnam 1967a), that the entire universe is a giant computing machine (Zuse 1970), or that computation is somehow the building block of the physical world (e.g., Fredkin 1990; Wheeler 1982). Pancomputationalism took on a life of its own within certain physics circles, where it became known as digital physics or "it from bit" (Wheeler 1990) and—after the birth of quantum computation—transmuted into the view that the universe is a quantum computer (Lloyd 2006).[3]

6.4 Physical Computation

All this talk of computation, especially in psychology and philosophy of mind, raised the question of what it takes for a physical system to perform a computation. Is the brain a computing system? Is the mind the software of the brain? What about other physical systems: which of them perform computations? For these questions to be answerable, some account must be given of what it takes for a brain or other physical system to perform a computation.

One early and influential account of physical computation asserted that a physical system performs a computation just in case there is a mapping between the computational state transitions that constitute that computation and the physical state transitions that the system undergoes (Putnam 1967a). The main appeal of this *simple mapping account* is that it's simple and intuitive. Eventually it became clear that it's far too weak.

The main problem with the simple mapping account is that, without further constraints, it's too easy to map a computation to a series of physical state transitions. Ordinary physical descriptions define trajectories with undenumerably many state transitions, whereas classical computational descriptions such as Turing machine programs define trajectories with denumerably many state transitions. If nothing more than a mapping is required, any denumerable series of state transition can be mapped onto any undenumerable series of state transitions. Thus, by the simple mapping account, any computation is implemented by any ordinary physical system (Putnam 1988). This result is now known as *unlimited pancomputationalism*; it provides a reductio ad absurdum of the simple mapping account of physical computation.

To avoid unlimited pancomputationalism, several theorists introduced constraints on which mappings are acceptable. Perhaps the most popular constrain is a causal one. According to the *causal account* of physical computation, only mappings that respect the causal relations between the computational state transitions—including those that are not instantiated in a given computation—are acceptable (Chalmers 2011; Chrisley 1994; Scheutz 1999). In other words, mapping individual computations onto individual physical state space trajectories

[3] Independently, some logicians and mathematicians showed that large classes of physical or computational systems could be simulated more or less exactly by Turing machines (e.g., Gandy 1980; Rubel 1989); others looked for possible exceptions (e.g., Pour-El and Richards 1989; Cubitt et al. 2015).

is insufficient. What is also needed is that the mapping be set up so that the physical states that correspond to the computational states stand in appropriate causal relations with one another, mirroring the causal relations implicitly defined by the computational description.

The main advantage of sophisticated mapping accounts, such as the causal account, is that they remain close to the simplicity of the mapping account while avoiding unlimited pancomputationalism. The main disadvantage of these accounts is that they remain committed to *limited* pancomputationalism—the view that everything performs some computation or another. Limited pancomputationalism is in line with von Neumann's sweeping interpretation of the Church–Turing thesis, but it is in tension with a common way of understanding the computational theory of cognition. Specifically, it is in tension with the common view that computation is a special kind of process with unique resources to explain cognition.

According to many theorists (e.g., Fodor 1968b, 1975), cognition is explained computationally in a way that other physical processes are not. Specifically, the computational explanation of cognition has to do with the manipulation of representations. Another popular family of accounts of physical computation—*semantic accounts*—is tailored to the perceived needs of this representational version of the computational theory of cognition. According to semantic accounts, physical computation is the manipulation of representations—or, more precisely, the manipulation of certain kinds of representation in appropriate ways (Cummins 1983; Fodor 1975; Pylyshyn 1984; Shagrir 2006).

The greatest advantage of semantic accounts is that they can avoid all forms of pancomputationalism. In fact, semantic accounts restrict physical computation to those physical systems that manipulate the right kinds of representation in the right way—presumably, only cognitive systems and artificial computing systems qualify.

Semantic accounts also come with a disadvantage: they make no sense of any (alleged) computational system that does not manipulate representations. It turns out that, following computability theory, there is no difficulty in defining computations that manipulate meaningless letters. It is just as easy to define a Turing machine that manipulates uninterpreted marks as it is to define one that manipulates meaningful symbols. It is equally easy to program a digital computer to sort meaningless marks into meaningless orders as it is to program it to manipulate meaningful symbols in meaningful ways. Semantic accounts make no room for this. Therefore, they are inadequate accounts of computation in general.

Later we will see that the difficulties of both mapping and semantic accounts led to the most recent accounts of physical computation: mechanistic accounts. But first we need to prepare the terrain by questioning some of the assumptions that are built into the traditional understanding of physical computation.

6.5 The Scope of the Church–Turing Thesis

Given the hype about computability discussed in Section 6.3, eventually scholars began to investigate its history and proper scope. One of the earliest, rigorous investigations of this sort was Guglielmo Tamburrini's PhD dissertation (Tamburrini 1988), which was soon followed by a series of incisive articles by his advisor Wilfried Sieg (e.g., Sieg 1994, 2009). Such investigations reconstructed the intellectual context of Church and Turing's work—which, as I mentioned above, was the foundations of mathematics. They also clarified the interpretation and conceptual foundation of Turing's results. They showed that the Church–Turing thesis, properly understood, is about what mathematical problems can be solved by following algorithms—not about what the mind in particular or physical systems in general can do. The latter questions are independent of the original Church–Turing thesis and must be answered by other means.

With this more rigorous and restrictive understanding of the Church–Turing thesis in the background, Jack Copeland (2000) argued that we should distinguish between a mechanism in the broad, generic sense and a mechanism in the sense of a procedure that computes a Turing-computable function. A mechanism in the broad, generic sense is a system of organized components—this is the notion behind the new mechanism in philosophy of science, which I will discuss in the next section. Whether a mechanism in the broad, generic sense computes a Turing-computable function or a Turing-*un*computable function is independent of the original Church–Turing thesis.

Copeland pointed out that, at least in mathematics, there are hypothetical machines that are more powerful than Turing machines—they compute Turing-uncomputable functions (Turing 1939). Copeland called a machine that is computationally more powerful than a Turing machine a *hypercomputer*. Finally, Copeland suggested that whether the brain or any other physical system is an ordinary computing system or a hypercomputer is an empirical question—it is possible that brains are hypercomputers or that artificial hypercomputers can be built. Given Copeland's argument, the mind may be mechanistic in the generic sense even if it does something that is Turing-uncomputable.

What about the view that something is mechanical if and only if it is computable by some Turing machine? This kind of view relies on the notion of "mechanical" used in logic, whereby algorithms are said to be mechanical procedures. Since Turing machines are formal counterparts to algorithms, any procedure that is not computable by Turing machines is ipso facto not mechanical in that sense. As we've seen, this reasonable conclusion was fallaciously used to imply that the human mind is mechanical (in an unspecified sense) if and only if it is equivalent to a Turing machine. Copeland dubbed this kind of inference—from the Church–Turing thesis to conclusions about whether minds are mechanistic or

Turing-computable—the Church–Turing fallacy (Copeland 2000, more on this in Chapter 10).

Parallel to and independently of Copeland's argument, a literature on physical hypercomputation began to emerge within the foundations of physics. Various philosophers and physicists proposed hypothetical means by which exotic physical systems may be able to compute Turing-uncomputable functions. A prominent example involves a receiver system traveling towards a huge, rotating black hole while receiving signals on the results of a Turing machine's computation performed by a Turing machine orbiting the black hole. Given the way the black hole distorts space-time, the Turing machine may be able to complete an infinite number of operations while the receiver system that launched the Turing machine traverses a finite time-like trajectory; this allows the receiver to compute a Turing-uncomputable function, at least for one value of the function (Hogarth 1994; Etesi and Németi 2002). Even though there is little indication that hypercomputation is feasible in practice, this literature as well as Copeland's argument undermine the idea that Turing machines are a general model of mechanisms. Still, none of this directly challenges (limited) pancomputationalism in a broader sense. It may still be that every physical system is computational—however, now some physical systems may compute functions that no Turing machine can compute.

6.6 The Rise of the New Mechanism

Around the time that Copeland and others were defending the physical possibility of hypercomputation, another group of philosophers of science revived the view that constitutive explanation—explanation of what a system does in terms of what its organized subsystems do—is provided by mechanisms. These philosophers became known as *the new mechanists*. They argued that many special sciences, including biology and neuroscience, explain phenomena by uncovering mechanisms. What they mean by "mechanism" is what Copeland meant as well as what I mean in this book: an organized system of components such that the components' activities, standing in certain organizational relations, constitute a phenomenon.[4]

A few points about the new mechanism are relevant here.

First, the notion of mechanism articulated by the new mechanists is grounded in the explanations provided by the special sciences. The new mechanists scrutinized scientific practices in sciences such as molecular biology and neuroscience. They argued that such sciences explain phenomena mechanistically. This raises

[4] E.g., Bechtel 2008; Bechtel and Richardson 2010; Craver 2007; Glennan 1996, 2017; Glennan and Illari 2017; Illari and Williamson 2012; Krickel 2018; Machamer, Darden, and Craver 2000.

the question of whether other disciplines—including, say, computer science and computer engineering—explain mechanistically.

Second, the new mechanists differed slightly in the way they conceptualized mechanisms. For example, some talked about mechanisms performing operations, others of mechanisms performing activities, yet others of mechanisms performing functions. Those differences don't matter for our purposes. What matters is that none of these theorists conceptualized mechanisms in terms of computation. The view examined in previous sections, whereby a mechanism is something whose behavior can be modeled by a Turing machine, was absent from the new mechanism. Mechanisms are simply systems of concrete components organized to perform operations, activities, or functions. (In Piccinini 2017b, I argue that these formulations are essentially equivalent anyway; cf. Illari and Williamson 2012.)

Third, even when mechanists talked about functions, they were not talking about the kind of mathematical functions of a denumerable domain that may or may not be Turing computable. Instead, they were talking about functions as causal roles. In this sense of function, the function of the heart is to pump blood, and the function of a drill is to make holes. Even with respect to functions, this notion of mechanism is independent of the notion of computation.

Fourth, many new mechanists were hostile to teleology. When they talked about functions, they typically understood them in a nonteleological way, as something a mechanism *does* rather than something a mechanism is *for* (cf. Craver 2001). But the functions of mechanisms can also be understood teleologically, as what something is *for*. The notion of teleological function may be combined with mechanisms to yield *functional* mechanisms, that is, mechanisms with teleological functions (Chapter 3). Functional mechanisms in this teleological sense are a subset of all mechanisms.

6.7 The Mechanistic Account of Digital Computation

The new mechanists argued forcefully that many special sciences explain by uncovering mechanisms. The most obvious candidate sciences were molecular biology, physiology, neuroscience, and perhaps the study of engineered artifacts.

If all of these disciplines explain mechanistically, the question arises of how computer science and computer engineering explain. A closely related question is how computational explanation works—not only in computer science and engineering but also in psychology and neuroscience. A mechanistic answer to these questions is the starting point of the mechanistic account of physical computation.

I began defending a version of the mechanistic account in the early 2000s (Piccinini 2003a; although at that time I called it the "functional account") and spent much of the subsequent decade developing and extending it (2015). Others followed suit with their own versions, sometimes tailoring it to specific sciences or varieties of computation (Kaplan 2011; Fresco 2014; Milkowski 2013; Duwell 2017). Mechanistic accounts explicate physical computation in terms of a special kind of mechanism.

A first observation is that computer science and computer engineering have their own special domain. They study *computational* systems. By contrast, other sciences—with the exception of cognitive science and neuroscience—generally do not focus on computational systems. Thus, if computing systems are mechanisms, there seems to be something distinctive about them—something that makes them the proper domain of specific disciplines. Accordingly, an important task of the mechanistic account is to say what distinguishes computing mechanisms from other mechanisms.

This task is made difficult by the contested boundaries between computing systems and noncomputing systems. Some theorists take a very restrictive attitude: only systems that manipulate representations according to rules and such that their manipulations are caused by representations of the rules count as genuinely computing systems (Fodor 1975; Pylyshyn 1984). Other theorists take a very liberal attitude: every physical system performs some computation or another (Chalmers 2011; Scheutz 1999; Shagrir 2006). Yet other theorists fall somewhere between these two extremes.

To establish reasonable boundaries, I used the following strategy. First, identify a set of paradigmatic examples of physical computing systems (digital computers, calculators, etc.) and a set of relatively uncontroversial examples of noncomputing systems (digestive systems, hurricanes, etc.). Second, identify what distinguishes the computing systems from the noncomputing systems. Finally, use the resulting account to decide any boundary cases. My first stab at a mechanistic account covered only digital computation (including hypothetical hypercomputers), precisely because digital computing systems such as computers and calculators are generally accepted as computational and because they fall under Turing's mathematical theory of computation (Piccinini 2007a, 2008a, b).

To construct a mechanistic account of digital computation, the key notion was that of a digit, which is a concrete counterpart of the abstract notion of letter (or symbol) employed by computability theorists. As we saw above, a Turing machine is said to manipulate strings of *letters* (symbols) according to a set of instructions. A digit is a physical macrostate that corresponds to the mathematical notion of letter. It comes in finitely many types that the system must be able to reliably distinguish and manipulate. Like letters, digits can be concatenated into strings such that the physical system that manipulates the digits can differentiate between digits based on their place within a string and treat each digit differently depending on its place within the string. Given these preliminaries, I argued that physical computing systems are functional mechanisms whose function is manipulating strings of digits according to a rule defined over the digits. The rule is just the mapping from input strings of digits and internal states to output strings of digits.

This account has several virtues.

First, it makes computation an objective property of physical systems. Whether a physical system has functions, whether it manipulates strings of digits, and whether it does so in accordance with rules defined over the digits are objective properties of the system, which can be empirically investigated. Unlimited pancomputationalism is thus avoided.

Second, only a fairly restricted class of physical systems counts as computational. This way, limited pancomputationalism is avoided. By being thus restricted, the mechanistic account fulfills the needs of robust computational theories of cognition, according to which computation is a special kind of process that has something specific to do with cognition.

Third, the mechanistic account does not assume that computation requires representation. It shares this virtue with mapping accounts. This is a virtue because nothing prevents us from building computing systems that do not manipulate representations in any interesting sense.

Fourth, the mechanistic account comes with a clear and compelling notion of computational explanation, which fits the practices of the relevant sciences. That is, computational explanation is a species of mechanistic explanation. It is the species of mechanistic explanation that appeals to components that manipulate strings of digits in accordance with rules defined over the digits.

Fifth, the mechanistic account is the first to explain miscomputation. Computer scientists and engineers work hard to test the reliability of computer circuits, debug computer programs, and fix computers. In other words, they fight miscomputation. Mapping accounts have a hard time making any sense of that—after all, there is always a mapping between any physical state transition and some computation or another. Semantic accounts are not much better off. In any case, no one attempted to make sense of miscomputation until mechanistic accounts came along. According to mechanistic accounts, there are several possible sources of miscomputation. A system may miscompute because it was poorly designed (relative to the goals of the designers), or poorly built (relative to the design plan), or misused. If a system is programmable, it can also miscompute because it was programmed incorrectly (relative to the programmer's intentions).[5]

Sixth, the mechanistic account can be used to provide an illuminating taxonomy of computing systems (Piccinini 2008a, c).

Based on this initial mechanistic account, I argued that so-called analog computers and neural networks that have no digital inputs and outputs are not

[5] For more on miscomputation, see Piccinini 2015, 148–50.

computing mechanisms properly so called (Piccinini 2008a, c). This turned out to be overly restrictive.

6.8 The Mechanistic Account of Computation in the Generic Sense

Clearly, there are important differences between digital computers and analog computers. Whereas digital computers manipulate strings of digits, analog computers manipulate variables that can be continuous. Whereas digital computers can follow precise instructions defined over digits, analog computers (typically) follow systems of differential equations. Whereas digital computers compute functions defined over denumerable domains, analog computers (typically) solve systems of differential equations. Whereas digital computers can be universal in Turing's sense, Turing's notion of universality does not apply to analog computers. Whereas digital computers produce exact results, analog computers produce approximate results. These differences may be seen as evidence that analog "computers" are so different from digital computers that they shouldn't even be considered computers at all (Piccinini 2008a).

And yet, those same differences also point at deep similarities. Both digital and analog computers are used to solve mathematical problems. Both digital and analog computers solve mathematical problems encoded in the values of physical variables. Both digital and analog computers solve mathematical problems by following some sort of procedure or rule defined over the variables they manipulate. Both digital and analog computers used to compete within the same market—the market for machines that solve systems of equations. (Digital computers eventually won that competition, in part because they do much more than solving systems of equations.) These similarities are serious enough that many scientific communities found it appropriate to use the same word—"computer"—for both classes of systems. Maybe they are onto something; philosophers may be better off figuring out what that is rather than implying that scientists are misguided.

A similar point applies to other unconventional models of computation, such as DNA computing and quantum computing. In recent decades, computer scientists have investigated models of computation that depart in various ways from mainstream electronic digital computing. These models employ chemical reactions, DNA molecules, or quantum systems to perform computations—or so say the people who study them. It would be valuable to have an account of physical computation general enough to cover them all. Notice that neither supporters of mapping accounts nor supporters of semantic accounts ever addressed this sort of problem.

In light of the above, we need a notion of physical computation broader than digital computation—a notion of computation in a generic sense that covers digital computation, analog computation, and other unconventional kinds of computation. Someone might be tempted to simply conclude that everything computes (limited pancomputationalism). That is unhelpful for several reasons. To begin with, pancomputationalism erases all boundaries between the domain of computer science and every other scientific domain. More importantly, pancomputationalism per se does not elucidate physical computation—it does not tell us what it takes for a physical system to perform a computation. Finally, pancomputationalism does not tell us what distinguishes a physical system that performs a digital computation from one that performs an analog computation or any other kind of computation. It would be better if there were a way to characterize what digital computing systems have in common with analog computers and other unconventional computing systems as well as what distinguishes all of them from other physical systems.

A helpful notion is *medium independence*, which was used by Justin Garson (2003) to characterize the notion of information introduced by Edgar Adrian (1928) in neurobiology. Garson argued that information is medium-independent in the sense that it can be carried by a physical variable (such as spike trains) regardless of the physical origin of the information (which may be light, sound waves, pressure, etc.).

Medium independence in this sense is stronger than multiple realizability (Chapter 2). If a property is medium independent, it is also multiply realizable, because it's realizable in different media. The converse does not hold: a multiply realizable property may or may not be medium independent. For example, the property of being a mousetrap is multiply realizable but not medium independent: any mousetrap must handle the same medium—mice!

A similar notion of medium independence can be used to characterize all the vehicles of computation, whether digital or not. Specifically, all physical computing systems—digital, analog, or what have you—can be implemented using physically different variables so long as the variables possess the right degrees of freedom organized in the right way and the implementing mechanisms can manipulate those degrees of freedom in the right way. In other words, computational vehicles are macroscopic variables defined in a medium-independent way. This, then, became my mechanistic account of computation in the generic sense: a physical computing system is a functional mechanism whose function is to manipulate medium-independent variables in accordance with a rule defined over the variables (Piccinini and Scarantino 2011; Piccinini 2015).

This more general mechanistic account of physical computation inherits the virtues of the mechanistic account of digital computation. In addition, it covers analog computers and other unconventional models of computation. Thanks to its generality, the account covers the notion of computation used by computational neuroscientists without assuming that it must be digital—or analog, for that matter.

6.9 Kinds of Computation: Digital, Analog, and More

6.9.1 Digital Computation

As I'm using the term, digital computation is the kind that was analyzed by the classical mathematical theory of computation that began with Alan Turing and other logicians. Classical computability theory is a well-established branch of mathematics that lies at the foundation of computer science. Digital computation encompasses a widely used class of formalisms such as recursive functions and Turing machines. It gives us access to the fundamental notion of universal computer: a digital computer that can compute any digital computation that can be spelled out by an algorithm (Davis et al. 1994).

Digital computation may be defined both abstractly and concretely. Roughly speaking, abstract digital computation is the manipulation of strings of discrete elements, that is, strings of letters from a finite alphabet. Here we are interested primarily in concrete computation, or physical computation. Letters from a finite alphabet may be physically implemented by what I call *digits*. As I said above in section 6.7, to a first approximation, concrete digital computation is the processing of strings of digits by a functional mechanism according to general rules defined over the digits. Let us consider the main ingredients of digital computation in more detail.

The atomic vehicles of concrete digital computation are digits, where a digit is a macroscopic state (of a component of the system) whose type can be reliably and unambiguously distinguished by the system from other macroscopic types. To each (macroscopic) digit type, there correspond a large number of possible microscopic states. This means that digits are variably realizable (Chapter 2). Artificial digital systems are engineered to treat all those microscopic states in the same way—the one way that corresponds to their (macroscopic) digit type. For instance, a system may treat 0 volts plus or minus some noise in the same way (as a "0"), whereas it may treat 4 volts plus or minus some noise in a different way (as a "1"). To ensure reliable manipulation of digits based on their type, a physical system must manipulate at most a finite number of digit types. For instance, ordinary computers contain only two digit types, usually referred to as "0" and "1." Digits need not mean or represent anything, but they can. For instance, numerals represent numbers, while other digits (e.g., "|," "\") need not represent anything in particular.

Digits can be concatenated (i.e., ordered) to form sequences or strings. Strings of digits are the vehicles of digital computations. A digital computation consists in the processing of strings of digits according to rules. A rule for a digital computation is simply a map from input strings of digits, plus possibly internal states, to output strings of digits. Examples of rules that may figure in a digital computation include addition, multiplication, identity, and sorting. To be a bit more precise, for each digital computing system, there is a finite alphabet out of which strings of digits can be formed and a fixed rule that specifies, for any input string on that alphabet (and for any internal state, if relevant), whether there is an output string defined for that input (internal state) and which output string that is. If the rule defines no output for some inputs (internal states), the mechanism should produce no output for those inputs (internal states).

Digits are unambiguously distinguishable by the processing mechanism under normal operating conditions. Strings of digits are ordered pluralities of digits, i.e., digits such that the system can distinguish different members of the plurality depending on where they lie along the string. The rules defining digital computations are defined, in turn, in terms of strings of digits and internal states of the system, which are simply states that the system can distinguish from one another. No further physical properties of a physical medium are relevant to whether its states implement digital computations. Thus, digital computations can be implemented by any physical medium with the right degrees of freedom.

To summarize, a physical system is a digital computing system just in case it is a mechanism whose function is manipulating input strings of digits, depending on the digits' type and their location on the string, in accordance with a rule defined over the strings (and possibly certain internal states of the system).

The notion of digital computation here defined is quite general. It should not be confused with three other commonly invoked but more restrictive notions of computation: classical computation in the sense of Fodor and Pylyshyn (1988), algorithmic computation, and computation of Turing-computable functions (see Figure 6.1).

Let's begin with the most restrictive notion of the three: classical computation. A classical computation is a digital computation that has two additional features. First, it is algorithmic: it proceeds in accordance with effective, step-by-step procedures that manipulate strings of digits and produce a result within finitely many steps. Second, it manipulates a special kind of digital vehicle: language-like vehicles. Thus, a classical computation is a digital, algorithmic computation whose algorithms are sensitive to the combinatorial syntax of the strings (Fodor and Pylyshyn 1988).

A classical computation is algorithmic, but the notion of algorithmic computation—digital computation that follows an algorithm—is more inclusive, because it does not require that the vehicles being manipulated be language-like.

Any algorithmic computation, in turn, can be performed by some Turing machine (i.e., the computed function is Turing-computable). This is a version of the Church–Turing thesis. But the computation of Turing-computable functions

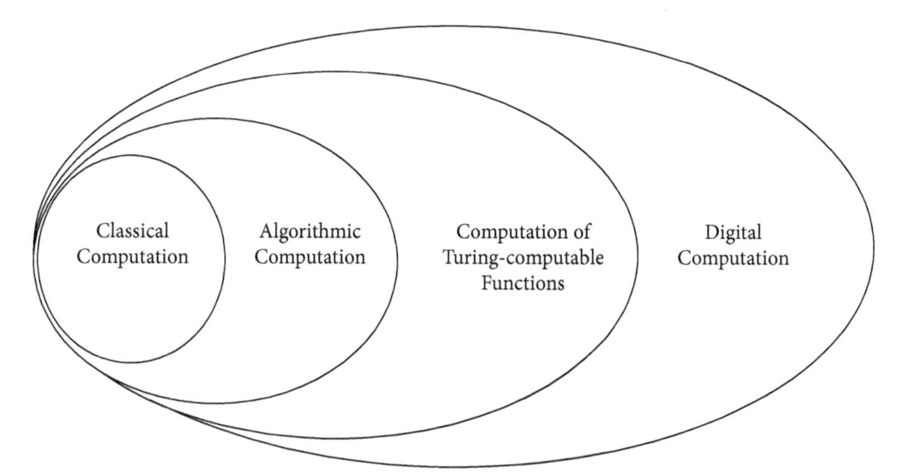

Figure 6.1 Types of digital computation and their relations of class inclusion.

need not be carried out by following an algorithm. For instance, many neural networks compute Turing-computable functions (their inputs and outputs are strings of digits, and the input–output map is Turing-computable), but such networks need not have a level of functional organization at which they follow the steps of an algorithm for computing their functions; there may be no functional level at which their internal states and state transitions are discrete.

Here is another way to draw the distinction between algorithmic digital computation and digital computation more generally. Algorithmic digital computation is fully digital—digital every step of the way. Fully digital systems, such as McCulloch–Pitts networks (including, of course, digital computers), produce digital outputs from digital inputs by means of discrete intermediate steps. Thus, the computations of fully digital computing systems can be characterized as the (step-by-step) algorithmic manipulations of strings of digits. By contrast, digital computations more generally are only input/output (I/O) digital. I/O digital systems, including many neural networks, produce digital outputs from digital inputs by means of irreducibly continuous intermediate processes. Thus, the computations of merely I/O digital computing systems cannot be characterized as step-by-step algorithmic manipulations of strings of digits.

Finally, the computation of a Turing-computable function is a digital computation, because Turing-computable functions are by definition functions of a denumerable domain—a domain whose elements may be counted—and the arguments and values of such functions are, or may be reduced to, strings of digits.[6] But it is equally possible to define functions of strings of digits that are not

[6] Some cognitively oriented computer scientists have suggested a version of the Computational Theory of Cognition (CTC) according to which cognition is computable (e.g., Shapiro 1995; Rapaport

Turing-computable, and to mathematically define processes that compute such functions. Some authors have speculated that some functions over strings of digits that are not Turing-computable may be computable by some physical systems (e.g., Copeland 2000). According to the present usage, any such computations still count as digital computations. It may well be that only the Turing-computable functions are computable by physical systems—whether this is the case is an empirical question that cannot even be properly posed without the distinctions we are drawing. In any case, the computation of Turing-uncomputable functions is unlikely to be relevant to the study of cognition (more on this in Chapter 10). Be that as it may—I will continue to talk about digital computation in general.

Many other distinctions may be drawn within digital computation, such as hardwired vs. programmable, special purpose vs. general purpose, and serial vs. parallel (cf. Piccinini 2015, chapters 11 and 13). These distinctions play an important role in debates about the computational architecture of cognitive systems.

6.9.2 Analog Computation

Analog computation is often contrasted with digital computation, but analog computation is a vaguer and more slippery concept. The clearest notion of analog computation is that of Pour-El (1974). Roughly, abstract analog computers are systems that manipulate variables that can be continuous—variables that can vary continuously over time and take any real values within certain intervals. The functional relations between such variables are typically specified by differential equations. A major function of analog computers is solving certain systems of differential equations by instantiating the relations between continuous variables specified in the equations.[7]

1998). This view appears to be a specific version of digital CTC according to which the computations performed by cognitive systems fall in the class of Turing-computable functions.

[7] The class of functions of a real variable whose values can be generated by a general-purpose analog computer are the differentially algebraic functions, i.e., solutions to algebraic differential equations (Pour-El 1974; see also Lipshitz and Rubel 1987; and Rubel and Singer 1985). Algebraic differential equations have the form $P(y, y', y'', \ldots y^{(n)}) = 0$. Here, P is a polynomial function with integer coefficients and y is a function of x. Some algebraic differential equations can be classified as "universal" to the extent that a solution of the equation can be used to approximate any continuous function of a real variable over the interval $0 \leq t < \infty$, up to an arbitrary degree of accuracy. General-purpose analog computers with as few as four integrators can generate outputs which, corresponding to this type of "universal" equation, can approximate any continuous function of a real variable arbitrarily well (Duffin 1981; Boshernitzan 1986). Like analog computation, neural firing can be described using coupled nonlinear ordinary differential equations. But whether neural firing constitutes analog computation depends on more precise similarities between the properties of neural firing and those of analog computation. As I shall argue in Chapter 13, there are important disanalogies between the two.

Analog computers can be physically implemented, and physically implemented continuous variables are a different kind of vehicle than strings of digits. While a digital computing system can always unambiguously distinguish digits and their types from one another, a concrete analog computing system cannot do the same with the exact values of (physically implemented) continuous variables. This is because continuous variables can take any real values but there is a lower bound to the sensitivity of any physical system. Thus, it is always possible that the difference between two portions of a continuous variable is small enough to go undetected by the system. From this it follows that the vehicles of analog computation are not strings of digits and analog computations (in the present, strict sense) are a different kind of process than digital computations.

Nevertheless, analog computations are only sensitive to the differences between portions of the variables being manipulated, to the degree that they can be distinguished by the system. Any further physical properties of the media implementing the variables are irrelevant to the computation. Like digital computers, therefore, analog computers operate on medium-independent vehicles.

There are other kinds of computation besides digital and analog computation. One increasingly popular kind, quantum computation, is an extension of digital computation in which digits are replaced by quantum states called qudits (most commonly, binary qudits, which are known as qubits). Suffice it to conclude that computation in the generic sense includes digital computation, analog computation, quantum computation, and more (Figure 6.2).

6.10 Two Ways of Explaining Computation Mechanistically

Oron Shagrir and Lotem Elber-Dorozko (2019; see also Shagrir 2017; cf. Coelho Mollo 2018) have raised an important question for the mechanistic account of physical computation. According to the mechanistic account, computational explanation is mechanistic. That is, we explain computations by decomposing them into subcomputations, each of which is performed by computing components which, when organized in the right way, give rise to the whole computation.

This strategy iterates: subcomputations can be decomposed into sub-sub-computations, until we reach atomic computations performed by primitive computing components, which in ordinary digital computers are logic gates. So far, this sounds like ordinary mechanistic explanation—except that both the functions and the structures that constitute the mechanisms are defined in a medium-independent way. That's why computational properties are realizable by any physical medium with appropriate degrees of freedom.

Once we reach primitive computing components, we are done with computational explanation, but typically we are not done with mechanistic explanation. We can identify classes of concrete physical microstates (e.g., voltages) that realize

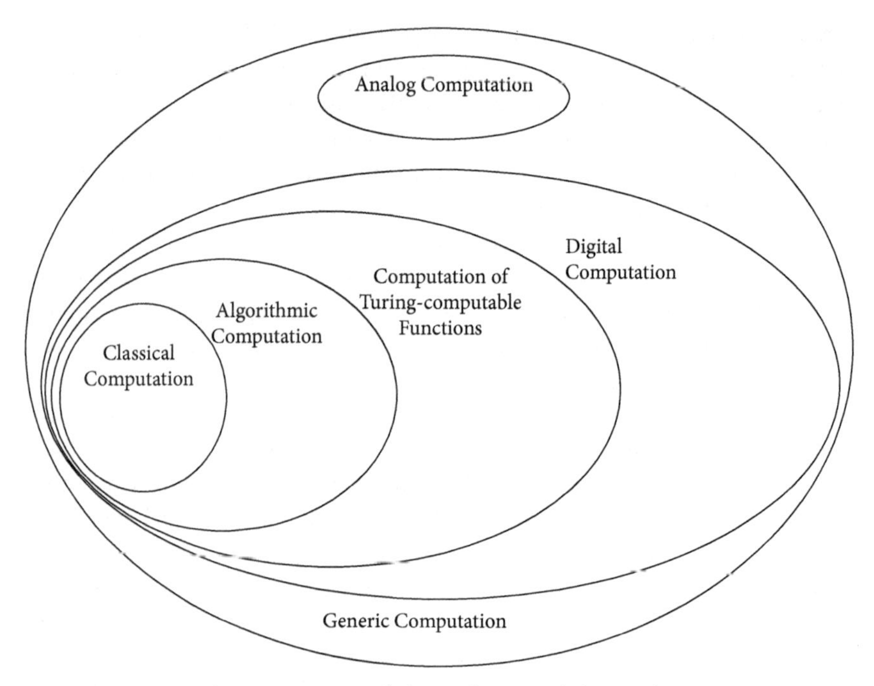

Figure 6.2 Types of computation and their relations of class inclusion.

the computational inputs, internal states, and outputs; we can also identify concrete physical structures (e.g., electronic circuits) that perform the relevant operations. This level is what is often called technological *implementation*. It provides a mechanistic explanation of the operation of primitive computing components. Since it's medium-*dependent*, it is not computational.

Now that we have medium-dependent components, however, we can reconstruct the whole mechanism in medium-dependent terms. Instead of describing it as a system of mere logic gates, we can describe it as a system of electronic circuits. If, instead, we had established that the logical operations are performed by relays, we could reconstruct the whole system as a system of relays. Ditto for any other technology with appropriate degrees of freedom organized in the right way. The point is that we now have two kinds of decomposition of a computational system: a medium-*independent* decomposition and a medium-*dependent* one. Which of the two is the genuinely mechanistic decomposition? Which of the two provides the full mechanistic explanation of the system's computational capacities?

Answer: both.

Which mechanistic explanation is appropriate depends on the explanandum. As I mentioned in Chapter 2, capacities (functions) can be described more narrowly or more broadly. We can ask how to track time, how to track time with a certain degree of precision, or how to track time using sunlight and shades.

We can ask how to cook, how to cook using a microwave oven, or how to cook using a convection microwave oven. Each functional specification puts different constraints on the explanans. The resulting class of mechanistic explanations—explanations that could possibly perform the function in question—is correspondingly broader or narrower. Broadly described functions are multiply realizable by a larger class of mechanisms than narrowly defined functions.

The same point applies to computing systems. If the explanandum is a pure computational capacity—say, performing prime factorization over strings of binary digits—then an algorithm is the first explanatory level. If we'd like to understand how the algorithm is executed, we also need a description of the processor and memory, perhaps all the way down to the logic gates. These are purely computational yet complete mechanistic explanations of the explanandum at the desired levels—they are medium independent and compatible with different technological implementations.

In contrast, if the explanandum also includes aspects of a technology—say, performing prime factorization in an integrated circuit—then that aspect of the technology must also be included in the explanation. As we decompose the capacities of the system, we must decompose them into subcapacities of circuits built out of the relevant technology. The result is still a complete mechanistic explanation of the explanandum. But since the explanandum includes aspects of the medium, it's not a purely computational, medium-independent explanation. It's a hybrid, computational-technological mechanistic explanation.

We can get even more specific—there are many ways of building integrated circuits. We can ask how to perform prime factorization using integrated circuits that have a certain clock cycle, or circuits that use complementary metal-oxide-semiconductor (CMOS) technology, or are built out of specific semiconductors. These restrictions on the functional specification further constrain the explanandum, giving rise to mechanistic decompositions that depend progressively more on specific physical media, and include a narrower and narrower range of possible mechanisms.

In conclusion, both kinds of mechanistic decomposition—medium-independent and medium-dependent—provide fully adequate mechanistic explanations. They provide explanations of different capacities—medium-independent and medium-dependent capacities, respectively. If the function to be explained is defined purely computationally, then the resulting decomposition is purely computational and therefore more inclusive; if the function is defined in terms of a specific medium, then the resulting decomposition is hybrid and therefore less inclusive. More or less general functional specifications generate more or less inclusive decompositions.

6.11 Internalism vs. Externalism

Another question that keeps coming up is whether physical computation should be individuated solely based on intrinsic properties of a system (internalism) or also in terms of relational properties (externalism). There was a time when computational individuation was thought to be paradigmatically internalistic in a slightly different sense—that is, based solely on causal powers rather than broad representational content (Fodor 1980). That time was followed by a long debate predicated on the semantic account of computation.

According to externalists (e.g., Burge 1986), computational states are individuated by their broad representational content, which depends in part on certain semantic relations between the system and its environment. Others replied that there is a viable notion of narrow content, which is entirely intrinsic to a representational system. If there is a viable notion of narrow content, then internalism about computational states is viable (e.g., Segal 1991; Egan 1999).

This debate shifted when Oron Shagrir (2001) pointed out that the intrinsic properties of a physical system are always consistent with multiple computational individuations, depending on how the microstates of the system are grouped into equivalence classes. This is now known as the problem of computational indeterminacy (Fresco et al. 2016). In Shagrir's example, the states of a tristable system can be grouped and labeled in several different ways, leading to different computational individuations.

A caveat. The problem of computational indeterminacy is *not* that the same macrostates can be labeled in different ways. For instance, consider a simple bistable device taking two input types and yielding two output types. If the device returns type 1 outputs if and only if both inputs are of type 1, and it returns type 0 outputs in all other cases, this is called an AND gate. If the labels are switched, this same device is called an OR gate (Sprevak 2010). Some have suggested that this multiplicity of computational individuations is a problem for the mechanistic account, because the mechanistic account seems unable to determine whether the device is an AND or an OR gate.

On the contrary, this is no problem for the mechanistic account. The mechanistic account assigns a computational identity to a physical system based on its mechanistic properties, up to relabeling the state types. Relabeling state types is a purely observer-dependent maneuver. It does not alter how the system works or what it's capable of. In practice, when we study artificial systems, we are free to label them as we please. Needless to say, we label them in a way that suits our purposes, which corresponds to what we build them to do. When we study natural computational systems, however, we must try to discover a system's computational functions and structures and figure out how the structures perform the functions. How we label the functions makes no difference as long as our labeling scheme attaches to the correct way of grouping the system's microstates into macrostates—that is, classes of functionally equivalent microstates.

The real problem of computational indeterminacy is finding the correct grouping of microstates. Shagrir argued that the only way to select the correct computational individuation of a system is to find what the microstates semantically represent outside the system. Since this external semantic interpretation leads outside the intrinsic properties of the system, Shagrir defended externalism about computational individuation. In response, I accepted that externalism is needed but rejected semantic individuation. Instead of using semantics, I argued that we can also select the correct computational individuation of a system by finding the broad computational function performed by the system (Piccinini 2015, chap. 3).

Subsequent commentators have defended nonsemantic internalism about computational individuation (Coelho Mollo 2018; Dewhurst 2018b; Tucker 2018). They contend that the mechanistic structure intrinsic to a physical system is enough to individuate the computations it performs. Tucker adds that intrinsic individuation is *needed* to identify the computation *actually* performed by a system, as opposed to the computation it has the teleological function to perform. Tucker grants that the function the system ought to perform, as a matter of its teleological function, is individuated externalistically. He insists, however, that the computation the system actually performs is individuated internalistically, and we need this internalistic individuation to obtain an adequate account of miscomputation: when the function a system actually performs is different from the function the system ought to perform, a miscomputation occurs.

As they are presented, these neo-internalist proposals do not solve the problem of computational indeterminacy. Consider an arbitrary physical system, with its indefinitely many microstates and transitions between microstates. Microstates can be grouped into equivalence classes in indefinitely many ways, and the transitions between microstates will be correspondingly grouped into indefinitely many equivalence classes. Out of all of these possible groupings, which grouping is the correct one? In other words, which grouping corresponds to the computation performed by the system? None of the internalist proposals in the literature answer this question. This is not surprising, because by restricting themselves to what's intrinsic to the system, they lack resources with which to solve the problem. Nevertheless, by pushing back against externalism, they help clarify an important way in which computational explanation depends on wide functions.

Again, the situation is different for natural versus artificial systems. Since we design and build artificial systems, we design and build them to compute certain functions. Therefore, we engineer artificial computing systems to have certain stable macrostates and undergo certain transitions between macrostates. We get to define the correct equivalence classes between microstates as we please and, if we are successful, we engineer them to be that way. We define the function the system ought to compute as a function from macrostates to macrostates. If the system computes a different function—if the wrong macrostates show up at the end of a sequence of transitions between macrostates—a miscomputation occurs. Both the functions that a system ought to compute and the functions it actually computes are individuated in relation to us—designers, builders, and users of the devices.[8]

In contrast, when we study natural systems we can't *define* the macrostates and transitions between macrostates. We have to discover them. The way we discover them is the same way we discover the functions of any other functional mechanism. Here is a brief rational reconstruction. We begin by identifying the capacity of interest, the structures that perform the capacity, and the organization that allows the structures to perform the capacity. In this case, the capacity is cognition—understood to include governing behavior—and the structure is the nervous system. Cognition is a broad function of the nervous system. When we examine the nervous system more closely, we discover that it cognizes by transducing different sources of physical information into internal signals—spike trains—and processing those signals.

Assuming for the sake of the argument that processing neural signals is a type of computation, computing is a narrower function of the nervous system than cognizing. But computing is still broader than the nervous system itself, because discovering that spikes are the computational vehicle requires studying the transduction process at sensory receptors as well as the neuromuscular junction to observe that sensory receptors produce spikes and spikes cause muscle fibers to contract. Finally, we look for equivalence classes between spike trains and transitions between such equivalence classes, which correspond to the specific computational functions of neural systems. If a system computes a different function—if

[8] If we are unsuccessful at designing or building the system, there is no fact of the matter as to which function the system we build ought to compute. There are just the functions we want it to compute and the functions it computes under the circuits' normal operating conditions. This situation is reminiscent of strong skepticism about which function is computed by a physical system (Kripke 1982; Buechner 2018). Even here, however, we can check whether the circuits are operating properly independently of which function they are computing, because we have independent criteria for both. On that basis, we take the mathematical function that we intend the machine to compute as our standard, conclude that the circuit designers or builders made a mistake, and recall our product from the market. At any rate, in this case the lack of a normative function is philosophically harmless. By contrast, when design and construction are executed correctly, we can test the circuits and collect evidence that they are functioning correctly, and this gives us a benchmark for establishing both which function the system ought to compute and which function it is actually computing. More precisely, we have evidence that the circuits we are using function correctly under physical conditions C, and we take correct functioning under conditions C as a defeasible assumption. On that basis, we can tell that tokens of type X are supposed to compute F under conditions C by observing such tokens under conditions C and seeing that they succeed at computing F. Then we compare a particular run of a particular token of type X to F and the physical conditions of the machine to C. If the conditions match C and the output matches F, the machine correctly computed F. If not, and the machine still yielded a computational output, there was a miscomputation. (Strictly speaking, a malfunction could still lead to the correct output, and we could discover the malfunction by looking at other functions as well as at the mismatch between current physical conditions and conditions C. This is unlikely to happen in real life.) This is just a sketch of an answer to strong skepticism about physically computing a function. Doing justice to strong skepticism would require a more detailed treatment.

the wrong equivalence classes of spike trains show up as outputs, if the system behaves incorrectly—a miscomputation occurs.

Needless to say, identifying broad functions requires looking at the relation between a mechanism and its context. It cannot be done by looking solely at the intrinsic properties of a system. And identifying miscomputations is parasitic on identifying computational functions. Therefore, insofar as we are looking for explanations of the capacities of natural biological systems, computational individuation remains externalistic.[9]

6.12 Is All Computational Explanation Mechanistic?

I have defended a mechanistic account of physical computation. Physical computation is the processing of medium-independent vehicles by a functional mechanism in accordance with a rule. I've also argued that computation can occur without information processing and vice versa. Nevertheless, typical computing systems process information, and typical information processors are computing systems. Computation and information processing are performed by mechanisms, which can be defined in a medium-independent way because computation and information processing can be defined in a medium-independent way.

Perhaps the biggest point of contention surrounding the mechanistic account of physical computation is whether it adequately subsumes all forms of computational explanation. Carl Craver and I (Piccinini and Craver 2011) argued that it does because functional analysis—the traditional alternative to mechanistic explanation—is just a sketch of a mechanism. Others responded that, at least in cognitive science or cognitive neuroscience, there are forms of computational explanation that are not mechanistic (Chirimuuta 2014; Shagrir 2010a; Shapiro 2016; Weiskopf 2011b). The most common reason they give is that such forms of computational explanation abstract away from some aspects of the mechanism that carries out the computation. In the next chapter I will argue that, when the proper roles of abstraction within mechanistic explanation are appreciated, constitutive explanation—including computational explanation—remains mechanistic.

[9] Is there anything wrong with externalism? As far as I can tell, there is no downside to externalism in the present sense. This innocuous form of externalism is not to be confused with the kind of externalism according to which the belief that water is wet is of a different type on Earth versus Twin Earth simply because "water" supposedly refers to different substances on Earth versus Twin Earth, even though the subject can't tell the difference (Putnam 1975a; Fodor 1980).

7

Mechanistic Models of Cognition

From Autonomy to Integration

7.1 Causal Explanation: Etiological, Contextual, and Constitutive

This chapter extends the account of mechanisms and mechanistic explanation that I developed in previous chapters. It argues that constitutive explanation, including computational explanation, is mechanistic.

Science explains phenomena. By "phenomenon," I mean either a system doing something or a system's capacity to do something. That's what we try to explain. I am interested in *causal* explanation—explanation that situates phenomena within the causal structure of the world (Salmon 1984; Woodward 2003; Craver 2007). Causal explanation has three aspects.

A first aspect is the extrinsic causes of a phenomenon—the factors, external to the phenomenon itself, which bring it about. For example, why did the plant grow? Because you planted a seed in fertile soil at a suitable temperature within a hospitable atmosphere under enough sunlight and watered it regularly. The seed, soil, temperature, atmosphere, sunlight, and water are the relevant causes. They make a difference to the phenomenon. Manipulating them is a way of altering the phenomenon. This is *etiological* explanation.

A second aspect of causal explanation is what a phenomenon contributes to a larger phenomenon or system that contains it. For example, what role does the plant play within the ecosystem? It generates oxygen, which many organisms breathe; it provides a home for some organisms; it is food for others. The plant makes a difference to the broader system that contains it. Manipulating the plant is a way of altering the larger system. This is *contextual* explanation (Craver 2001).

A third aspect of causal explanation is the causal structure that is intrinsic to a phenomenon itself—the internal factors that bring the phenomenon about from the inside out. If a phenomenon involves a system's capacity, function, activity, or process, this aspect of causal explanation involves suitably organized sub-capacities, sub-functions, sub-activities, or sub-processes that constitute the phenomenon. For example, how does the plant grow from a seed? To answer this question, we need to find out what the components of the seed are, what each of them does, and how they are organized. The seed has an outer coat that protects and preserves it, an endosperm that provides nutrients, and a cell called zygote.

Neurocognitive Mechanisms: Explaining Biological Cognition. Gualtiero Piccinini, Oxford University Press (2020).

DOI: 10.1093/oso/9780198866282.001.0001

When enough water enters the seed, the zygote takes nutrients from the endosperm and divides until it forms an embryo. The embryo sprouts—first a root that absorbs water and nutrients from the soil, and then a stem with leaves that take sunlight to transform water and carbon dioxide into carbohydrates, generating oxygen as byproduct. The parts of the plants and their activities make a difference to plant development. Manipulating them is a way of altering plant development, and manipulating plant development is a way of altering them. This is *constitutive* explanation.

Constitutive explanation includes realization and adds internal causal structure. As we have seen, realization is composition plus parthood between properties: a plurality of properties **Q** realizes a property P if and only if those **Q** belong to the component parts that compose the object possessing P and P is an aspect of **Q** (Chapter 2). Constitutive explanation adds the causal relations between a mechanism's components. By doing so, constitutive explanation explains how a mechanism's higher-level property comes about.

The three aspects of causal explanation go hand in hand. Knowing the etiology of a phenomenon helps us understand internal structure and vice versa. Knowing etiology also helps us understand contribution to larger systems and vice versa. Finally, knowing internal structure helps us understand contribution to larger systems and vice versa. The three kinds of explanation are inseparable in practice and mutually constrain one another.

The task of understanding a phenomenon and its place in the causal structure of the world can be approached in many ways. One task is simply to find out what a system does under various conditions. To do that, you either observe the system and see what it does or intervene on the system and observe how it responds. To describe the phenomenon systematically and succinctly, you may build a *phenomenological model*—a representation of what the system does under various conditions. The main purpose of phenomenological models is to describe and predict, *not* to explain.

A second task is to identify the causal relations the phenomenon is involved in. As we've seen, there are three kinds of such causal relations: between the phenomenon and its external causes (etiological explanation), between the phenomenon and a larger system that contains it (contextual explanation), and between the phenomenon and causes internal to it (constitutive explanation). To identify causes, you can use statistical techniques, derive causal conclusions from background theories and models that are independently established, or intervene directly on the system. When you intervene on the system carefully enough that you can rule out other causes, you may conclude that your intervention caused the effect. To help yourself, you can also build a *causal model*—a representation of the causal relations that affect the phenomenon. By representing causal relations, causal models describe, predict, and also explain the phenomenon by identifying what makes a difference to the phenomenon itself, which is also what you'd have to manipulate if you wanted to alter the phenomenon.

When a phenomenon is very complex, involving many variables, it can be extraordinarily difficult to find its etiology, causal role, and causal structure. Cognition is a perfect example. Cognition involves an indefinite number of capacities, an indefinite number of conditions under which these capacities are exercised, and an indefinite number of contexts in which such capacities play a role. Thousands of researchers have been studying cognition for centuries and they've made a lot of progress, but there is plenty more that remains mysterious.

Complex and mysterious phenomena must be approached piecemeal, from multiple directions. One strategy is to articulate the phenomenon—to identify different cognitive capacities and describe them under various conditions, and to establish regularities about them. A second strategy is to alter the variables that surround the phenomenon and see what (if any) difference they make. A third strategy is to observe how the phenomenon impacts various systems that contain it to identify what is most significant about the phenomenon itself. A fourth strategy is to form hypotheses about which internal variables might be involved in some aspect of the phenomenon and test these hypotheses by intervening on select variables to see if they make a difference to the phenomenon. A fifth strategy is to identify components of the system that exhibits the phenomenon and study how they contribute to the phenomenon. These same strategies can be iterated for the components of the system and what they do.

All these strategies constrain one another in ways that individual researchers may or may not be aware of, and they can be combined in any way we please. For example, suppose we are studying the cognitive capacity to solve problems. We may build a theory of the internal variables that affect problem solving. Our theory may posit internal variables representing the state of the problem and a mechanism that searches for a solution through such representations, without saying much about precisely which components of the system compose the posited mechanism or which of their states realize those variables (Newell and Simon 1972). Such a theory involves identifying problems of interest that the system can solve, articulating problem-solving behavior, establishing generalizations about it through experimentation and modeling, and proposing a hypothesis about the inner working of the system. It does not involve specifying which concrete components and which of their states, if any, correspond to the posited mechanism.

Just because we may choose to ignore some aspects of the system that constrain our explanation of a phenomenon, it doesn't follow that those constraints go away. In our example, just because we choose to ignore the concrete components of the system and the way they may or may not compose the mechanism we posit, it doesn't follow that our explanation is complete. In fact, until we identify the components of the mechanism and the component states that realize the variables we posit, it's difficult to be sure that we've found the *actual* explanation. We have a more or less plausible hypothesis; a *how-possibly* explanation. We may be able to confirm aspects of our hypothesis by manipulating internal variables, but we probably won't know whether the posited mechanism is the actual one until we identify its components and show that they work in the way we hypothesized.

By the same token, we may build a theory to the effect that neurocognitive activity is digital computation without worrying too much about the ways in which the activities of individual neurons combine to cause the various capacities that constitute cognition (McCulloch and Pitts 1943; Chapter 5, this volume). Such a theory involves identifying relevant aspects of experimentally established neuronal behavior, making assumptions, idealizations, and simplifications about it, and proposing a complex hypothesis about how the activities of individual neurons combine to constitute complex cognitive capacities, without working out the details of how that goes.

A sweeping theory like that is a great achievement that can push a field forward for decades. It can also engender complacency. It might give the impression that, whether we work out the details or not, we already know that the system must work in the way we hypothesized—i.e., as a digital computer. We might feel that as long as we posit some form of digital computation to explain some cognitive capacity, our explanation is in the right ballpark because the components of the system are digital computing devices that, if suitably organized, can perform any digital computation. This might even be seen as a justification for positing complex internal digital computations without worrying about how they are implemented (as in Newell and Simon 1972). Nevertheless, not only is our explanation incomplete—it is only as plausible as the assumptions and simplifications in the underlying theory.

Of course, we may also purposefully limit the scope of our inquiry. We may decide that we are only interested in the causal role of certain high-level holistic variables (as in low-dimensional dynamical systems approaches), or in a few select internal variables that affect a task (as in Bayesian psychology). What we obtain is *partial* constitutive explanations. If we engage in this enterprise, we may be satisfied with an explanation that invokes select variables and no explicit mechanism. We may even identify variables that actually play a crucial causal role without understanding the underlying mechanism. We may tell ourselves that that's all we wanted and conclude our inquiry there.

Nevertheless, we would gain further explanatory depth if we also identified the relevant components, their functions, and their organization, and showed that their functions, correctly organized, constitute the explanandum phenomenon. This is full-blown constitutive explanation. This is *mechanistic* explanation.

Mechanistic explanation does not limit itself to identifying hypothetical internal variables. It tells us which components play a role, which role they play, and how such roles, taken collectively, constitute the phenomenon under various background conditions. This gives us a deeper understanding of the system. It's the kind of understanding that allows us to take the system apart, put it back together, or build another one like it. It's the kind of understanding that allows us to break the system in selective ways and fix it when it's broken. Nothing less than a mechanistic explanation gives us this depth of understanding (cf. Dretske 1994).

To summarize where we've gotten, fully understanding what a system does requires fitting it in the causal structure of the world. It requires understanding its external causes (etiological explanation), the role it plays within larger systems (contextual explanation), and its mechanism (constitutive explanation). Understanding the mechanism requires identifying the components, their relevant functions, and their organization.

As I argued in Chapter 1, what the system as a whole does is an aspect of what its organized components do. Thus, mechanistic explanation involves identifying which aspects of the organized, working parts are the phenomenon. Needless to say, this requires studying and understanding both the whole system and its components. Iterating this process through the upward composition and downward decomposition of systems within systems and their mechanisms, and integrating the mechanistic explanations at multiple levels, yields *multilevel* mechanistic explanation.

Multilevel mechanistic explanation involves abstraction at all levels because, at each level, we need to identify the *relevant* causes—the ones that make the most difference to the behavior of the whole—and abstract away from the irrelevant ones. Multilevel mechanistic explanation also involves idealization because, most of the time, we cannot possibly capture all the relevant causal details in our models on pain of the models becoming mathematically and computationally intractable.

The rest of this chapter expands on this account of multilevel mechanistic explanation and argues that it does justice to the roles of abstraction in constitutive explanation.

7.2 Functional Analysis

When the cognitive revolution occurred in the 1940s and 1950s, single cell recording was in its infancy, and there was no way to image the brain during cognitive tasks. Therefore, there was hardly enough information to integrate models at the levels of neurons, circuits, networks, and systems with models of cognitive functions. To explain cognition, it was not unreasonable to either build relatively speculative theories about how neural computations might give rise to cognitive functions (e.g., McCulloch and Pitts 1943), or build computational explanations of cognitive functions divorced from information about the brain (e.g., Newell and Simon 1972).

In this context, it made sense to justify the rise of cognitive psychology and its modeling styles in such a way that cognitive psychology was not required to take information about brains into account. The priority was to legitimize the new science of cognition in the face of behaviorist resistance. Integrating psychology and neuroscience was secondary and could wait.

A helpful step towards justifying cognitive psychology was taken by John von Neumann in response to McCulloch and Pitts's work. In their 1943 paper, McCulloch and Pitts assert that neural nets—more precisely, neural nets as idealized and simplified by their theory—are computationally equivalent to Turing machines (without tape).

Presumably confident that McCulloch and Pitts's assertion was correct, von Neumann called it a "result" and boldly claimed that it guaranteed the success of (digital) computational psychology (von Neumann 1951, pp. 22–3). Von Neumann was one of the most distinguished and influential scientists of his day—often praised as a genius. He was also one of the few with the expertise to understand and evaluate his claim about McCulloch and Pitts's "result." Many influential scientists took him at his word. They went about developing cognitive models for solving cognitive tasks, confident that the neural details would work themselves out. As von Neumann assured them, McCulloch and Pitts had proven so.

Soon thereafter, philosophers followed up with rationalizations as to why psychologists could theorize about and explain cognition while ignoring neuroscience. Their main doctrine is that psychological explanation is functional analysis. Roughly speaking, functional analysis is the analysis of a function into sub-functions and the order in which the sub-functions must be performed. In other words, a functional analysis looks like a flow chart or computer program for performing the original function. The original function is the explanandum; the flow chart is the explanans.

Many argued that psychologists could by and large ignore neuroscience because neuroscience is about mechanisms, whereas functional analysis is distinct and autonomous from mechanistic explanation (e.g., Fodor 1965, 1974; Cummins 1975, 1983; for context and analysis, see Piccinini 2004b). Functional analysis is allegedly *distinct* because it abstracts away from the components that carry out the functions. While mechanistic explanation refers to components and their organization, functional analysis refers only to the functions, without any commitment to components. Or, if functional analysis posits components at all, those components are black boxes—individuated solely by the functions they perform.

Functional analysis is also allegedly *autonomous* from mechanistic explanation, in the sense that there are no direct constrains between higher-level explanations and lower-level ones (Fodor 1965; Cummins 1983).[1] The reasons for autonomy are a bit murky. One alleged reason is multiple realizability, which supposedly entails traditional anti-reductionist functionalism, which supposedly entails that higher-level properties are distinct from lower level properties and therefore can only be investigated by sciences at higher levels. As I've argued in previous chapters, none of these entailments hold. On the contrary, multiple realizability is consistent with mechanistic explanation (Chapter 2), and functionalism is best understood within a mechanistic framework (Chapter 4). Higher levels and lower levels directly constrain one another, because higher levels are an aspect of their realizers (Chapter 1). To understand higher levels, we ought to understand lower levels and what aspects thereof the higher levels are. In light of this, I'll set multiple realizability aside and consider other alleged reasons for autonomy.

Another alleged reason is that cognition—or at least an important aspect of cognition—is carried out by a general-purpose, stored-program digital computer (e.g., Fodor 1968b, 1975; Cummins 1983). At any rate, classical cognitive psychological theories often posited something along similar lines (e.g., Miller, Galanter, and Pribram 1960; Newell and Simon 1976). It turns out that all general-purpose computers are computationally equivalent to one another in the sense that they can compute the same functions, though not necessarily at the same speed. Therefore, according to this view, the job of the cognitive psychologist is to figure out which programs are running on the cognitive computer when a given task is executed. Since the machinery is the same for all cognitive functions anyway, looking at the machinery is not going to illuminate how specific functions are performed. This is somewhat reminiscent of von Neumann's claim that, given McCulloch and Pitts's result, we need not worry about how the digital computations performed by the brain are neurally implemented (von Neumann 1951, 23). They are implemented somehow.

I will argue in Chapter 10 that von Neumann's argument is unsound, so the autonomy of psychology cannot be derived from it. The argument for the autonomy of psychology from the view that cognition is carried out by a general-purpose, stored-program digital computer is equally unsound. For a general-purpose, stored-program digital computer is a type of mechanism (Chapter 6 and Piccinini 2015). So this is a mechanistic hypothesis, which ought to be tested. If cognition is carried out by such a mechanism, this mechanism ought to be found in the organ of cognition—presumably, the nervous system. Where is the general-purpose processor? (Or processors, if there are many?) Where are the memory registers? Even the specific programs are stable states of memory registers, which ought to be identified. This version of the Computational Theory of Cognition constrains what we ought to find in the nervous system, and

[1] I rejected other notions of autonomy in Chapter 2, Section 2.10.

what we actually find in the nervous system constrains the plausibility of this hypothesis. Needless to say, this hypothesis is far from confirmed. At any rate, the autonomy of psychology does not follow from it—not the strong autonomy that was derived from it.

A third reason for autonomy is that the vocabulary of cognition, computation, and representation is proprietary to psychology. Neurons can't be said to compute, represent, or cognize—only cognitive systems can. Cognition is within the domain of psychology, not neuroscience.[2] This is pure dogmatism. It ignores that the very Computational Theory of Cognition (CTC) was initially justified on the basis of the all-or-none law of nervous activity, among other neurophysiological results (Chapter 5). That such a justification was shaky is beside the point. The point is that CTC was initially grounded in neurophysiology. In contemporary terms, CTC is the view that nervous systems perform computations over representations and that such neural computations explain cognition. There is no good reason to deny that nervous systems may compute, represent, and cognize. There is no way to properly assess CTC without considering evidence about nervous systems.

7.3 Mechanistic Integration

Functional analysis is neither distinct nor autonomous from mechanistic explanation. Consider that, right under the autonomists' noses, circa 1980s, cognitive neuroscience arose. Needless to say, cognitive neuroscientists had no problem saying that neural systems represent, compute, and cognize. They had no problem constructing mechanistic explanations of how neural systems perform cognitive functions. What gives? To answer that question, let's take a step back.

When autonomists talk about functions, what do they mean? "Function" can mean many things. At the very least, "function" can refer to mathematical functions, causal roles, teleological functions (Chapter 3), and even teleological functions of computing systems (Chapter 6). Typically, when autonomists talk about functions, they mean causal roles within a complex system. As I've argued in Chapter 3, cognitive and computational systems are better understood using teleological functions rather than mere causal roles, but this subtlety makes little difference here. In any case, there remains an important distinction: functions performed by complex systems in general versus functions performed by computing systems.

[2] Here are a couple of examples:
[I]n the language of neurology..., presumably, notions like computational state and representation aren't accessible (Fodor 1998, 96).
[E]xplanation in terms of distinctively psychological representational notions is, as far as we now know, basic and ineliminable (Burge 2010, 27).

Let's begin with complex systems in general. Here, it should be particularly easy to see that functional analysis is neither distinct nor autonomous from mechanistic explanation. Instead, functional analysis boils down to mechanistic explanation.

Consider someone who claims that we can understand how internal combustion engines work by functional analysis alone, and that such functional analysis is distinct and autonomous from the mechanistic explanation of internal combustion engines. What would the functional analysis of an internal combustion engine even look like? Believe it or not, one of the earliest defenses of functional analysis actually appeals to internal combustion engines as an example (Fodor 1965; cf. Piccinini 2004b). Paraphrasing in modern terms, Fodor alleges that the same component can be described as either a camshaft or a valve lifter. The first is a structural kind term, the second a functional kind term. Functional kinds can be used to construct functional analyses, which are distinct and autonomous from the mechanistic explanations that employ structural kinds. Or so Fodor says. Except that nothing could be further from the truth.

A valve lifter is not a black box; it is a concrete component that lifts valves. Lifting is a physical operation that requires physical force. A valve is also not a black box but a concrete component that requires to be lifted to let fuel into the pistons. Thus, "valve lifter" is not a purely functional kind term that can be used to construct a purely functional analysis in abstraction from structural information. I'm not aware of any putative examples of a purely functional analysis of an internal combustion engine.

Of course, you can program a computer simulation of an internal combustion engine. Mechanical engineers do it all the time. These are representations of engines and their functioning. They represent concrete structures, their functions, and their behavior. Such representations do not purport to reproduce the processes they model. In a computational simulation of an internal combustion engine there is nothing that actually burns fuel or generates motive power.

If you attempt to construct a purely functional analysis of an internal combustion engine, you face two problems: first, you can't do it without considering the concrete components of the engine; second, you can't describe what you find within engines without describing concrete components and their structural properties. The result is nothing less than a mechanistic explanation, although possibly a slightly more general mechanistic explanation than one that employs the term "camshaft." The point generalizes to any ordinary (i.e., noncomputational) complex system, whose functions are medium *dependent*.

In conclusion, functional analyses of ordinary complex systems are neither distinct nor autonomous from mechanistic explanations. A "functional analysis" is just a mechanistic explanation that abstracts away from one particular aspect of the mechanism—e.g., that in a certain kind of engine the valves are lifted by camshafts.

Is autonomy any more plausible when it comes to the functional analysis of computing systems? Here is where the mechanistic account of physical computation comes in handy (Chapter 6). I argued that computational functions are medium independent, which makes them importantly different from medium-dependent functions. But computing systems are still mechanisms, to be understood using mechanistic explanation. Their components may be defined in a medium-independent way, but they are still concrete components defined at an appropriate level—like valve lifters, but medium independent. Accordingly, a functional analysis of a computing system is still neither distinct nor autonomous from its mechanistic explanation. Insofar as we can abstract away from information about the components of the system, we are providing either a partial explanation—a sketch of a mechanism—or a more general mechanistic explanation. This is not necessarily a defective explanation. But it's not a distinct *kind* of explanation. Since it's not distinct, there's nothing distinct for it to be autonomous *from*.

The bottom line is the same as with any complex system. Higher-level properties are aspects of their realizers. Explaining a phenomenon requires not only identifying which higher-level variables are causally relevant, but also which concrete components are involved in manipulating the variables and how the components are organized to produce the phenomenon. If you know the realizer, you have a direct constrain on any realized property—the realized property is an aspect of its realizer, so it must be contained within its realizer. And if you know the realized property, you have a direct constrain on its realizers—they must all contain the realized property as an aspect. This is multilevel explanatory integration understood within an egalitarian ontology (Chapter 1).

The only difference between explaining medium-*dependent* systems such as internal combustion engines and explaining medium-*in*dependent systems such as computers is that when we deal with medium-independent phenomena, the components themselves can be defined in a medium-independent way (although they don't have to; they can also be defined in a medium-dependent way if we so choose; cf. Chapter 6, Section 6.10). Sure, we can limit ourselves to identifying higher-level variables or black boxes and their functions. The result is not an explanation that is distinct and autonomous from a mechanistic explanation—it is a partial mechanistic explanation.

After various authors defended the integration of psychology and neuroscience along these lines (Piccinini and Craver 2011; cf. Bechtel 2008; Kaplan 2011; Kaplan and Hewitson unpublished), others objected. They argue that some explanations are noncausal (Chirimuuta 2018), some causal explanations are nonmechanistic and claiming otherwise trivializes the notion of mechanism (Chirimuuta 2014; Rescorla 2018; Weiskopf 2017; Woodward 2017; van Eck 2018), some nonmechanistic explanations are autonomous from mechanistic ones (Knoll 2018; Weiskopf 2017; Rescorla 2018; Rusanen and Lappi 2016;

Shapiro 2016; Polger and Shapiro 2016), mechanistic integration neglects the role of abstraction in explanation (Weiskopf 2011b; Barberis 2013; Barrett 2014; Levy 2014; Levy and Bechtel 2013; Chirimuuta 2014; Ross 2015; Shapiro 2019), mechanistic integration neglects the difficulty of integrating different levels (Stinson 2016; Sullivan 2016), or integrating psychology and neuroscience is impossible because they make different idealizations (Hochstein 2016).

While many good points have been made that can be incorporated in a nuanced understanding of mechanistic integration, I will now argue that the basic integrationist picture stands. *A full causal explanation of a higher-level phenomenon includes a constitutive explanation, and a full constitutive explanation includes a mechanism.* By the same token, integrating psychology and neuroscience requires piecing together their explanations to describe multilevel mechanisms.

Since most of the objections to mechanistic explanation appeal to the role of abstraction within explanation, I now turn to this crucial topic. I will provide an explicit taxonomy of legitimate kinds of abstraction within constitutive explanation. I will argue that abstraction is an inherent aspect of adequate mechanistic explanation. Mechanistic explanations—even ideally complete ones—typically involve many kinds of abstraction and therefore do not require maximal detail. Some kinds of abstraction play the ontic role of identifying the specific complex components, causal powers, and organizational relations that produce a suitably general phenomenon. Therefore, abstract constitutive explanations are both legitimate and mechanistic. With that conclusion in place, I will address other concerns about integration.

7.4 The Requirement of Maximal Detail

Some constitutive explanations are *mechanistic*: they attribute the explanatory sub-phenomena to *structural components* (as opposed to black boxes). Some constitutive explanations are more *abstract* than others. That is, some constitutive explanations include *fewer (causally relevant) details* than others about their explanandum, their explanantia, or both. Question: is abstraction within constitutive explanation compatible with mechanistic explanation, or does it lead to nonmechanistic forms of explanation?[3]

To answer this question, let's distinguish between two kinds of role that abstraction could play within explanation: epistemic and ontic. This distinction maps onto epistemic and ontic conceptions of explanation.

According to the epistemic conception, explanations are *representations* that meet certain epistemic requirements. Events are explained by subsuming them

[3] This section and the following two are a revised descendant of Boone and Piccinini 2016a, so Worth Boone deserves partial credit for what is correct here.

under appropriate representations. For example, we explain why someone developed lung cancer by pointing out that they smoked and that people who smoke have a certain probability of developing lung cancer. Thus we explain the lung cancer by subsuming this particular instance under a statistical generalization. The epistemic conception easily accommodates abstract explanation. We can represent the world more or less abstractly, so we can give more or less abstract explanations.

According to the ontic conception, by contrast, the adequacy conditions for explanations are facts in the world. To explain a phenomenon is to identify the explanatory facts themselves. In our example, there are biological facts that explain why someone develops lung cancer from smoking (e.g., carcinogens accumulate and are metabolized into forms that covalently bond with DNA to form DNA adducts). These facts provide the adequacy conditions of our explanatory descriptions: there is still a sense in which representations explain, but only insofar as they denote the explanatory facts.

Although the ontic and epistemic conceptions are sometimes pitched against each another, they need not be. Explanatory facts can be described; explanatory representations explain to the extent that they describe explanatory facts. When it comes to causal explanations, causal representations explain to the extent that they describe relevant causes. The causes, which are in the world, explain in the ontic sense; the descriptions, which describe the causes, explain in the epistemic sense. That being said, the question remains of whether explanation in the ontic sense can accommodate explanatory abstraction.

The ontic conception of mechanistic explanation has been interpreted as implying that the more relevant details an explanatory description includes, and hence the less abstract it is, the better it explains. After all, the world is maximally concrete. The more relevant details we include, the more completely we denote the relevant facts or events. Therefore, the less abstract a model is, the better it approximates an ideally complete mechanistic explanation. Call this the *requirement of maximal detail*.[4]

The requirement of maximal detail implies that abstract constitutive explanations—since they omit relevant details—are partial, inferior, or inadequate. Critics have balked at these implications and defended abstract explanations (Barberis 2013; Barrett 2014; Levy 2014; Levy and Bechtel 2013; Chirimuuta 2014; Weiskopf 2011b, 2017; Rescorla 2018). As these critics see things, proponents of the ontic conception are committed to the requirement of maximal detail and must therefore reject abstract constitutive explanations. These critics conclude

[4] Some statements by some mechanist philosophers (e.g., Kaplan 2011, 347; Craver 2007, 114) are sometimes interpreted as implying the requirement of maximal detail, although their authors reject the charge (Craver and Kaplan 2018). For a parallel debate in neuroscience, see Markram 2006; Markram et al. 2015; and Eliasmith and Trujillo 2014.

that either explanation is solely epistemic (not ontic) or abstract explanations are not mechanistic precisely because they fail the requirement of maximal detail (or both).

This debate bears on at least five important questions. First, there is the proper characterization of constitutive explanation itself—whether mechanistic explanation requires maximal detail, whether abstract constitutive explanations are legitimate, and whether abstract constitutive explanations are mechanistic or are, instead, distinct and autonomous from mechanistic explanations.

A second question is whether explanation is epistemic or ontic. If the requirement of maximal detail holds and abstract explanations are legitimate, this appears to challenge the ontic conception. Some critics respond by rejecting the ontic conception. Another response is to assign abstraction merely *epistemic* roles: abstraction is a feature of our limitations in *representing* the world (cf. Craver 2014). I will defend a third option: abstraction plays ontic as well as epistemic roles.

A third related question is whether and when multiple realizability occurs (cf. Chapters 2 and 4). If abstract constitutive explanations are legitimate, abstract explanations may denote higher-level properties that are multiply realizable—or at least variably realizable—because they are invariant under transformations in the details they abstract from. If such abstract explanations are illegitimate or incomplete at best, then multiple realizability may be ruled out. The status of multiple realizability has important consequences for the metaphysics of mind: if mental states are multiply realizable, mental states are realized by, but not identical to, brain states; if mental states are not multiply realizable, they might be identical to brain states (Polger and Shapiro 2016).

A fourth related question is the proper characterization of computational explanation. Computational explanation is paradigmatically abstract.[5] If abstraction leads to inadequate explanations, then computational explanations are inadequate. If abstraction does not lead to inadequate explanations but does lead outside mechanistic explanation, then computational explanation may be adequate but not mechanistic (cf. Haimovici 2013; Shagrir 2010a). If abstraction in explanation is both legitimate and compatible with mechanistic explanation, however, computational explanation may well be both legitimate and mechanistic (cf. Chapter 6).

A fifth related question is whether higher-level explanations, which abstract away from relevant lower-level details, are eliminable, reducible to, irreducible to, or can be integrated with lower-level explanations. If higher-level explanations are

[5] Once again, by "abstraction" I mean omission of details. Some authors claim that computational explanation is abstract in that it involves abstract objects. I don't mean that at all—I don't even believe in abstract objects. In any case, abstract objects are beside the point. Our topic is constitutive explanations of concrete physical systems, including physical computational systems, which exist in spacetime and have causal powers.

distinct and autonomous from lower-level explanations, they are irreducible to lower-level explanations. If abstract explanations are illegitimate, they should be eliminated. If they are legitimate and mechanistic and there is no multiple realizability, they may be reducible to lower-level explanations. If they are legitimate and there is multiple realizability, they may be integrated with lower-level explanations by forming multi-level mechanistic explanations.

I will answer these questions by providing an explicit taxonomy of legitimate kinds of abstraction within constitutive explanation. I will reject the requirement of maximal detail and argue that abstraction is an inherent aspect of adequate mechanistic explanation. Mechanistic explanations—*even ideally complete ones*—typically involve many kinds of abstraction and do not require maximal detail. Some kinds of abstraction play the *ontic* role of identifying the specific complex components and properties that produce a suitably general phenomenon. Therefore, abstract constitutive explanations are both legitimate and mechanistic.

Before proceeding, a brief reminder of the framework I'm working with. Complex systems contain mechanisms that come in levels of functional organization. Mechanisms (level 0) *compose* larger mechanisms (level +1) and *are composed by* components, which are in turn (smaller) mechanisms (level -1). Mechanisms have *properties*. The properties of mechanisms (level 0), when such mechanisms and their properties are suitably organized, *realize* the properties of the larger mechanism that contains them (level +1). By the same token, the properties of a mechanism (level 0) are *realized by* the properties of its suitably organized components (level -1). The properties of a mechanism considered as a whole (level 0) are *aspects* of the properties of its organized components (level -1) (Chapters 1 and 2).

At each level of organization, mechanistic models articulate how the relevant properties of the relevant components, suitably organized, produce the phenomenon (cf. Bechtel and Richardson 2010; Craver 2007; Glennan 2017; Zednik 2019). Ontically, a phenomenon is constitutively explained by the mechanism that produces it and its relevant properties. An *ideally complete mechanistic model* is an articulation of the mechanism that produces a phenomenon and its relevant properties at all levels of organization, including an articulation of the compositional relations between the whole mechanism, its components and their properties and organization, their sub-components and properties and organization, etc. (cf. Craver 2006, 360).

In the next section, I list several distinct epistemic roles that abstraction plays within this mechanistic framework. These epistemic roles are relatively uncontroversial and consistent with a range of views of the nature of mechanistic explanation. In the subsequent section I outline several distinct *ontic* roles that abstraction plays within the mechanistic framework; these ontic roles had not been previously articulated explicitly. Since abstraction plays these ontic roles, even ideally complete mechanistic models involve abstraction.

7.5 Epistemic Roles of Mechanistic Abstraction

Mechanistic abstraction plays a number of epistemic roles. These roles are dictated by our epistemic interests and limitations rather than any ontic norms of mechanistic explanation. Such abstract explanations are thus incomplete in the relevant sense but may nonetheless be indispensable to science. Thus, incomplete abstract explanations may be explanatorily adequate in many contexts. The adequacy of such explanations may shift, however, as scientific knowledge and methods evolve.

Two types of abstraction play epistemic roles: abstraction of one or more levels of organization among others, abstraction of one or more aspects of one level among others. These types of abstraction are needed if we are interested in some specific aspect of the phenomenon at the expense of others, or because including more details makes the explanation intractable, or simply because we are ignorant of the details we omit.

Epistemically motivated abstractions are often mixed with *simplifications* and *idealizations*. Simplifications and idealizations introduce features into a description or model that distort or misrepresent the target system. Although simplification and idealization are not my main topic, it is important to realize that, along with abstractions, simplifications and idealizations are part and parcel of mechanistic models as well as scientific descriptions in general (cf. Wimsatt 2007, chap. 6; Weisberg 2013, chap. 6; Rohwer and Rice 2013; Milkowski 2015; Potochnik 2017, chap. 2).

One obvious role for abstraction in mechanistic explanation is that many details of a system may as yet be unknown. In such cases, mechanistic explanations take the form of mechanism *sketches*. In mechanism sketches, black boxes or filler terms may be included to highlight the incompleteness of a mechanistic model and motivate future research.

Even when certain details are known, there are other nontrivial epistemic reasons to omit them from a mechanistic model. In the first place, mechanistic explanations are typically offered at one or a few levels of organization among others. Including all levels of a mechanism in an explanation may be unfeasible for several reasons: not enough is known about some levels (as I just said), too much information would be needed to describe all the levels and it would be unfeasible to include it all, or much of the additional information is irrelevant to the levels of interest. Even when multiple levels of mechanistic organization are included in a model, typically the included levels are only some of the levels of organization of the entire multilevel mechanism. Other levels are omitted. Thus multilevel mechanistic models—the ones actually developed in practice, as opposed to ideally complete ones—typically employ this form of abstraction.

A second epistemic role for abstraction comes from considerations of solvability and tractability in mathematical and computational models. Mechanistic explanations are often presented in the form of (interpreted) mathematical or computational models—e.g., the Hodgkin–Huxley model of the action potential. Typically, such models become intractable or difficult to manipulate if they include too many details about their target systems. Instead, mathematical and computational models are typically constructed not only by abstracting away from many details of the target system, but also by replacing those details with simplifications and idealizations that distort or misrepresent the target system. Nevertheless, these models retain crucial epistemic roles (e.g., for prediction and discovery) and often retain explanatory value to the extent that they denote components, causal powers, or organizational relations of target systems that are relevant to the explanandum phenomenon.

A third epistemic role for abstraction is due to our interest in one component or property of a mechanism at the expense of other components or properties. The result is another type of mechanism sketch, or partial (i.e., elliptical) mechanistic explanation. Consider what it takes to explain why, under special conditions, a mechanism functions differently than it normally does. Explaining a deviation from normal functioning typically requires simply demonstrating some aberrant feature of the mechanism, while omitting the details relevant to normal function (van Eck and Weber 2014). For instance, to explain why certain patients have left-side hemineglect (roughly: inattention to and unawareness of the left side of visual space) it may be enough to point out that such patients suffered damage to the contralateral (right-side) cortical areas responsible for spatial attention, without describing the whole mechanisms involved in normal spatial attention and consciousness.

Interim conclusion: our epistemic interests and limitations often demand abstractions, namely, abstraction from unknown details, abstraction from some levels of a mechanism in favor of one or more levels, abstractions in the service of mathematical and computational tractability, and abstraction from some aspects of one level in favor of other aspects of that level.

7.6 Ontic Roles of Mechanistic Abstraction

The epistemic roles of abstraction are consistent with a range of views on the nature of mechanistic explanation, including the requirement of maximal detail. Contrary to the requirement of maximal detail, however, even ideally complete mechanistic explanations require abstraction.

Within an ideally complete mechanistic explanation, abstraction serves two interrelated roles: (1) to explain a phenomenon at a particular degree of generality, and (2) to identify all and only the components, properties, and organizational relations that constitute the mechanism at different levels. To fulfill these roles, two types of abstraction must be performed: (i) abstraction to sufficiently general types of components, properties, and organizational relations; and (ii) abstraction from lower levels of organization to higher levels of organization.

To a first approximation, I construe types as classes of components, property instances, or relation instances whose members resemble one another. I leave open whether property types are tropes that resemble one another, universals, or classes of particulars. Since I'm discussing causal explanation, as shorthand, I will say that a type of property is a class of property instances that share a plurality of causal powers, a type of component is a class of components that share relevant properties, a type of organizational relation is a class of relation instances that are similar to one another, and a regularity is a manifestation of a type of mechanism (component types, property types, and organizational relation types).

At face value, the requirement of maximal detail entails that all explanations are explanations of token events; obtaining any level of generality requires at least some omission of the particular details of token events (cf. Ross 2015). But explananda can themselves be regularities. For instance, the link between smoking and lung cancer (explanandum) is itself a regularity in the world, which we explain by situating it within other causal regularities—namely, the processes by which carcinogens are metabolized into forms that covalently bond with DNA to form DNA adducts. On the ontic conception, we thus explain such regularities in the same way we explain token events—both are perfectly legitimate explananda; they simply differ in their scope or generality.[6]

Explanandum phenomena can be more specific or more general in many ways. We may wish to explain how a particular rat navigates a particular maze at a particular time (a particular event), how a particular rat navigates in general, rat navigation more generally, rodent navigation, mammalian navigation, and so forth. Neural recordings from a particular rat engaged in a particular behavior at a particular time may be part of the evidence for discovering explanations of all of the above explananda, typically in combination with other sources of evidence. Depending on which phenomenon we aim to explain, and how general it is, we need to invoke more specific or more general types of components, properties, and organizational relations.

Generalizing beyond token events requires isolating target phenomena that are stable across variations in those events. For example, memory researchers may implant electrodes in the hippocampi of specific rats and gather recordings while those rats navigate a specific maze at a specific time. But those researchers are

[6] Abstraction and generality are distinct notions. Abstraction does not always increase generality, as can be seen in the examples discussed in the previous section. But generality requires abstraction: for any explanation to have scope beyond token events, some idiosyncratic details of those token events must be omitted.

typically not merely interested in explaining how the particular rats in the test population navigated those particular mazes at those particular times; rather, they more often seek to generalize their results to conclusions about rat hippocampi in general, or rodent hippocampi more generally, or even mammalian hippocampi.

If we are explaining a particular navigation event, we are free to appeal to idiosyncratic features of that event. If, by contrast, we are explaining rat navigation in general, we have to appeal to features shared by the relevant mechanisms present in normal rats. In such cases, we target a mechanism type, as opposed to a mechanism token (cf. Machamer, Darden, and Craver 2000, 15). This requires abstracting away from the idiosyncratic details of particular rats and particular navigation events and specifying the relevant property types, common to all tokens of the relevant component types, which, when appropriately organized, produce the phenomenon at the specified degree of generality—here, rat navigation.

Explaining still more general phenomena like mammalian navigation requires abstracting away from the specifics of rat physiology to find relevant similarities across the class of organisms that feature in the explanandum. This may require identifying an even more restricted plurality of properties common to the more general type of mechanism that is shared by different sorts of mammals. These abstractions succeed to the extent that they track stable features of those progressively more general types of system. Those stable features delimit different, and progressively more general, *types* of mechanism. Identifying and explaining different mechanism types requires omitting the idiosyncratic details of more specific (less abstract) types of mechanism in order to reach a description that is general enough to denote the relevant features that the more specific types have in common.

Beyond a certain degree of generality, explananda may no longer share enough relevant features to constitute a unified phenomenon that can properly be a target of explanation. Alternatively, the systems that are responsible for a phenomenon may no longer share relevant features, or the relevant features that are shared may no longer productively explain the supposed target phenomenon. In other words, there may be cases in which no relevant shared causal powers can be identified; in such cases there is no relevant mechanism type to be identified and no unified phenomenon to be explained.

As stated above, an ideally complete mechanistic explanation of a phenomenon includes all and only the details that produce the explanandum at *all* levels of organization. This requires both omitting details that are irrelevant to producing the phenomenon and including all the relevant features at all relevant levels of organization. Moving from one level to a higher level requires another kind of abstraction. The details of lower-level mechanism *types*, which implement the level under consideration, are omitted. As a result, the properties being described at the higher level are only *some aspects* of the properties of the lower levels. They are precisely the higher-level properties I've been harping on (Chapters 1 and 2).

Consider the rat hippocampus and its components. The properties of the rat hippocampus taken as a whole are only some of the causal powers of the neurons and other components that make up the rat hippocampus, even when the rat hippocampus' neurons and other components are taken to be organized into the rat hippocampus itself. For instance, the rat hippocampus' neurons send neurotransmitters to one another under various conditions, whereas the rat hippocampus taken as a whole lacks that property. Instead, the hippocampus as a whole tracks the location of a rat in its environment. This function of the hippocampus may well be stable over massive amounts of variation in the properties of the individual neurons that compose the system.

When we go up a level in our mechanistic explanations (e.g., from neurons to hippocampi), we abstract away from some of the properties of the components in order to single out the specific properties that produce the phenomenon at the higher level. For instance, we single out the properties that are specific to hippocampi as wholes, as opposed to their organized components. In other words, we identify the most specific difference makers—the aspects of the world that explain the phenomenon at the higher level.

This kind of ubiquitous abstraction tracks one form of multiple realizability between levels of mechanistic organization, because the same higher-level property, which operates at one level, may be embedded in different realizers (Chapter 2). For example, the property that all functioning corkscrews share (lifting corks out of bottlenecks) may be an aspect of different realizers (lifting corks out of bottlenecks by screwing into the cork and directly pulling it out, by leveraging it out using an arm placed on one side of the bottle, by indirectly pulling it out with two levers attached to a rack and pinion, and so on) depending on which type of lower-level mechanism is being used to generate the original power (cf. Shapiro 2004).

A related but distinct ontic role for abstraction in mechanistic explanation consists in isolating features that are shared by mechanisms that occur within radically different systems and may even occur at different levels of organization. This form of analysis yields a kind of mechanism *schema* (Machamer, Darden, and Craver 2000). Mechanism schemata involve deliberate omissions of details, capturing a mechanistic structure that many different systems have in common. As above, a schema denotes specific properties that are found in many different systems, where they may be embedded in different realizers—thus providing a distinct form of multiple realizability.

Graph-theoretic models provide a common way of representing mechanism schemata in neuroscience (cf. Levy and Bechtel 2013; Zednik 2019). Such models are often used to represent *types* of neural circuits that crop up in different parts of the nervous system. The particular microstructural details—the types of neurons, neurotransmitters, receptors, etc.—are omitted in order to isolate organizational similarities that are invariant over these details.

For instance, lateral inhibition can be represented by a simple set of input nodes with inhibitory projections towards the neighbors of their corresponding output nodes (Figure 7.1). When many input nodes have overlapping receptive fields, those that receive the greatest stimulation inhibit the neighbors of their corresponding output nodes, creating a "winner-take-all" processing stream. This form of circuitry is found in many different neural regions and is particularly ubiquitous in peripheral sensory processing—e.g., in the retina and in the somatosensory periphery—because it allows greater precision in transduction of sensory signals.

Such patterns in neural circuitry offer a distinct but similar form of generality to that found in moving from features of particular rat hippocampi to rat hippocampi in general, or rodent or mammalian hippocampi. In the case of hippocampi, identifying the relevant mechanism type that explains the phenomenon of interest (rat navigation, rodent navigation, or mammalian navigation) relies heavily on the structural similarity (i.e., the shared components) of those different systems; whereas in the case of lateral inhibition, identifying the relevant mechanism type relies more on organizational similarity of the different systems.

Interim conclusion: abstraction plays two ontic roles within mechanistic explanation, namely, to explain a phenomenon at a suitable degree of generality and to identify all and only the features of the mechanism at all levels of organization. Two kinds of abstraction are needed to fulfill these roles: abstraction to general *types* of component, property, and organizational relation, and abstraction from lower to higher levels of organization.

7.7 From Autonomy to Inter-level Integration

Contrary to what is often asserted or implied, mechanistic explanation does not require maximal detail. Multi-level mechanistic explanation—even *ideally*

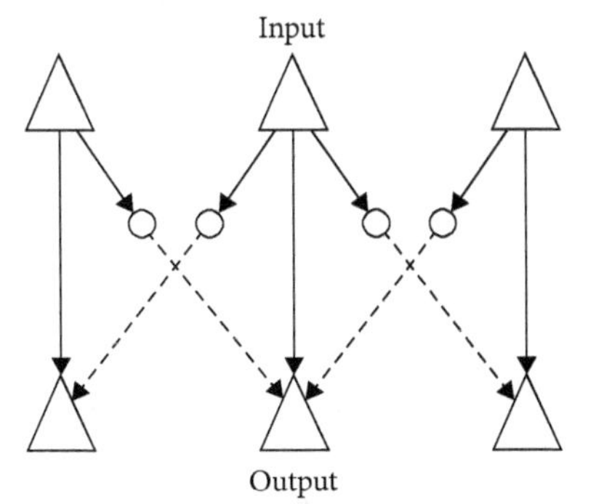

Figure 7.1 Graph-theoretic representation of a lateral inhibition circuit. Solid lines represent excitatory connections, dotted lines represent inhibitory connections.

complete mechanistic explanation—mandates several legitimate kinds of abstraction. Of course, actual models are not ideally complete explanations: they contain abstractions as well as simplifications and idealizations that play epistemic roles in addition to ontic roles. Such models still provide (possibly partial) mechanistic explanations.

Sometimes mechanistic explanations are schemata that cover many specific instantiations, sometimes they are schemata that cover many kinds of realizing mechanisms, sometimes they are schemata that cover similar mechanistic structures in different kinds of systems and possibly at different levels of organization, sometimes they are either sketches or schemata that involve deliberate omissions of known details, and sometimes they are sketches that omit unknown details. These types of abstraction are essential to mechanistic explanation both in practice and in principle.

With these conclusions as background, I will now address some other recent objections to the integrationist approach.

One objection is that some explanations are noncausal but "distinctively mathematical" (Lange 2017, 2018; Chirimuuta 2018). For instance, I have exactly two biological children because I had one, then I had another, and 1 child + 1 child = 2 children. That 1 + 1 = 2 is a mathematical fact that, given the premises, necessitates the explanandum. This does not affect my argument for two reasons. First, the integrationist framework I am defending is explicitly restricted to *causal* explanation. Therefore, strictly speaking, any noncausal explanation lies outside its scope. Second, qualities are part and parcel of causal explanation. For example, solids leave marks on soft surfaces that depend on their shape. A square solid leaves a square mark, while a round solid leaves a round mark. The shape of the solid is a qualitative, mathematically described property, which necessitates certain causal consequences. Deriving necessary causal consequences from qualities is integral to causal explanation. Therefore, insofar as distinctively mathematical explanations amount to deriving causal consequences that are necessitated by mathematically described qualities, they are not an alternative to causal explanation but an integral aspect of it (cf. Andersen 2016; Craver and Povich 2017). From now on, I will set distinctively mathematical explanation aside to focus exclusively on causal explanation.

A cluster of objections revolves around the claim that some explanatory models do not describe mechanisms (Chirimuuta 2014; Knoll 2018; Rescorla 2018; Rusanen and Lappi 2016; Shapiro 2016, 2019; van Eck 2018; Weiskopf 2017; Woodward 2017). Often, these explanations are said to be nonmechanistic because they abstract away from some information about components. In some cases, these explanations avoid the internal structure of a mechanism altogether, focusing on features of the phenomenon itself, its extrinsic causes, or its role within a broader context. If they focus on describing the phenomenon itself and what it correlates with, they are descriptive or phenomenological models, not explanations. Aside from that, addressing this cluster of objections requires the distinction between kinds of causal explanation that I introduced in Section 7.1.

As we saw then, some causal explanations are etiological: they provide the extrinsic causes of a phenomenon. Therefore, by definition they are not constitutive explanations and they are not an alternative to mechanistic constitutive explanation for the simple reason that they answer a different explanatory question.

Other explanations are contextual: they identify the role a phenomenon plays within a larger containing system. An especially salient example of contextual explanation is adaptationist or optimality considerations, to the effect that a system's capacity satisfies certain external constraints more or less well and that's why it's the way it is. Contextual explanations can contribute to the mechanistic explanation of higher-level systems, but they don't have to. If they don't, then they are not constitutive explanations, but they are also not an alternative to mechanistic constitutive explanation, because they are not constitutive explanation at all.

Finally, there are constitutive explanations, which identify the causal structure—intrinsic to a system—that produces a phenomenon given appropriate conditions. Here, nothing forces us to always look for the full mechanism. We can restrict our attention to some causally relevant variables, sub-functions, or structures. Nevertheless, the full mechanism at the relevant level of organization is what gives us the deepest understanding—a full explanation. It answers the largest number of what-if-things-had-been-different questions. It identifies all the difference makers at that level. It gives us the most control over the phenomenon, including how to take the system apart, how to build another like it, and how to fix it if it's broken. It's what the search for constitutive explanation ideally strives for (see also Povich 2015, 2018, 2019; Wajnerman Paz 2017).

Some might object that some constitutive explanations abstract away from the "physical" or "implementational" details of the mechanism. Therefore, they are not mechanistic in the interesting sense (cf. Weiskopf 2017). This objection equivocates. In one sense, every concrete mechanism is physical, and every mechanism implements a phenomenon. But in another sense, "physical" and "implementational" are used to refer to lower-level details that may well be irrelevant to explaining a higher-level phenomenon. Furthermore, the phenomenon itself may be defined more specifically or more generally. The more generally the phenomenon is defined, the broader the class of mechanisms that are covered by its explanation, the more lower-level details must be abstracted away. This is especially obvious for computational explanation, where a mechanistic explanation may be general enough to apply to both biological and artificial systems.

Mechanistic explanations explain phenomena. To explain a phenomenon, we need to identify level(s) of mechanistic organization that are relevant to producing the phenomenon. This includes all and only the causal factors—intrinsic to the phenomenon and the system that exhibits it—that produce the phenomenon. Finding such a mechanism will always abstract away from myriad lower-level details—i.e., "physical," "implementational" in one sense. Far from leading outside the mechanistic framework, this leads to the mechanistic level that is relevant to explaining the phenomenon.

A critic might reply as follows: what if it's difficult to identify specific roles for specific components? Perhaps the system that produces the phenomenon is merely aggregative (like a pile of sand), or the components are so tightly coupled with one another that it's difficult to decompose the phenomenon into subfunctions (cf. Weiskopf 2017; Woodward 2017). No matter. These are specific mechanistic considerations pertaining to specific classes of systems. They point at mechanistically trivial phenomena (aggregative phenomena), relatively complex ones, or—if worse comes to worse—lack of mechanistic explanation for lack of adequate knowledge. It doesn't follow that we should stop searching for mechanisms. At any rate, my main concern is the science of cognition. The evidence we have and the experience of cognitive neuroscience suggests that explaining cognition mechanistically is difficult but possible and much progress has been made in the relatively short time that cognitive neuroscience has been around.

A separate concern pertains to factors that may lie outside the spatial boundaries of the system and yet be relevant to individuating, constraining, or even causally explaining the phenomenon (cf. Hochstein 2016; Knoll 2018). For example, cognitive systems respond to distal objects, which may be represented within the system. The representations may be individuated by what they represent outside the system, and yet they may be causally relevant to explaining the system's behavior. Or cognition may involve the use of smartphones, which lie outside the boundaries of the organism. Does this lead outside mechanistic explanation? No, because nothing forbids mechanisms from being individuated broadly, by reference to entities outside the system, or crossing the boundaries between organisms and their environment. If the system that exhibits a phenomenon includes organisms that interact with smartphones, then the mechanism for that phenomenon may include components of both organisms and smartphones.

But what if explanations at different levels make different idealizations? Won't that prevent them from being integrated into multilevel mechanistic explanations? Hochstein (2016) argues that some psychological theories explain behavior by assuming that human beings make *rational* choices based on their propositional attitudes, whereas "neuroscientists instead treat neural mechanisms as neither rational nor irrational. The idealization of rationality is one that has no role to play in such neuroscientific theories" (Hochstein 2016: 5). Given that they make incompatible idealizations, Hochstein concludes, psychological and neuroscientific explanations cannot be integrated.

On the contrary, idealizations at different levels are no obstacle to integration. As I pointed out, idealizations along with abstractions are a necessary aspect of scientific modeling. They allow scientists to identify difference makers that contribute to explaining phenomena at different levels; such difference makers can then be linked together by causal and constitutive relations to form multilevel mechanistic explanations. As to rationality, it is not just an idealization. People as well as nonhuman animals do behave rationally to a degree, albeit within limits that have been empirically investigated by psychologists for decades. That's why rationality explains some aspects of behavior while irrationality explains others. Whatever degree of rationality biological agents exhibit is the product of, and can be constitutively explained by, neurocognitive mechanisms. Such mechanisms can be discovered and investigated. When enough is known about them, an integrated explanation of rationality will be available. Contrary to Hochstein's assertion, nothing prevents neuroscientists from investigating rationality. Cognitive neuroscience includes a vast literature on the mechanisms of decision-making and the degree to which they produce rational behavior not only in human beings but also in nonhuman animals (Glimcher and Fehr 2014).

Another concern is that integrating explanations at different levels—such as psychological and neuroscientific explanations—faces methodological challenges. It requires stable constructs across laboratories and areas of science, experimental protocols that measure real capacities rather than experimental artifacts, and taking into account the limits of reverse inference (Sullivan 2016). In addition, the integration of explanations at different levels is often only partial and may require future revisions (Stinson 2016). These challenges are real and worthy of consideration. Science is hard, and multilevel mechanistic integration is especially hard. Nevertheless, it's been happening for a while and for good reasons. Sullivan (2016) herself mentions the classic example of *cognitive maps*, which originated as a purely psychological posit but were eventually located in the *hippocampus*. This discovery gave rise to the project of investigating the neurophysiological mechanisms of spatial navigation (Bechtel 2016). Stinson (2016) discusses another beautiful example: the integration of classical cognitivist theories of attention, which were initially developed without reference to neural components, with the neuroscience of attention.

7.8 Explanatory Integration: Object and Properties, Parts and Wholes

I have defended the following picture of multilevel mechanistic explanatory integration.

First, abstract constitutive explanations are both legitimate and mechanistic; therefore, contrary to common assertions, abstract constitutive explanations are neither distinct nor autonomous from mechanistic explanations. All instances of mechanistic explanation involve mutual constraints between higher and lower levels. Omission of details presupposes some sufficient understanding of those details to know what is and is not relevant. In other words, explanations that deliberately omit known details are obviously not autonomous from those details. When details are unknown or it is unknown whether they are relevant, those details must be investigated and then either added or deliberately omitted in order to explain how the system actually performs a particular operation. Putatively nonmechanistic constitutive explanations are either *sketches* of mechanisms or mechanism *schemata* that describe aspects of one or more mechanistic levels. This does not make them inferior. It just makes them an indispensable step towards multilevel mechanistic explanation.

Second, some abstract explanations denote an important aspect of the objective causal structure of the world. Specifically, they denote a system's complex components, higher-level properties, and higher-level organizational relations that explain a phenomenon with a suitable degree of generality. This aspect of mechanistic abstraction is indispensable to explanation in the ontic sense—it should be included in any adequate ontic conception of explanation.

Third, multiple realizability is perfectly compatible with multi-level mechanistic explanation precisely because mechanistic explanation involves abstraction to specific higher-level properties that are shared by many different systems, which properties can be embedded in different realizers in different systems (Chapter 2).

Fourth, computational explanation involves abstraction while nonetheless remaining mechanistic (Chapter 6).

Fifth, abstract constitutive explanations at higher levels can be integrated with less abstract explanations at lower levels to form multilevel mechanistic explanations.

Sixth, multilevel mechanistic explanations can be partial in various ways. Various details can be omitted for different purposes.

Seventh, multilevel mechanistic explanations can include fewer or more levels (e.g., from neurons down to neurotransmitters, or also circuits, or also neural systems, or also the behaving animal, or also groups of animals, or also ecosystems, etc.).

Eighth, multilevel mechanistic explanations can be more general or more specific (that is, include broader or narrower classes of mechanisms), depending on how broad a phenomenon we are interested in.

Ninth, multilevel mechanistic explanation can be more focused on structure, on function, or on dynamics.

Nevertheless, an ideally complete multilevel mechanistic explanation must show which components of a system constitute a mechanism for the explanandum (at multiple levels), and which higher-level properties—that is, which aspects of the properties of the parts of a mechanism—are involved in a phenomenon. This requires studying both higher and lower levels and discovering how they fit together. It also requires, from the bottom up, abstracting away from the irrelevant lower-level details to identify the relevant higher-level details, and, from the top down, to explain what lower-level mechanism brings about the higher-level properties. This process must be repeated at all (relevant) levels. The result is a multilevel mechanistic explanation.

In the next chapter I will illustrate how this integrationist explanatory strategy is an important aspect of the cognitive neuroscience revolution.

8

The Cognitive Neuroscience Revolution

8.1 Is 'Cognitive Neuroscience' an Oxymoron?

The traditional framework of cognitive science included (aspects of) six disciplines: psychology, computer science, linguistics, anthropology, neuroscience, and philosophy. (There is also behavioral biology, which is not included in the traditional list but belongs there as well.) These disciplines were supposed to work together towards understanding cognition in accordance with a neat division of labor, to which many practitioners conformed. On one side stood psychology, with the help of computer science, linguistics, anthropology, and philosophy; on the other side stood neuroscience. Psychology etc. studied the functional or cognitive level, or—in Marr's terminology—the computational and algorithmic levels; neuroscience investigated the neural, mechanistic, or implementation level. Explanations at these two levels were considered distinct and autonomous from one another.[1]

This division of labor leaves no room for cognitive *neuro*science. Indeed, from this perspective, the very term "cognitive neuroscience" is almost an oxymoron, because neuroscience is supposed to study the mechanisms that implement cognition, not cognition proper. Yet cognitive neuroscience has emerged as the mainstream approach to cognition. What gives?

In this chapter, I argue that cognitive science as traditionally conceived is on its way out and is being replaced by cognitive neuroscience, broadly construed. Cognitive neuroscience is still an interdisciplinary investigation of cognition. It still includes (aspects of) the same disciplines—psychology, computer science, linguistics, anthropology, neuroscience, behavioral biology, and philosophy. But the old division of labor is gone, because the strong autonomy assumption that supported it has proven wrong.

The scientific practices based on the old two-level view (functional/cognitive/computational vs. neural/mechanistic/implementation) are being replaced by scientific practices based on the view that there are *many* levels of neurocognitive organization. No one level has a monopoly on cognition proper. Instead, different levels are more or less cognitive depending on their specific properties.

Explanations at different levels and the disciplines that study them are not autonomous from one another. Instead, the different disciplines contribute to the

[1] This chapter is a revised descendant of Boone and Piccinini 2016b, so Worth Boone deserves partial credit for most of what is correct here.

Neurocognitive Mechanisms: Explaining Biological Cognition. Gualtiero Piccinini, Oxford University Press (2020).

DOI: 10.1093/oso/9780198866282.001.0001

common enterprise of constructing multilevel mechanistic explanations of cognitive phenomena. In other words, there is no longer any meaningful distinction between cognitive psychology and the relevant portions of neuroscience—they are merging to form cognitive neuroscience. Or so I will argue.

I have two primary, closely related goals. The first is to bring to fruition the framework developed in previous chapters to explicate how contemporary cognitive neuroscience explains cognition, contrasting it with traditional cognitive science. The second is to soften any residual resistance to the mechanistic integration of psychology and neuroscience. I proceed as follows. After recapitulating the received view of explanation in cognitive science (Section 8.2), I briefly indicate why traditional responses to the received view fail to square with cognitive neuroscience (Section 8.3). I then articulate a framework of multilevel neurocognitive mechanisms (Section 8.4) and the levels that constitute them (Section 8.5). I conclude by highlighting three important aspects of cognitive neuroscience that illustrate the integrationist mechanistic framework: the incorporation of experimental protocols from cognitive psychology into neuroscience experiments, the development and evolution of functional neuroimaging, and the movement toward biological realism in computational modeling (Section 8.6). One important consequence of the picture I advance is that neither structures nor functions have primacy in cognitive neuroscience. The upshot is that explanation in cognitive neuroscience is multilevel, mechanistic, computational, and representational.

8.2 Cognitive Science as Traditionally Conceived

The *cognitive revolution* of the 1950s is most often contrasted with the behaviorist program it supplanted. By contrast with behaviorism's methodology and metaphysics, which is widely assumed to reject the positing of cognitive states and processes, cognitive science explicitly posits internal cognitive states and processes to explain intelligent capacities. An important motivation for this approach came from the analogy between cognitive systems and digital computers.

Computers possess internal states and processes that contribute to their capacities, some of which—playing chess, solving problems, etc.—are capacities that require intelligence in humans. Since it's patently legitimate to explain a computer's capacities in terms of its internal states and processes, cognitive scientists argued that it is equally legitimate to explain human cognition in terms of internal states and processes. More importantly, the internal states and processes of computers are representations and computations, which are typically considered cognitive notions. Thus, the argument continues, it is legitimate to explain human cognition in terms of computations and representations. Indeed, in this tradition cognition is often *identified* with some form of computation—more specifically, some form of *digital* computation over representations (e.g., Newell and Simon 1976; Anderson 1983; Johnson-Laird 1983; Pylyshyn 1984).

This focus on the contrast between behaviorism and cognitive psychology often obscures some of the substantive commitments that came out of the cognitive revolution. At all stages of Western history, available technology has constrained the analogies used to think about the operations of the human mind and body. For instance, water technologies—pumps, fountains, etc.—provided the dominant analogy behind the ancient Greek concept of the soul—the "pneuma"—and the humor theories that dominated Western medicine for 2000 years (Vartanian 1973); the gears and springs of clocks and wristwatches played a similar role for early mechanist thinking during the enlightenment (e.g., La Mettrie's *L'Homme Machine*, 1748); hydraulics for Freud's concept of libido; telephone switchboards for behaviorist theories of reflexes; and so on (cf. Daugman 1990). It is no coincidence that the cognitive revolution co-occurred with the advent of computers.

Whenever technology guides thinking about the human mind or body, the analogy may be taken too far. While it may be true that cognition involves transitions between internal states analogous to computations of some kind, the commitments of traditional cognitive science go far beyond this basic point. Specifically, the analogy between cognition and computation was taken to imply that cognition may be studied *independently* of the nervous system (Chapter 7). The main rationale for this autonomy is that digital computers—more specifically, universal, program-controlled digital computers—reuse the same hardware for the different programs (i.e., software) they execute. Each particular program explains a specific capacity, while the hardware remains the same. By the same token, a widespread assumption of traditional cognitive science was that the brain is a universal, program-controlled digital computer; therefore, cognition can be studied simply by figuring out what programs run on such a computer, without worrying over the details of the wetware implementation of those programs (e.g., Fodor 1975; Newell and Simon 1976; Pylyshyn 1984). Additionally, many who thought of the brain simply as some kind of digital computer (without assuming that it is universal and program-controlled) nonetheless agreed that cognition could be explained independently of neuroscience (e.g., Cummins 1983).

A close ally of this computer analogy and its rationale for the autonomy of psychology is the view that psychological explanation is different in kind from neuroscientific explanation. According to this view, psychological explanation captures cognitive functions and functional relations between cognitive states and capacities, whereas neuroscientific explanation aims at the structures that implement cognitive functions. The two types of explanation are supposed to place few constraints on one another with the upshot that each can proceed independently from the other.

The resulting picture of cognitive science is that psychology studies cognition in functional terms, which are autonomous from the noncognitive mechanisms studied by neuroscience. Aspects of this two-level picture can be found in the writings of many philosophers of cognitive science. Here are a couple of classic examples:

> The conventional wisdom in the philosophy of mind is that psychological states are functional and the laws and theories that figure in psychological explanations are autonomous. (Fodor 1997, 149)
>
> We could be made of Swiss cheese and it wouldn't matter. (Putnam 1975b, 291)

These philosophers were trying to capture what cognitive scientists were doing at the time. And while cognitive scientists were perhaps less explicit about the two-level picture, something similar to this view can be found in many landmark works that came out during the heyday of classical cognitive science (e.g., Newell and Simon 1976; Newell 1980; Marr 1982; Anderson 1983; Johnson-Laird 1983; Pylyshyn 1984).

8.3 Traditional Responses to Cognitive Science

This traditional two-level picture of cognitive science fails to capture explanation in contemporary cognitive *neuro*science. Cognitive neuroscience strives to explain cognition on the basis of neural mechanisms and thus involves integration, not autonomy, between psychology and neuroscience. After the cognitive revolution, the mechanistic integration of psychology and neuroscience amounts to another paradigm shift: the cognitive *neuro*science revolution. Cognitive neuroscience requires a different way of thinking about levels, cognitive explanation, representation, and computation. The resulting explanatory framework, multilevel neurocognitive mechanisms, is what I aim to articulate in this chapter.

In seeking an account of explanation in cognitive neuroscience, let's see how the reductionist strategy that I discussed in Chapter 1 applies to neuroscience. I will briefly argue that reductionism and its close cousin, eliminativism, don't capture the kind of integration found in cognitive neuroscience. These arguments will motivate a positive proposal (Section 8.4).

One traditional alternative to autonomy is to eliminate the theoretical constructs posited by psychology *in favor of* the theoretical constructs posited by neuroscience. The model for such eliminativism is the past elimination of theoretical constructs, such as epicycles, phlogiston, or the ether, from past scientific theories. Just as those theoretical posits were eventually eliminated from our scientific theories of, respectively, planetary motion, combustion, and the transmission of radiation through space, so the theoretical posits of psychology, such as the language-like mental representations posited by classical cognitive psychology, should be eliminated in favor of posits that are more amenable to neuroscience (P.M. Churchland 1981; P.S. Churchland 1986).

If eliminativism is construed radically enough—that is, as the literal elimination of any science of cognition other than neuroscience—it offers a partial solution to the problem at hand. That problem is to understand how the disciplines that study cognition fit together and how cognition ought to be explained. If any discipline other than neuroscience is eliminated, the first half of the problem is solved: since the other disciplines no longer exist, we don't need to worry about how they fit together with neuroscience. But this radical construal is hardly a solution to the most interesting part of the problem—how to explain cognition.

Contemporary cognitive neuroscience explains cognition in terms of neural computation over neural representations (more on this below). If the eliminativist approach implies that cognition itself—and all "cognitive" theoretical posits, such as representation, computation, or information processing—should be eliminated or at least deflated (cf. Ramsey 2007), then we are faced with a solution that is antithetical to cognitive neuroscience.

Another alternative to traditional (two-level) cognitive science is to *reduce* psychological theoretical posits to neuroscientific theoretical posits. The models for this reductionist strategy come from examples from physics, such as the reduction of classical thermodynamics to statistical mechanics or the reduction of Newton's theory of gravitation to a special case of Einstein's theory of General Relativity. The main difficulty for this reductionist approach in cognitive neuroscience is that, even assuming that it works for some physical theories (which has been debated), psychological and neuroscientific explanations lack the appropriately general mathematical formulations to be conducive to such reductions (cf. Cummins 2000).

Nevertheless, some have argued that when we can intervene on molecular structures in the brain and affect some cognitive behavior, the specific molecular events "directly explain" the behavioral data; we thereby reduce the relevant cognitive capacity to the relevant molecular events, and we thereby obviate the need for intermediate levels of explanation (cf. Bickle 2003, 2006). The main problem with this form of reductionism is that specific molecular events are at best very *partial* explanations of cognitive phenomena. It is one thing to correlate specific molecular events with cognitive phenomena via some specific intervention, or to establish that they are causally relevant; to actually explain a cognitive phenomenon requires a lot more.

As I argued at length in Chapter 7, lower-level events such as molecular events are only relevant to higher-level phenomena to the extent that they occur within specific neural structures in specific ways, and locating the relevant neural structures and ways of occurrence requires more than purely molecular neuroscience. In addition, even identifying a molecular event within a neural structure that contributes to a cognitive behavior falls short of a full explanation. A full explanation requires identifying how molecular events contribute to relevant neural events, how relevant neural events contribute to circuit and network events, how those in turn contribute to relevant systems-level events, and finally how the relevant systems, appropriately coupled with the organism's body and environment, produce the behavior. These intermediate links in the causal-mechanistic chain are crucial to connecting molecular events to cognitive phenomena in a way that is fully explanatory. And identifying these intermediate level structures and their causal contributions requires going well beyond molecular neuroscience (cf. Craver 2007; Bechtel 2008).

In spite of their respective limitations, both eliminativism and reductionism put pressure on the received view of cognitive science—most helpfully, by pointing out that cognitive scientists who ignore neuroscience do so at their peril and by pushing towards the integration of psychology and neuroscience. But neither eliminativism nor reductionism offers a satisfactory framework for explanation in *cognitive* neuroscience: the former insofar as it neglects cognition altogether; the latter because it offers only partial explanations that lack the necessary contextual factors provided by intermediate levels of mechanistic organization.

8.4 Multilevel Neurocognitive Mechanisms

Cognitive neuroscience stands in stark contrast to the traditional two-level picture of cognitive science. Broadly, cognitive neuroscience is the scientific study of how neural activity explains cognition and the behavior it gives rise to. Cognitive neuroscientists study nervous systems using many techniques at many levels. They study how cortical areas and other neural systems contribute to various cognitive capacities, how the operations of the neural subsystems that compose them (columns, nuclei) contribute to the capacities of neural systems, how networks and circuits contribute to their containing systems, how neurons contribute to networks and circuits, and how sub-neuronal structures contribute to neuronal capacities. Analyzing systems across such varied levels involves coordinating techniques ranging from molecular neuroscience and genetics to neurophysiology, neuroimaging, mathematical analysis, computational modeling, and a wide range of behavioral tasks.

Cognitive neuroscience thus strives to explain cognitive phenomena by appealing to and analyzing (both separately and conjointly) multiple levels of organization within neural systems. In previous chapters, I defended *multilevel mechanisms* as the right explanatory framework for thinking about the relations between levels of organization. A multilevel mechanism is a system of components in which the organized capacities of the components constitute (and thus mechanistically explain) the capacities of the whole. Note that it may take multiple subcapacities organized in specific ways to bring about specific capacities. In this section, I argue that cognitive neuroscience aims to discover multilevel *neurocognitive* mechanisms.

Multilevel neurocognitive mechanisms have a nested structure: at any level, each component of the mechanism is in turn another mechanism whose capacities are explained by the organized capacities of *its* components; and each whole mechanism is itself a component that contributes to the capacities of a larger whole. This multilevel nested structure tops off in the capacities of whole organisms and their interactions with other organisms, which are studied by social neuroscience and neuroeconomics; it bottoms out in structures—such as the atoms that compose neurotransmitters—that fall outside the disciplinary boundaries of cognitive neuroscience.

Cognitive neuroscience is not the only science that explains mechanistically, but it is one of the few whose mechanisms perform computations over representations. In Chapter 12, I will articulate what neural representations are. Briefly, a vehicle carries *semantic information* about a source just in case it reliably correlates with the state of the source. For instance, the spike trains generated by neurons in cortical area V1 reliably correlate with the presence and location of edges in the visual environment; thus, they carry semantic information about the presence and location of edges in the visual environment. But correlation alone is insufficient for representation.

A vehicle *represents* a source just in case it has the teleological *function* of carrying information about the source. For a vehicle to have such a function, the information it carries must be usable by some part of the system in which it is embedded. The information is usable by the system to the extent that it's causally relevant to other operations of the system. In our example, the spike trains generated by neurons in V1 have the function of carrying information about the visual environment because this information is usable by downstream areas for further processing of visual stimuli—i.e., it is causally relevant to the operations of those downstream areas. Thus, in the relevant sense, V1 neurons represent the presence and locations of the edges with which they correlate.

Finally, a system performs *computations* just in case it manipulates vehicles in accordance with rules that are sensitive to inputs and internal states and are defined in terms of differences between different portions of the vehicles it manipulates (Chapter 6). Which computations are performed by a system depends on its specific mechanistic properties—its component types, its vehicle type, and the rules it follows. That is, computation is defined nonsemantically based on the mechanistic properties of the system and the vehicles it manipulates. Although computation can occur in the absence of representation, processing representations is a form of mechanistic computation.

A distinctive feature of neural systems is that they pick up information from the environment and organism, transmit it through the system via appropriate signals (neural representations), and process such signals in conjunction with pre-existing representations and in accordance with rules (neural computation) in order to generate further signals that can regulate the organism's behavior. This appeal to representation and computation distinguishes mechanistic explanations in cognitive neuroscience from mechanistic explanations in many other sciences.

The above account of computation is diametrically opposed to persistent views of computation that draw a stark contrast between computational and mechanistic explanations.[2] Such views maintain that computations are *abstract* or *mathematical* in a way that evades mechanistic explanation. While it's true that computation can be mathematically characterized, the physical computations performed by nervous systems (and artificial computers, for that matter) are functions performed by concrete mechanisms.[3] Like other functions, information processing via neural computation is performed by mechanisms—specifically, *neurocomputational* mechanisms. With this said, an important caveat is that computing mechanisms, like all mechanisms, can be characterized with different degrees of abstraction (Chapter 7). This is an integral aspect of multilevel mechanistic explanation.

A much-discussed example, which is particularly relevant to the present context, is the Hodgkin–Huxley model of the action potential (Hodgkin and Huxley 1952). The Hodgkin–Huxley model explains the voltage profile of the action potential in terms of a neural membrane's changing voltage conductivity. Lower-level mechanistic details about how changes in membrane permeability arise were omitted from the model, initially because they were unknown (Hodgkin and Huxley 1952, 541), but also later because this omission affords the model greater generality (Schaffner 2008; Levy 2014; Chirimuuta 2014, 141). The Hodgkin–Huxley model has been described as nonexplanatory (Bogen 2005), as providing a nonmechanistic explanation (Weber 2005, 2008), and as a mechanism sketch because it omits information about the role of ion channels in allowing membrane permeability (Craver 2008). None of these characterizations fully hit the mark. Rather, the HH model is an example of a mechanism sketch that evolved into a mechanism schema—it explains a phenomenon (the action potential) at one mechanistic level (changes in membrane conductivity) while abstracting away from lower mechanistic levels (ion channels and their components).

As the preceding example illustrates, mechanistic explanations—particularly those that involve computations and representations in the sense outlined above—are often presented in the form of (interpreted) mathematical or computational

[2] A recent example: "My key claim is that the use of the term 'normalization' in neuroscience retains much of its original mathematical-engineering sense. It indicates a mathematical operation—a computation—not a biological mechanism" (Chirimuuta 2014, 124).

[3] Not all *mathematical* models in cognitive neuroscience ascribe computations to the nervous system; only those that explain phenomena through computations performed by the target systems do so.

models. Typically, such models become analytically insoluble and computationally intractable if they include too much detail about their target systems. As such, issues relating to solubility and tractability provide another motivation for the exclusion of detail from models of neurocomputational mechanisms. Issues regarding tractability are ubiquitous in computational neuroscience given the plethora of biological detail that could potentially figure into modeling scenarios.

For instance, one controversial but extremely common assumption among computational neuroscientists is that individual neurons can be treated as integrating a linear sum of dendritic inputs, and then initiating an action potential when that sum reaches a threshold. The dynamics of actual neurons are more complex than this assumption suggests, which in turn has led to the development of more complex models. For example, Waxman (1972) provides an alternative model, which introduces nonlinearities into the branching regions of the dendritic (input) and axonal (output) trees, rather than treating those regions as, respectively, collecting and distributing charge linearly. But the basic, linear treatment of dendritic input integration has been a powerful tool in a wide variety of modeling contexts. One explanation for the success of these simplified modeling strategies is that they capture important aspects of neural responses that are adequate for some epistemic purposes.

One clear takeaway is that an important skill of mathematical and computational modelers is to capture the main features of the systems they study that are needed to explain phenomena of interest, often by introducing appropriate idealizations and simplifications. Those idealizations and simplifications allow modelers to represent systems to desired degrees of approximation while maintaining mathematical solubility or computational tractability (cf. Humphreys 2004; Piccinini 2007b; Winsberg 2010; Weisberg 2013). Computational or information processing explanations often require these forms of detail omission. What is crucial to appreciate here is that computation and information processing do *not* lead outside the multilevel mechanistic framework but are instead best seen as a special case of it.[4]

Relatedly, as a matter of methodology, we are often interested in one aspect (some components or capacities) of a mechanism at the expense of other aspects (other components or capacities). This is one type of mechanism *sketch*, or partial (elliptical) mechanistic explanation. Special cases of this type of mechanism sketch are descriptions of computational (the function computed and why it is adequate to the task, cf. Shagrir 2010b) and algorithmic (the computational operations and representations) levels of a system, which omit details about the components that

[4] This is not to say that all analyses of neural computation or information-processing are mechanistic. Some focus only on the information content and efficiency of a neural code without saying anything about the processing mechanisms (Dayan and Abbott 2001, xiii; Chirimuuta 2014, 143ff). These models are not especially relevant here because they do not provide the kind of constitutive explanation that is the present topic.

carry out the algorithm.[5] There is certainly value to such approaches in cognitive neuroscience, particularly in the context of discovery. Marr (1982) argued that a neural "algorithm" is "likely to be understood more readily by understanding the nature of the problem being solved than by examining the mechanism (and the hardware) in which it is embodied" (Marr 1982, 27). Marr's arguments have often been cited in defense of autonomist views of cognitive science.[6] I have a different take, consistent with seeing computational and algorithmic accounts (in Marr's sense) as mechanism sketches (or schemata to the extent that underlying details are known but deliberately omitted).

Understanding the capacities of a system often requires looking "up" to situate the system within some higher-level mechanism or environmental context as much as looking "down" to understand how those capacities are implemented by the lower level components, their capacities, and organization. In addition, more may be known about the mechanistic or environmental context of a system than its components and their operations. In such cases, investigators may be forced to constrain their models primarily by examining the problem being solved rather than the components and their operations, even though the likely result of such a method is a "how-possibly" model that falls short of explaining how the system *actually* works. Much of Marr's work belongs in this how-possibly category. We certainly face some of the same problems in contemporary cognitive neuroscience, but the field has developed to the point where integration, rather than autonomy, is the appropriate framework. The computational-level descriptions Marr and others sought are best construed as a valuable step along the way to integrated multilevel mechanistic explanations. It is no longer enough to simply home in on ways in which problems *might* be solved in the brain; contemporary cognitive neuroscience aims to understand how those problems are *actually* solved in the brain.

8.5 Neurocognitive Levels

A primary motivation behind the traditional autonomist picture of cognitive science is the idea that functions can be understood independently from the structures that perform them. Therein lies the putative distinction between the "functional" level, which is cognitive, representational, and computational, and the "structural" level, which is noncognitive, mechanistic, and implementational.

[5] In computational neuroscience, the term "algorithm" and its cognates are often used for any computational process, whether or not the process is fully or even partially digital. This should not be confused with the use of "algorithm" in computer science and computability theory, where it usually means a sequence of fully digital computational operations.

[6] Bechtel and Shagrir (2015) is a good entry into the extensive literature on Marr's levels, including how they might fit within a mechanistic framework.

The present account of multilevel neurocognitive mechanisms adopts a different notion of neurocognitive levels, which undermines this traditional dichotomy.

Contrary to the received view, there is no single "functional," "cognitive," or "representational/computational" level of explanation, standing in opposition to a single (or even multiple) "structural," "neural," or "implementational" level(s).[7] In this section I analyze each of these concepts from the perspective of neurocognitive mechanisms in order to highlight how the integrationist mechanistic framework improves upon traditional autonomist and reductionist views.

In the first place, every level of a multilevel mechanism is both functional *and* structural, because every level contains structures performing functions. This stands in contrast to traditional views that maintain that structural analyses and functional analyses are distinct and autonomous from one another. Traditional reductionists—e.g. type identity theorists (Smart 1959)—strove to identify mental types with lower-level physical types. As a result, they may be interpreted as focusing on structural properties at the expense of functional properties, relegating the latter to "second order states" of physical types (Smart 2007). Traditional functionalists do the opposite: they give primacy to functional properties at the expense of structural properties (e.g., Putnam 1967a; Fodor 1968a). This is a somewhat unorthodox way of characterizing these views, so some brief unpacking is in order.

Classical reductionist views of the mind–brain relation, specifically type identity theorists, look to identify higher-level kinds (e.g., mental kinds, like "pain") with corresponding physical kinds. These reductive views were developed in contrast to dualism: the view that the mind and brain are distinct kinds of substance. Dualists have a notoriously difficult time specifying the means by which these distinct substances interact; type identity theorists dissolve this problem. To say that water is H_2O just is to identify a putatively higher-level kind with a lower-level physical kind—an arrangement of atoms. With such an identity in hand, it is illegitimate to wonder how water and H_2O interact. In a similar vein, type identity theorists argued that mental kinds, like pain, are identical to some neural kinds, like C-fibers firing. This identification dissolves dualistic concerns about how mental states interact with bodily states. What is noteworthy for present purposes is that the defining features of the kinds that figure into higher-level explanations just are the lower-level physical features common to instances of those kinds. This identification with lower-level physical features downplays the role of functional features of those higher-level kinds. Water is not individuated by its ability to quench thirst, nourish plants, etc., nor pain with its role in avoidance of noxious stimuli, protecting injured body parts,

[7] This point is reminiscent of Lycan's underappreciated critique of "two-levelism," which was pointing in the right direction (Lycan 1990).

etc. Instead, both kinds are identified with particular physical types, which possess causal powers that are, incidentally, associated with these functions.

Classic functionalist views turn this story on its head: the defining features of higher-level kinds, according to such views, are their functional features while the structural features are incidental. The crux of this disagreement between functionalists and reductionists turns on multiple realizability—i.e., the claim that the same function can be realized in different ways. Carburetors provide a classic example: they are defined by their function in internal combustion engines (mixing fuel and air); they retain this function over variations in the stuff they're made of (e.g., cast iron, zinc, aluminum, plastic) and the types of engine they're found in (e.g., car, motorcycle, lawn mower). Putnam's original arguments for autonomy in the 1960s were based on this insight (e.g., Putnam 1967a). Fodor took up the torch and used multiple realizability to argue for the general autonomy of the special sciences from lower-level sciences—physics, in particular (Fodor 1968a, 1974, 1997). The idea behind these arguments is that, while higher-level states are token identical to particular lower-level physical states, there is no single lower-level physical kind for the higher-level states to be identified with. Rather, when higher-level kinds are realized, the underlying physical kinds will form unruly disjunctions (e.g., cast iron OR zinc OR aluminum OR plastic); the only thing tying this disjoint set of physical features together is the higher-level kind itself (e.g., the function, "mixing fuel and air"). Thus, nothing is added to the higher-level analysis by looking at its realizers.

Neither of these approaches adequately captures the main thrust of work in cognitive neuroscience, because that work is aimed at understanding the complex interplay *between* structure and function at multiple levels of organization. The integrationist mechanistic framework adequately captures this aspect of cognitive neuroscientific explanations; in the integrationist mechanistic framework, functions constrain structures and vice versa. Functions *cum* contextual factors—i.e., mechanistic context—constrain the range of structures that can perform those functions. Similarly, structures *cum* contextual factors determine the range of functions those structures can perform. In this framework, neither structures nor functions have primacy over the other; neither can explain cognition without the other.

Any given structure is only capable of performing a restricted range of functions. For an everyday example, consider again the functions that can be associated with water. Structural facts about the chemical composition of water both enable and restrict its ability to perform certain functions. For instance, the facts that water is liquid at physiological temperatures and is composed of hydrogen (positively charged) and oxygen (negatively charged) make it capable of dissolving ionic compounds into ions essential for normal cell function. Contextual factors—like ambient temperature and available compounds—combine with structural factors to determine the appropriate range of functions. In cognitive neuroscience, similar observations abound. For instance, neurons have a refractory period, during which they cannot fire. This refractory period restricts a neuron's maximum firing rate to about 1,000 Hz, which in turn limits the kinds of code by which the brain can encode and transmit information. The structural properties that determine the recovery period of a neuron—blocks that prohibit influx of sodium ions through voltage-gated channels—limit the encoding and signaling functions neurons can perform.

In the other direction, any function can only be performed by a restricted range of structures. For an everyday example, reconsider the carburetor. While it's true that carburetors can be made from different materials, the appropriate materials are severely restricted once mechanistic context and desired function are considered. A plastic carburetor from a lawn mower engine will cease to function as a carburetor in the context of a Ford F150 engine. The function and the context in which the function is embedded determine the range of structures that can implement that function. As a cognitive neuroscience example, consider the function of storing information long term in a read/write, addressable form similar to the way memory works in digital computers (Gallistel and King 2009). Fulfilling this function requires memory registers whose states persist over a sufficiently long time, which must be able to communicate appropriately with the processing components; it also requires a system of addresses that are stored in memory components and manipulated by an appropriate control structure. None of this comes for free by positing a certain function; for a functional hypothesis to prove correct, the structures that perform that function within the nervous system must be identified.

The upshot is that cognition cannot be explained without accounting for the ways in which structures constrain functions and vice versa. In the long run, the mutual constraints between structures and functions lead cognitive psychologists and neuroscientists to look to each other's work to inform their explanations. At any given level of organization, the goal is to identify both which structures are in play and which functions are performed. The more we know about functions and the context in which they are embedded, the more we can formulate sensible hypotheses about which structures must be involved. Similarly, the more we know about structures and the contextual factors that influence them, the more we can formulate sensible hypotheses about which functions they perform. The best strategy is to investigate *both* structures *and* functions simultaneously. As I will illustrate in the next section, this is the main driving force between the merging of neuroscience and cognitive psychology into cognitive neuroscience.

Building on these observations about the relations between structures and functions, similar points can be made about *implementation* (or realization): there is no single implementation (or realization) level. Every system of organized components implements (realizes) the capacities of the whole it composes (Chapter 2). Every capacity of a whole is implemented (realized) by its organized components. Implementation is thus relative to level. In other words, every level of a multilevel mechanism implements the level above it and is implemented by the level below it. The only exceptions occur at the (somewhat arbitrary) boundaries of cognitive neuroscientific inquiry—e.g., the whole behaving organism need not implement anything, at least as far as cognitive neuroscience is concerned.

Relatedly, every level of a neurocognitive mechanism is *neural*—or more precisely, every level is either (at least partially) composed of neurons or is a component of neurons.[8] The fact that neurocognitive mechanisms "bottom out" in components of neurons is a contingent feature of the disciplinary boundaries of cognitive neuroscience. The crucial point for present purposes, in terms of deviation from the classical autonomist view, is that there is no "nonmechanistic" level of explanation to be added to the mechanistic ones.

Whether a level of a neurocognitive mechanism is *representational* or *computational* depends on whether it contains representations or performs computations in the relevant sense (Chapters 12 and 13). Many cortical areas and other large neural systems contain representations and perform computations, so they are representational and computational. Many of their components (columns, nuclei) also contain representations and perform more limited computations over them; the computations they perform are component processes of the computations performed by their containing systems. Therefore, large neural components are representational and computational, and the same holds for their components (e.g., networks and circuits). Again, the computations performed by smaller components are constituents of the larger computations performed by their containing systems, and that is how the computations of their containing systems are mechanistically explained. At a still lower level, the response profiles of some single neurons reliably correlate with specific variables and it appears to be their *function* to correlate in this way; if this is correct, then they are representational in the relevant sense. Whether individual neurons perform computations over these representations is a matter of debate that can be left open.

Sub-neuronal structures may or may not contain representations and perform computations depending on the extent to which they satisfy the relevant criteria.[9] At some point, we reach explanations that are no longer computational but instead are purely biophysical. Here certain biophysical mechanisms explain how certain neural systems register and transmit information. The purely biophysical level is reached when our explanation of the processes no longer appeals to medium-independent features of the vehicles—which in the case of neural

[8] For convenience, I'm ignoring that glial cells might also play a role in cognition. Whether they do does not affect the thrust of my argument.

[9] For example, Gallistel (2017a, b) speculates that some classical digital computations occur within neurons, by manipulating appropriately structured molecules.

vehicles are mostly spike frequency and timing—in favor of the specific biophysical properties of neurons, such as the flow of specific ions through their cell membranes. These purely biophysical (and lower) levels are no longer representational and computational in the relevant sense.

Finally, whether a level of a neurocognitive mechanism is *cognitive* depends on whether and how it contributes to a cognitive capacity. Given the account in the previous section, to the effect that explaining cognitive capacities involves neural computation and representation, a neurocognitive level is cognitive depending on the extent to which the components of that level perform computations over representations in a way that is relevant to explaining some cognitive capacity. As above, the lowest-level neural computations are explained purely biophysically. In some simple organisms, these simple computations may be sufficient to explain the organism's behaviors. In more complex organisms, however, these simple computations combine with other simple neural computations to constitute higher-level neural computations, which in turn constitute still higher-level neural computations, and so on, until we reach the highest-level neural computations, which explain cognitive capacities.

An example of such an explanatory strategy would be the following sketch of an account of vision. Individual cells in V1 selectively respond to particular line orientations from the visual scene. Several of these cells in conjunction form an orientation column, which provide the basis for edge detection in the visual scene. These orientation columns combine to constitute V1, which computes the boundaries of visual objects. V1 then operates in conjunction with downstream parietal and temporal areas to constitute the different "streams" of visual processing and visual object representation.[10]

The resulting framework for explaining cognition is a mechanistic, computational, and representational version of homuncular functionalism, whereby higher-level cognitive capacities are iteratively explained by lower-level cognitive capacities until we reach a level at which the lower-level capacities are no longer cognitive in the relevant sense (Attneave 1961; Fodor 1968b; Dennett 1978a; Lycan 1981, Chapter 4). The rise of cognitive neuroscience illustrates how this framework has developed and been applied (and continues to develop and be applied) in scientific practice. In the next section, I highlight three aspects of cognitive neuroscience that demonstrate the development and application of this framework: the incorporation of experimental protocols from cognitive psychology into neuroscience experiments, the evolution of functional neuroimaging, and the movement toward biological realism in computational modeling in cognitive science.

[10] I am not committed to the adequacy of this particular explanation of visual processing, just to its exemplifying the explanatory strategy of multilevel neurocognitive mechanisms that I am explicating here.

8.6 How Cognitive Neuroscience Exhibits Multilevel Mechanistic Integration

Cognitive neuroscience emerged most forcefully in the late 1980s, although neuroscientists attempted to explain cognition well before that. Prior to the 1980s, cognitive science and neuroscience had developed largely in isolation from one another. Cognitive science developed between the 1950s and the 1970s as an interdisciplinary field comprised primarily of aspects of psychology, linguistics, and computer science. In linguistics, this involved the development of generative grammars aimed at explaining the syntax structuring human linguistic behavior. In psychology, researchers began developing information processing accounts aimed at explaining capacities like problem solving and memory. In computer science, researchers began developing computational models, involving discrete state-transitions, in order to model psychological capacities like reasoning and problem solving. The development of cognitive science accelerated through the 1960s and 1970s, with these approaches proceeding on their own terms with little contact with neuroscience. The lack of contact with neuroscientific evidence contributed to a significant underdetermination of cognitive scientific hypotheses by available evidence (cf. Anderson 1978).

Meanwhile, neuroscience developed as an interdisciplinary field investigating both normal and abnormal functioning of the nervous system. Neurophysiological investigations had been carried out since at least the late 19th century, at a time when neuroscience and psychology were seen as disciplines that should be integrated (e.g., Freud 1895/1966; James 1890/1983). But the term "neuroscience" was only coined in the 1960s with the development of new techniques for investigating the cellular and molecular levels of nervous systems and for relating those investigations to systems and behavioral levels. As a result, early neuroscience illuminated candidate structures for implementing cognitive functions, but it did so with little connection to functional context, thereby making limited progress towards explaining cognitive functions.

Throughout the development of both fields in the 1960s and 1970s, neuroscience and cognitive science dealt with domains with a great degree of overlap. In principle, they could have merged; in practice, they tended to exclude one another. Conceptual motivation for this exclusion was rooted in views already discussed: autonomist commitments (both implicit and explicit) among practicing cognitive scientists versus reductionist commitments among many practicing neuroscientists. Meanwhile, practical motivation that reinforced the exclusion was rooted in the pace of early developments that shaped both fields. In neuroscience, techniques for investigation at the cellular and molecular level developed at a pace that outstripped and overshadowed work at the systems level. In cognitive science, rapid developments in computer science and artificial intelligence in the 1970s provided a computational framework in which processes were decomposed into specific operations performed on symbolic (language-like) structures. This framework fostered a gulf between cognitive analyses and neural analyses because there was no obvious way for these symbolic computational units to be realized in neural tissue.

The differences between the fields began to abate in the 1980s. Bechtel (2001) cites two chief contributors: the need for more sophisticated behavioral protocols in neuroscience and the related development of functional neuroimaging techniques.

The former developed naturally as neuroscience researchers shifted focus toward determining specific functions performed by recently discovered cellular and molecular structures, attempting to link those structures to particular behaviors. In order to draw these links and target higher-level functions, neuroscientists needed more sophisticated behavioral protocols. Cognitive psychologists had developed relatively sophisticated behavioral protocols in order to obtain informative data from a limited range of dependent variables. At the time, most experiments in cognitive psychology involved inference to some cognitive hypothesis based on patterns in two dependent variables: characteristic patterns of error human subjects exhibited on some task (error rate), and the typical amount of time taken for those subjects to perform the task (reaction time).

As neuroscientists began to shift their explanatory ambitions, they ran up against the same limited range of dependent variables. Rather than reinventing the wheel, they began incorporating behavioral protocols from cognitive psychology and applying those protocols to experimental setups in which neural activity could be monitored in both humans and model organisms. These more sophisticated behavioral protocols allowed neuroscientists to form and test hypotheses about the contributions of neural structures to cognitive functions.

This disciplinary shift demonstrates how, in practice, functions constrain structures: sophisticated behavioral protocols provided the functional context necessary to constrain the search for the structures involved in performing those functions. Of course, many of these protocols were subsequently revised in a give-and-take between the incoming physiological data and the existing functional models that motivated the protocols. But with the integration of these techniques and protocols, the underdetermination of structure-function mapping became a tractable empirical issue rather than a conceptual one.

Another main contributor to the practical integration of psychology and neuroscience was the development of functional neuroimaging techniques, which allow measurement of physiological changes in large neural structures in response to performance of particular tasks. The first functional neuroimaging technique to be developed was Positron Emission Tomography (PET). PET involves injecting a radioactive tracer into a subject's bloodstream, which can then be imaged as it decays to illuminate blood flow to different brain regions. In a seminal study, Fox et al. (1986) used PET to measure hemodynamic response in particular brain areas during different cognitive tasks—their results correlated sensory and motor tasks with increased blood flow in primary sensory and motor areas, respectively.

Prior to the development of neuroimaging, the primary way to attribute specific cognitive functions to neural systems and thereby to relate neural activity to behavior (in humans) was through the study of behavioral deficits resulting from lesions due to some form of traumatic brain injury. While these lesion studies remain an integral part of cognitive neuroscience to this day, there are a number of potential confounding factors involved in extrapolating from these data to brain function in nonpathological cases (see, e.g., Kosslyn and Van Kleeck 1990). Brain imaging assuages some of these concerns and, as a result, the early research into the applications of PET for functional brain imaging set the stage for the explosion of research in cognitive neuroscience precipitated by the development of even more powerful and less invasive imaging techniques like functional Magnetic Resonance Imaging (fMRI).

The ability to correlate activity in different brain regions with specific tasks has improved our ability to map cognitive functions to neural structures. Underdetermination problems remain, as cognitive functions cannot be simply read off of tasks and functional neuroimaging still has limits on spatial and temporal resolution that place corresponding limits on fine-grained attribution of functions to lower-level neural structures (cf. Roskies 2009). Nonetheless, these neuroimaging techniques provide valuable data to constrain structure-function mapping by situating putative functions within mechanistic context.

This mechanistic context needs to be supplemented by further modeling in order to provide fully integrated explanations of how cognitive phenomena relate to neural activity. The recent trend toward model-based fMRI studies, in which models from computational neuroscience are incorporated into traditional fMRI experimental designs, demonstrates one way in which these integrated explanations are approached (e.g., O'Doherty et al. 2007; Turner et al. 2017; cf. Povich 2019). These model-based imaging techniques illustrate the multilevel mechanistic framework. At a relatively coarse-grained level, neuroimaging allows identification of the cortical and subcortical networks that are active in particular tasks. In order to determine more precisely the functions performed by these intermediate-level networks, researchers look to modeling efforts in computational neuroscience that are highly constrained by the neurophysiology of the particular regions involved (more on this below). This strategy facilitates integration between different mechanistic levels and in so doing allows more precise identification of the functions involved in cognitive processes and the specific structures performing those functions. The proliferation of neuroimaging studies over the past two decades and, in particular, the current trend toward model-based approaches provide further evidence that cognitive neuroscience is indeed concerned with the complexities of structure-function mapping.

The evolution of computational modeling in cognitive science also exemplifies the shift from autonomist cognitive science to cognitive neuroscience. After McCulloch and Pitts (1943) introduced the first model that explicitly targets neural computation, three main modeling research programs developed. First, there is classical symbolic computationalism, which strives to explain cognitive capacities in terms of symbolic computation in putative autonomy from neuroscience (e.g., Newell and Simon 1972; Anderson 1983). Second is connectionism, which strives to explain cognitive capacities in terms of neural network computations, though such neural networks are artificial models that are minimally (if at all) constrained by what is known about actual neural systems (e.g., Rosenblatt 1962; Feldman and Ballard 1982; Rumelhart et al. 1986). The third modeling research program is computational neuroscience, which explains cognitive capacities by building models of neural systems that are explicitly constrained by neuroanatomical and neurophysiological evidence (e.g., Hodgkin and Huxley 1952; Caianiello 1961; Stein 1965; Knight 1972; Wilson and Cowan 1972). The critical difference is that while classicist and connectionist models cannot be mapped onto neural structures in any direct way, models from computational neuroscience target specific neural structures and form hypotheses about the specific functions they perform and how those functions contribute to cognitive behaviors (cf. Kaplan 2011). Thus, computational neuroscience models exhibit the integration of functions and structures that characterizes cognitive neuroscience.

For much of their history, these three traditions developed largely independently from one another. Classical computationalism gained a solid footing in the 1970s and was often based on the idea that the brain is a universal, program-controlled digital computer. In this modeling paradigm, cognitive processes were seen as the software that is implemented on such computers and thus can be studied independently from the hardware/wetware implementing the software. But this analogy between natural cognitive systems and digital computers is problematic for two reasons. First, whether nervous systems are universal, program-controlled digital computers is an empirical question, which cannot be settled independently of neuroscience. More importantly, evidence from neuroscience suggests that neural computation, in the general case, is in fact *not* a form of digital computation (Chapter 13). Since digital computation is a necessary (though insufficient) condition for a system to be a universal, program-controlled digital computer, current best evidence suggests that nervous systems are in fact not such systems.

In the 1980s, connectionism re-emerged and challenged the hardware/software analogy in favor of "neurally inspired" network models (Rumelhart and McClelland 1986, 131). But typical neo-connectionist psychology was not grounded in known neural processes and mechanisms. Connectionists made largely arbitrary assumptions about the number of neurons, number of layers, connectivity between neurons, response properties of neurons, and learning methods. Connectionist psychology made such assumptions in order to model and explain psychological phenomena. Since these assumptions were not grounded in neuroscience, connectionists were developing a different take on the computer analogy, replete with their own commitment to the autonomy of psychology from neuroscience. Thus, while connectionism pushed in the right direction, it fell short of actually integrating psychology and neuroscience. From the point of view of cognitive neuroscience, this kind of connectionism was more on the side of classical, autonomist cognitive science than it was on the side of neuroscience. As a result, both classical computationalism and connectionism foster models of cognitive systems that are largely disconnected from structural (neuroscientific) constraints (cf. Weiskopf 2011b).

While philosophers were captivated by the divide between classical computationalism and connectionism, computational neuroscientists developed powerful tools for modeling and explaining cognitive phenomena in terms of actual biological processes. They imported theoretical tools from mathematics and physics and took advantage of the exponentially increasing power of modern computers. By now, there are many highly sophisticated research programs developing detailed models of how specific neural structures perform cognitive functions at various levels of organization (e.g., Dayan and Abbott 2001; Ermentrout and Terman 2010; Izhikevich 2007; Eliasmith 2013).

The field has matured to a point where connectionism is disappearing as an independent research tradition, instead merging into computational cognitive neuroscience (compare O'Reilly and Munakata 2000 to O'Reilly et al. 2014). Most of the classicist research programs are also being shaped by this emergence of computational neuroscience. Many books could be written analyzing specific cognitive neuroscience explanations in detail. Luckily, some such books are already written (e.g., Bechtel 2008). I will just point out some recent pronouncements of some key figures in classical and connectionist modeling indicating that the field is undergoing a deep transformation.

Several influential early attempts at building classical cognitive architectures were based on *production systems* (Anderson 1983; Laird, Rosenbloom, and Newell 1987). Production systems model cognitive processes as software packages specifying a series of "if... then..." statements (rules) taking inputs to outputs. Initially, these quintessentially symbolic models were in no way constrained by neural data. Nevertheless, their proponents expressed great confidence: "Cognitive skills are realized by production rules. This is one of the most astounding and important discoveries in psychology and may provide a base around which to come to a general understanding of human cognition" (Anderson 1993, 1). More recently, work on these same cognitive architectures has evolved to respect multiple levels of computational organization that are constrained by evidence from neuroscience. A stark transition can be seen, in particular, in Anderson's work, where his initial ambitions for his ACT-R production system architecture as a univocal model for cognition have been replaced by the acknowledgement that hybrid architectures are more promising. In a more recent paper, Anderson et al. argue that "theories at different levels of detail and from different perspectives are mutually informative and constraining, and furthermore no single level can capture the full richness of cognition" (Jilk, Lebiere, O'Reilly, and Anderson 2008). Similarly, a relatively recent presentation by John Laird of his Soar architecture advocates constraint by evidence from neuroscience (as well as from psychology and AI): "we have found that combining what is known in psychology, in neuroscience, and in AI is an effective strategy to building a comprehensive cognitive architecture" (Laird 2012, 22).

Similar transitions can be seen in the works of other leading cognitive scientists. Stephen Kosslyn, a pioneer of mental imagery and the view that mental imagery involves a special, pictorial representational format, went from a traditional theory based primarily on behavioral data (Kosslyn 1980) to a thoroughly cognitive neuroscientific framework (Kosslyn 1994; Kosslyn, Thompson, and Ganis 2006). Kosslyn's trajectory is a good illustration of the process of deepening explanations via the investigation of underlying mechanisms (Thagard 2007), which is the hallmark of cognitive neuroscience. Kosslyn's early theory of mental imagery faced skeptical resistance from defenders of a nonpictorial alternative (Pylyshyn 1981). By appealing to fMRI and neuropsychological evidence, Kosslyn later gained widespread acceptance for his pictorial theory. The debate over the format of mental images is not entirely over, but the way to resolve it is not to reject neuroscientific evidence as irrelevant or insufficient (cf. Pylyshyn 2002, 2003). The way to resolve it is to learn even more about how the brain realizes and processes mental images.

An analogous shift from traditional cognitive science to cognitive *neuro*science can be seen in Anne Treisman's landmark work on attention (Treisman and Gelade 1980; Treisman 1996, 2009). James McClelland, who pioneered the neo-connectionist models that were developed autonomously from neuroscience (Rumelhart and McClelland 1986), subsequently co-founded the Center for the Neural Basis of Cognition (a collaboration between Carnegie-Mellon University and the University of Pittsburgh) and became a computational cognitive neuroscientist (e.g., McClelland and Lambon 2013). Michael Posner's authoritative treatment of the subtractive method employed in cognitive psychology (Posner 1976) became the basis for the rigorous use of neuroimaging methods, beginning with PET, that are the backbone of much cognitive neuroscience (Posner and Raichle 1994).

Along with the development of neuroimaging methods came new modeling methods specifically designed to model the functional organization of neural systems at the system level—the level described by neuroimaging methods. Examples include Granger Causality Analysis and Dynamic Causal Modeling.

Granger Causality Analysis is a statistical technique for reconstructing functional dependencies between neural structures such as brain areas from time-series data produced via neuroimaging. It is based on the notion of G-causality. Consider a prediction of the value of Y based on the value of a number of other variables. X contains *unique predictive information* about Y if and only if predictions of Y that include X as their prediction basis are more accurate than predictions that do not include X. Given this, X is a *G-cause* of Y if and only if X occurs before Y and X contains unique predictive information about Y. Granger Causality Analysis infers G-causal relations between neural structures based on time-series data yielded by imaging studies. Insofar as G-causation is a guide to causation simpliciter, Granger Causality Analysis helps researchers understand the functional relations between neural structures (Seth et al. 2015).

Unlike Granger Causality Analysis, Dynamic Causal Modeling does not extract causal relations directly from time-series data. Instead, it begins with differential equations that model, in a biologically realistic way, how certain brain regions or neuronal populations affect one another. It produces a family of such models and then adds a model of how the dynamics posited by the first family of models would affect measurements. Finally, it uses actual measurements obtained through neuroimaging studies to select the most plausible model within the initial family of models. Dynamic Causal Modeling was developed and is primarily used to infer the ways in which brain regions or neuronal populations affect one another and how their dynamics are modulated by contextual factors such as attention (Daunizeau et al. 2011).[11]

By identifying the higher-level functional organization of the nervous system, techniques such as Granger Causality Analysis and Dynamic Causal Modeling help integrate information about cognitive functions with information about neural networks, thus contributing to multilevel mechanistic integration.

Because we are still in the midst of this interdisciplinary shift toward the integration of psychology and neuroscience, it is easy to miss how significant it is. The old view of psychology as autonomous from neuroscience (as well as the faith in the reductionist program, from the other direction) has been effectively supplanted by a new framework where multilevel integration rules the day.

8.7 Neurocognitive Mechanisms

The cognitive neuroscience revolution consists in replacing the scientific practices stemming from the traditional two-level view of cognitive science with a fully integrated science of cognition. The traditional two-level view maintained a division of labor between the sciences of cognition proper (psychology, linguistics,

[11] An early discussion of systems-level dynamical modeling techniques such as Dynamic Causal Modeling may be found in Egan and Matthews 2006.

anthropology, AI, and philosophy) and sciences of implementation (neuroscience). This framework has fallen by the wayside as cognitive neuroscience has risen.

The old two-level picture fell apart for several reasons. First, new modeling and empirical techniques—including the emergence of neuroimaging methods—have provided more sophisticated ways to link cognitive capacities to the activities of specific neural systems. Second, the dubious assumptions about the nervous system that bolstered the received view, such as the assumption that the nervous system is a universal, program-controlled digital computer, simply have not panned out. Third, the received view of cognitive explanation, according to which there is one privileged cognitive level and one distinctive and autonomous explanatory style—functional analysis—was faulty.

Philosophers should take heed. In place of the eliminative/reductive and autonomist views of cognition, I have proposed the framework of integrated, multilevel, representational, and computational neural mechanisms as capturing the essence of successful explanation in cognitive neuroscience. Any discipline that studies cognition—including behavioral biology, which should be included in any list of the disciplines that study cognition—can fruitfully contribute to this project by characterizing one or more neurocognitive level(s) using its empirical and analytical techniques. In addition to avoiding the problems of the old two-level view, this framework also avoids the pitfalls of both reduction and elimination by retaining a role for organization within each neurocognitive level.

Much work remains to be done to more fully understand the implications, applications, and limitations of this framework. The first step is to articulate the Computational Theory of Cognition in a general form, of which McCulloch and Pitts's theory is a specific version. That is the task of the next chapter.

9
The Computational Theory of Cognition

9.1 Explaining Cognitive Capacities

In previous chapters I articulated a framework of multilevel mechanisms, which explain systems' capacities. When we explain certain capacities, we appeal to computations performed by the system. For example, smartphones—unlike, say, air conditioners—have the peculiar capacity of yielding the product of their inputs: if we open the calculator app and press appropriate buttons on a (well-functioning) smartphone in the appropriate order, the smartphone returns the product of the input data. Our most immediate explanation for this capacity is that, under those conditions, smartphones perform an appropriate computation—a multiplication—on the input data. This is a paradigmatic example of computational explanation. In Chapters 6 and 7, I argued that fully explaining a system's computing capacities requires mechanisms. In Chapter 8, I argued that cognitive neuroscience explains cognition mechanistically. I can now expand on what it takes to explain *cognitive* capacities.

Some systems exhibit cognitive capacities such as perception, memory, reasoning, emotion, language, planning, and motor control: capacities to respond to their environment in extraordinarily subtle, specialized, and adaptive ways, which require acquiring and integrating multiple sources of information within a specialized control organ. In explaining such cognitive capacities, we often appeal to cognitive states such as perceptions, memories, intentions, etc. We also recognize that, in earthly biological systems, the mechanisms that underlie cognitive capacities are primarily neural—no nervous systems, no cognition. I refer to the neural mechanisms that explain cognition as neurocognitive mechanisms. But it is difficult to connect cognitive processes to their neural realizers—to see how perception, memory, reasoning, and the like are realized by neural states and processes. This difficulty has haunted the sciences of cognition since their origin.

As we saw in Chapter 5, Warren McCulloch and Walter Pitts (1943) devised an ingenious solution: cognitive capacities have a computational explanation. This is the computational theory of cognition (CTC). Sometimes CTC is also called Computational Theory of Mind (CTM). I prefer to reserve CTM for the view that the whole mind, including phenomenal consciousness, is computational.

Neurocognitive Mechanisms: Explaining Biological Cognition. Gualtiero Piccinini, Oxford University Press (2020).

DOI: 10.1093/oso/9780198866282.001.0001

CTC leaves consciousness out. I will discuss CTM and its relation to CTC in Chapter 14.

Another way to put CTC is that cognition is a kind of computation. For simplicity, I will use these two formulations interchangeably unless otherwise noted. Most cognitive neuroscientists endorse some version of CTC and pursue computational explanations as their research program. Thus, when cognitive neuroscientists propose an explanation of a cognitive capacity, the explanation typically involves computations that result in the cognitive capacity.

There are many versions of CTC: *classical* versions, which often downplay what we know about neural systems (e.g., Newell and Simon 1976; Fodor 1975; Pylyshyn 1984; Gallistel and King 2009); *connectionist* versions, which pay lip service to it (e.g., Rumelhart and McClelland 1986); and *neurocomputational* versions, which are more or less grounded in what we know about neural systems (e.g., Churchland and Sejnowski 1992; O'Reilly and Munakata 2000). According to all of them, the nervous system is a computing system, and its cognitive capacities are explained by neural computations.

CTC has encountered resistance. Some neuroscientists are skeptical that the brain may be adequately characterized as a computing system in the relevant sense (e.g., Perkel 1990; Edelman 1992; Globus 1992). Some psychologists think computational explanation of cognitive capacities is inadequate (e.g., Gibson 1979; Varela, Thompson, and Rosch 1991; Thelen and Smith 1994; Port and van Gelder 1995; Johnson and Erneling 1997; Ó Nualláin, McKevitt and MacAogáin 1997; Erneling and Johnson 2005). And some philosophers find it implausible that everything about cognition may be explained by computation (e.g., Taube 1961; Putnam 1988; Mellor 1989; Bringsjord 1995; Dreyfus 1998; Harnad 1996; Penrose 1994; van Gelder 1995; Wright 1995; Horst 1996; Lucas 1996; Fetzer 2001). I'll address objections to CTC in Chapter 11. In this chapter, I focus on what CTC says and the evidence for it.

Digital computers are more similar to cognitive systems than anything else known to us. Computers can process information, perform calculations and inferences, and exhibit a dazzling variety of cognitive capacities, including guiding sophisticated robots. Computers and natural cognitive systems are sufficiently analogous that CTC is attractive to most of those who are searching for a mechanistic explanation of cognition. As a consequence, CTC has become the mainstream theory in psychology, neuroscience, and naturalistically inclined philosophy of mind. In some quarters, CTC is so entrenched that it seems commonsensical.

More than three quarters of a century after CTC's introduction, many take for granted that cognitive capacities are explained by neural computations. But if we *presuppose* that neural processes are computations before investigating, we turn CTC into dogma. For our theory to be genuinely empirical and explanatory, it needs to be empirically testable. To bring empirical evidence to bear on CTC, we need an appropriate notion of physical computation and a corresponding notion of computational explanation.

In Chapter 6, I argued that the appropriate notion of computation is mechanistic. To understand computational explanation in psychology, neuroscience, and computer science, the notion we need is that of functional mechanisms that perform computations. Computing systems are mechanisms whose teleological function is manipulating medium-independent vehicles in accordance with a rule. CTC should be understood to assert that neurocognitive systems are computing mechanisms in this sense.

Later in this chapter I will give a couple of reasons to conclude that CTC is correct. The first reason is that neural vehicles—spike trains—are functionally significant thanks to medium-independent properties and are processed by a functional mechanism in a rule-governed way. The second reason is that neural vehicles integrate information coming from disparate physical sources, and that's why they are medium-independent. Since this integrated information is processed in a rule-governed way, this is also a reason to conclude that neurocognitive processes are computations.

Under the broad umbrella of mechanistic computation in the generic sense, there are different kinds of computation. The best known is digital computation, which is the functional processing of strings of digits in accordance with a rule. Another important kind is analog computation, which is the functional processing of variables that may be continuous, primarily by means of integration. There are other kinds as well.

An important question for CTC is, which kind of computation explains biological cognition? In other words, is neural computation digital, analog, or a third kind? McCulloch and Pitts (1943) argued that it's digital, and many mainstream cognitive scientists accepted that view. A minority argued that it's analog computation (Gerard 1951; Lashley 1958, 539; Rubel 1985). In Chapter 13, I will argue that it's neither digital nor analog; neural computation is sui generis.

In the rest of this chapter, I will distinguish CTC from other theses it is sometimes conflated with. After that, I will give two arguments for CTC and clarify the debate between different versions of CTC.

9.2 Five Theses about Neurocognitive Mechanisms

Like mechanistic explanation in other fields, mechanistic explanation in cognitive neuroscience is about how different components and their capacities are organized to exhibit the activities of the whole. Unlike mechanistic explanation in other fields, mechanistic explanation in cognitive neuroscience focuses on the peculiar functions performed by neurocognitive mechanisms.

To understand what is distinctive about neurocognitive mechanisms, let's distinguish CTC proper from other related theses it is sometimes conflated with (e.g., Cao 2018).

(1) Neurocognitive systems are *input–output* systems.

This thesis holds that neurocognitive activity generates responses by processing stimuli. Inputs come into neurocognitive systems, are processed, and such processing affects the output. This is obviously true but says nothing about the nature of the inputs, outputs, and internal processes. In this sense, neurocognitive systems are analogous to any systems that generate outputs in response to inputs via internal processes. If the notion of input and output is liberal enough, any physical system belongs in the class of input–output systems. If so, then thesis (1) is equivalent to the trivial thesis that neurocognitive systems are physical. But even if the notion of input and output is more restricted—for example, by specifying that inputs and outputs must be of a certain kind—thesis (1) remains weak.

(2) Neurocognitive systems are *functionally organized* input-output systems.

Thesis (2) strengthens (1) by adding a further condition: neurocognitive activity has the teleological *function* of generating specific responses from specific stimuli. In other words, neurocognitive systems are functionally organized to exhibit certain capacities and not others. Under this stronger thesis, neurocognitive systems belong to the specific class of physical systems that are functional mechanisms (Chapter 3). Systems of this type include computers as well as systems that intuitively do not perform computations, such as engines, refrigerators, and digestive systems. This second thesis is stronger than the first but remains weak. Again, there is no reason to doubt it.

(3) Neurocognitive systems are *feedback control*, functionally organized, input–output systems.

Thesis (3) is stronger than (2) because it specifies a primary function of neural activity: feedback control. Feedback control consists in regulating a system's behavior based on what the system is doing. According to thesis (3), neurocognitive systems control the behavior of an organism in response to stimuli coming from the organism's body and environment, which stimuli include the effects of the neurocognitive system itself. Neurocognitive systems are in charge of bringing about a wide range of activities performed by organisms in a way that is sensitive to the state of both the organism and its environment. The activities in question include the cognitive functions as well as any functions involving the interaction between the whole organism and the environment, as in walking, feeding, sleeping, and also functions ranging from breathing to digesting to releasing hormones into the bloodstream. The specific function of neurocognitive activity—feedback control—differentiates the neurocognitive system from other functionally organized systems. Under thesis (3), neurocognitive systems belong in the class of feedback control systems together with, for instance, autopilot systems. Thesis (3) is not trivial and took serious scientific work to establish rigorously (Cannon 1932). Nevertheless, it is by now uncontroversial.[1]

(4) Neurocognitive systems are *information-processing*, feedback control, functionally organized, input–output systems.

Thesis (4) is stronger than (3) because it specifies the means by which neurocognitive systems exert their feedback control function. Those means are the processing of information. The term "information" is ambiguous between several senses. We can now see whether neurocognitive systems process information in any of the four senses of "information" that I reviewed in Chapter 6.

In the most basic sense, information is simply reduction of uncertainty and is closely related to thermodynamic entropy. Everything carries information and every physical process is an instance of information processing in this sense. This notion of information is too general to be directly relevant to theories of cognition. I will set it aside.

In a first relevant sense, "information" means mutual information in the sense of communication theory (Shannon and Weaver 1949). Mutual information is a measure of the statistical dependency between a source and a receiver. It is a property of a communication channel, not individual signals. It is experimentally well established that neural signals such as firing rates are statistically dependent on other variables inside and outside the nervous system. Thus, mutual information is often used to quantify the statistical dependence between neural signals and their sources and to estimate the coding efficiency of neural signals (Dayan and Abbott 2001, chap. 4; Baddeley et al. 2000). It should be uncontroversial that neurocognitive systems process signals that carry information in this sense, although this is still a fairly weak notion of information *processing*.

In a second relevant sense, "information" means natural semantic information, which is the informational analog of natural meaning (Grice 1957). As I use the term, a state of a signal carries natural semantic information *about* a state of a source to the extent that it raises the probability that the source is in that state. Natural semantic information is different from mutual information because it is a

[1] Of course, there are different kinds of feedback control mechanism, some of which are more powerful than others. Therefore, it remains to be determined which feedback control mechanisms are present or absent within the neurocognitive system. For some options, see Grush 2003. For an account of neural control within a mechanistic perspective, see Winning and Bechtel 2018.

property of individual signals and is a kind of *semantic* content, whereas mutual information is not a measure of semantic content. Nevertheless, the two senses are conceptually related because whenever there is mutual information between a signal and a source, there is by definition a reliable correlation between source events and receiver events. Thus, the receiver acquires natural semantic information *about* the source.

Once again, it is both evident to naïve observation and experimentally well established that neural variables carry natural semantic information about other internal and external variables (Adrian 1928; Rieke et al. 1999; cf. Garson 2003). Much contemporary neurophysiology centers on the discovery and manipulation of neural variables that correlate reliably—and often usefully—with variables external to the nervous system.

Given (3) and the complexity of neurocognitive systems' control functions, thesis (4) follows: in order to exert sufficiently complex control functions, the only efficient way to perform feedback control is to possess and process internal variables that correlate reliably with relevant external variables. These internal variables must be processed in a way that is sensitive to the natural information they carry. Thus, it should be uncontroversial that neurocognitive systems process information in the sense of processing signals based on the natural semantic information they carry.

This argument from (3) to (4), first presented by Piccinini and Bahar (2013), is closely related to the good regulator theorem, an important finding of control theory (Conant and Ashby 1970). The good regulator theorem says that, under broad assumptions, any control system that is maximally successful and simple must be a model that is isomorphic with the system it controls. Organismic traits need not be maximally successful and simple, but they must be good enough to get by. A consequence of the good regulator theorem is that any good enough control system must be a model that is at least approximately isomorphic (homomorphic) to the system it controls.

Control systems that rely on feedback, such as nervous systems, cannot rely on a fixed model of the system they control. Instead, they must build and actively modify their model in real time. Since nervous systems control a body within an environment, therefore, for neurocognitive systems to control organisms well enough, they must build a model of the body and environment of the organism and keep it constantly updated. Given the complexity of nervous systems' control functions, the only way neurocognitive systems can build and maintain such a model in order to exert feedback control is to collect and process natural semantic information about the system being regulated. Therefore, (3) plus the good regulator theorem entail (4).[2]

[2] Seth (2015, 8) writes as if the good regulator theorem entails, or at least strongly suggests, that the sole function of the brain is minimizing prediction error. That is not quite right. Klein (2018) points out

Notice that the processing of semantic information occurs as a matter of neurocognitive systems' control functions. Therefore, thesis (4) does not merely say that there is information in the system. It says that it is the function of the system to carry and process such information in the service of control functions. According to a tradition called Informational Teleosemantics, processing information in the service of a control function is what it takes to possess and process *representations* (Neander 2017). When a system has the function of carrying natural semantic information but fails to do so, it misrepresents. In Chapter 12, I will articulate Informational Teleosemantics in more detail and argue that neurocognitive systems process representations in this sense.[3]

In a third relevant sense, "information" means *non*natural semantic information, which is the informational analog of nonnatural meaning (Grice 1957). Nonnatural semantic information is the information carried by natural language and other signaling systems in which there may or may not be a correlation between a signal and what it carries information about. For example, the utterance "this is an armadillo" carries nonnatural semantic information about whatever "this" refers to, and it represents it as an armadillo. The represented object may be a *source* of the nonnatural information carried by a signal but it need not be. In fact, the object referred to may or may not be an armadillo and there may or may not be a represented object in the actual world: a signal may carry nonnatural information even when the objects it purports to be about don't exist.[4]

Since human neurocognitive systems process linguistic signals and other signals that carry nonnatural semantic information, they likely contain and process neural signals that carry nonnatural semantic information. That would help explain how they can process external signals that carry nonnatural semantic information. Still, nonnatural semantic information is a special case that poses special challenges. I will mostly set it aside in the rest of this book (for a theory of nonnatural representation, see Piccinini forthcoming).

The upshot of our discussion of thesis (4) is that there is at least one important and uncontroversial sense of information processing—the processing of signals that carry semantic information by a functional mechanism—in which neurocognitive systems process information.

that the good regulator theorem is consistent with ways to exert control that are not limited to minimizing prediction error. Klein also points out that structures outside a centralized control system can perform or at least facilitate control functions—e.g., the skin protects the body from much environmental perturbation, thereby facilitating homeostasis.

[3] Sometimes neurons themselves are said to constitute a representation. For instance, the motor cortex is said to carry a representation of the hand whether or not the motor cortex is firing. This notion of representation is analogous to the sense in which a map is a representation whether or not someone is looking at it and using it as such.

[4] This Gricean terminology is potentially misleading. I am not suggesting that there is anything nonnatural about "nonnatural" information. I am also not suggesting that nonnatural information depends on speakers' intentions, the way nonnatural meaning does according to Grice.

The vehicles most commonly employed by neurocognitive systems to carry information about the body and the environment are, of course, spikes produced by neurons. Typically, spikes are organized in sequences called *spike trains*, whose properties vary from neuron to neuron and from condition to condition. Spike trains from a single neuron are often insufficient to produce a functionally relevant effect. At least in large nervous systems, such as the human nervous system, in many cases spike trains from populations of several dozen neurons are thought to be the minimal processing units (Shadlen and Newsome 1998). Mechanistic explanation in neuroscience, at or above the levels that interest us here, consists in specifying how appropriately organized spike trains from different neuronal assemblies constitute the capacities of neural mechanisms, and how appropriately organized capacities of neural mechanisms constitute the brain's capacities—including its cognitive capacities.

From the perspective of neural function, then, the question of CTC becomes the following. Are neurocognitive processes—specifically, the processing of neuronal signals such a spike trains—computations? If so, which kind of computation are they?

(5) Neurocognitive systems are computing systems.

Thesis (5) is presupposed by any generic form of CTC. To evaluate CTC and assess its relation to (1)-(4) requires some precision about what counts as computation in physical systems such as neurocognitive systems. I explicated physical computation—independently of information processing—in Chapter 6. Physical computation is the functional processing of medium-independent vehicles in accordance with a rule. I can now outline the relationships between thesis (5) and its predecessors. As I already pointed out, each of (2), (3), and (4) is stronger than its predecessor.

While strictly speaking we could define computing systems without inputs, outputs, or both, I disregard them here because neurocognitive systems do have inputs and output. In addition, following Chapter 6, I assume that any computing system that performs nontrivial computations is functionally organized to perform its computations. With these assumptions in place, computing (5) entails having inputs and outputs (1) and being functionally organized (2). There is no entailment in the other direction. For there are plenty of systems—including functionally organized systems such as animals' locomotive, circulatory, and respiratory systems—that do not perform computations in any relevant sense. So (2) and a fortiori (1) do not entail (5).

Computing (5) entails neither feedback control (3) nor information processing (4). This is because computing systems need not perform any feedback-control or information-processing functions—an example would be a computer programmed to manipulate[5] meaningless digits according to some arbitrary rule.

[5] Terminological note: I use the terms "to manipulate" and "to process" interchangeably.

Feedback control (3) and information processing (4) do not entail computing (5); a system may exert feedback control functions by processing information without processing any medium-independent vehicles in accordance with a rule. For instance, a centrifugal governor is a control mechanism that regulates the speed of a steam engine by regulating the rate of steam that goes through a pipe. It does this by rotating at a speed that correlates with the speed of the engine and opening or closing a valve in the pipe accordingly. When the engine goes too fast, the governor closes the valve a bit, while the opposite happens when the engine goes too slow. This effect is obtained by transforming speed into intermediate physical variables and finally into the opening and closing of a valve by exploiting specific physical relationships between these variables. Thus, a centrifugal governor processes information in order to exert feedback control without ever transducing its input signals into medium-independent vehicles (cf. van Gelder 1995 who uses the term "Watt governor" for centrifugal governor).

The foregoing discussion clarifies how CTC, which presupposes thesis (5), relates to theses (1) through (4). I am now in a position to defend CTC.

9.3 Two Arguments for the Computational Theory of Cognition

Is cognition a type of computation? Here are two arguments.

9.3.1 The Argument from the Functional Organization of the Nervous System

1. Neurocognitive processes are functional manipulations of medium-independent vehicles in accordance with a rule.
2. Functional manipulations of medium-independent vehicles in accordance with a rule are computations (Chapter 6).
3. Therefore, neurocognitive processes are computations.

As I pointed out in Section 9.2, neurocognitive systems are functional mechanisms that have control and information processing functions. As a consequence, they manipulate vehicles not randomly but in a rule-governed way. The vehicles manipulated by neural processes include voltage changes in the dendrites, neuronal spikes (action potentials), neurotransmitters, and hormones. Spike trains are the primary means by which neurons send long-distance signals to one another as well as to muscle fibers. Therefore, I will focus primarily on spike trains. Current evidence suggests that the functionally relevant aspects of neural processes depend on dynamical aspects of the vehicles—most relevantly, spike rates and spike timing.

Early studies of neural signaling (Adrian and Zotterman 1926; Adrian 1928) demonstrated that nervous systems transduce different physical stimuli into a common internal medium. Spikes follow the *all-or-none* principle: neurons transmit a signal of a certain amplitude—a spike—above a certain threshold of input intensity, whereas below that threshold no signal is transmitted. Information about the stimulus is conveyed through changes in spike rates (*rate coding*). Much evidence suggests that precise spike timing can also play a role in information transmission (*temporal coding*). Indeed, spike timing may be critically important when stimuli must be processed more quickly than would be possible if the brain relied on spike rates (VanRullen et al. 2005). Examples of temporal coding can be found in the visual (Bair and Koch 1996; Meister and Berry 1999; Gollisch 2009) and somatosensory (Johansson and Birznieks 2004) systems. The relative importance and prevalence of temporal vs. rate coding remains an issue of intense debate (London et al. 2010).

Obviously, many aspects of spikes and their generation depends on the specific physical properties of the medium. For example, sending an action potential requires opening and closing ion channels, whose properties depend on the specific ions they allow flowing in and out of the cell membrane. Nevertheless, the upshot of the previous two paragraphs is that the aspects of spikes and spike trains that appear to be functionally relevant to cognition are primarily spike rates and spike timing. These features of spike trains are similar throughout the nervous system regardless of the physical properties of the stimuli they may encode (i.e., auditory, visual, somatosensory, etc.) and can be specified in abstraction from any physical medium. In fact, timing and rate are properties that can be realized either by neural tissue or by some other physical medium, such as a silicon-based circuit (cf. Craver 2010). That makes the vehicles of neurocognitive processes multiply realizable$_2$. Processes defined in terms of multiply realizable$_2$ inputs and outputs are medium independent (Chapter 2). Therefore, neurocognitive processes are medium independent. Analogous considerations apply to other vehicles manipulated by neurons, such as voltage changes in dendrites, neurotransmitters, and hormones.

Furthermore, it goes without saying that nervous systems process spike trains in accordance with rules. One particularly important rule is normalization, whose output is a ratio between the response of an individual neuron and the summed activity of a pool of neurons (Carandini and Heeger 2012). As Carandini and Heeger define it, normalization is a medium-independent operation found in many neural systems of many species, where it can be realized by many lower-level computations and can contribute to many broader computational and representational functions. Due to its ubiquity and versatility, Carandini and Heeger dub normalization a *canonical* neural computation.

Putting these points together, neurocognitive processes operate on medium-independent properties of their vehicles in accordance with appropriate rules, which is what we defined computation in a generic sense to be.

9.3.2 The Argument from Semantic Information Processing

1. Some neurocognitive systems integrate semantic information from different physical sources and process it in accordance with a rule.
2. Integrating and processing semantic information from different physical sources involves transducing it into medium-independent vehicles.
3. Processing medium-independent vehicles in accordance with a rule is computation (Chapter 6).
4. Therefore, some neurocognitive processes are computations.

The argument for generic CTC from semantic information processing begins with the observation made in Section 9.2 that cognition involves the processing of information in a rule-governed way, in at least three important senses. First, cognition involves processing stochastic signals; this is information in the non-semantic sense of mutual information between a source and a receiver. Second, cognition involves the processing of signals based on their raising the probability that their sources are in certain states (Adrian 1928; Rieke et al. 1999); I called this natural semantic information. Third, at least some forms of cognition, such as language processing, require manipulating linguistic signals, which carry nonnatural semantic information. Since biological cognition is carried out by neural processes, those neurocognitive processes manipulate semantic information. All these types of information processing serve cognitive functions that require them to be governed by rules—to yield certain kinds of information as output given certain kinds of information as input.

For those who identify computation and information processing, the fact that cognition involves information processing is enough to conclude that cognition involves computation. As I pointed out in Chapter 6, however, computation and information processing are best kept conceptually distinct, if nothing else because identifying them obliterates important distinctions between different notions of computation and information. In addition, strictly speaking, information can be processed by means other than computation.

Nevertheless, cognition involves information processing in a way that does entail computation. This is because the sources of information that neurocognitive systems integrate and process are physically disparate: light waves, sound waves, pressure, chemical concentrations, and more. Cognitively sophisticated organisms integrate information from these disparate sources by transducing the different sources into a common type of vehicle, which can correlate with all such sources. In biological cognition, this common vehicle is neuronal signals such as spike trains. Once information is transduced into a common vehicle, what matters for the purposes of information processing is the information the vehicles carry.

To allow the integration of physically disparate sources of information, the features of a vehicle that carry information must be medium-independent. In neurocognitive systems, as we've seen, such features include the frequencies and timing of the spikes. In other words, strictly speaking, the vehicles of information processing within neurocognitive systems must be medium independent. And computation in the generic sense is precisely the processing of medium-independent vehicles in accordance with a rule. Therefore, the kind of information processing we are interested in entails computation in the generic sense. (The converse does not hold because medium-independent vehicles need not carry any information.)

I conclude that cognition involves computation in the generic sense and neurocognitive systems perform computations in the generic sense. Can anything more specific be said about which kind of computation is involved?

9.4 Computation and Biological Cognition: Digital, Analog, or a Third Kind?

CTC takes different forms depending on which type of computation it invokes to explain cognitive capacities. The weakest and most general form of CTC is simply that cognition is computation of some kind. Stronger and more specific versions of CTC assert that cognition is either (a kind of) digital computation, (a kind of) analog computation, or computation of a third kind. Digital CTC, in particular, comes in many specialized versions that invoke different kinds of digital computation. A historically influential version of digital CTC is classical CTC, according to which cognition is (a kind of) classical computation.

While it's fairly safe to say that cognition involves computation in the generic sense and that nervous systems perform computations in the generic sense, it's much harder to establish that cognition involves a more specific kind of computation.

We've already encountered McCulloch and Pitts's argument that cognition involves digital computation because of the all-or-none nature of the action potential. This suggestion encountered mighty resistance and is now abandoned by most neuroscientists. I will explain why this is justified in Chapter 13.

Shortly after McCulloch and Pitts argued that brains perform digital computations, others countered that neural processes may be more similar to analog computations (Gerard 1951; see also Rubel 1985; Maley 2018). The evidence for the analog computational theory included neurotransmitters and hormones, which are released by neurons in degrees rather than in all-or-none fashion.

The claim that the brain is an analog computer is ambiguous between two interpretations. On the literal interpretation, "analog computer" is given Pour-El's (1974) precise meaning. The theory that the brain is an analog computer in this literal sense was never very popular. The primary reason is that although neural signals are a continuous function of time, they are also sequences of all-or-none signals.

On a looser interpretation, "analog computer" refers to a broader class of computing systems. For instance, Churchland and Sejnowski use the term "analog" so that containing continuous variables is sufficient for a system to count as analog: "The input to a neuron is analog (continuous values between 0 and 1)" (1992, 51). Under such a usage, even a slide rule counts as an analog computer. Sometimes, the notion of analog computation is simply left undefined, with the result that by "analog computer" one refers to some kind of presumably nondigital but otherwise unspecified computing system. The looser interpretation of "analog computer"—of which Churchland and Sejnowski's usage is one example—is not uncommon, but I find it misleading because it's easily confused with the stricter interpretation of analog computation in the sense of Pour-El (1974).

Not all arguments for specific versions of CTC are based on neuroscientific evidence. In fact, typical arguments for classical CTC make no reference to the brain. One argument for classical CTC begins with the observation that certain cognitive capacities exhibit productivity and systematicity—roughly speaking, they can generate an indefinite number of systematically related structured behaviors such as natural language sentences. The argument assumes that productive and systematic cognition requires the processing of language-like representations on the basis of their syntactic structure—that is, classical computation—and concludes that cognition is a kind of classical computation (Fodor and Pylyshyn 1988; Aizawa 2003). Like other arguments for specific versions of CTC, this argument has encountered resistance—either from people who deny that cognition involves the processing of language-like representations or from people who deny that the relevant processing of language-like representations requires a classical computational system.

Another argument for classical CTC is the argument from cognitive flexibility. Human beings can solve an indefinite number of problems and learn an indefinite range of behaviors. How do they do it? Consider digital computers—the most flexible of artifacts. Computers can do mathematical calculations, derive logical theorems, play board games, recognize objects, control robots, and much more. They can do this because they can execute different sets of instructions designed for different tasks. In Turing's terms, they approximate universal machines. Perhaps human beings are cognitively flexible because, like computers, they possess a general-purpose digital processor that executes different instructions for different tasks (Fodor 1968b; Newell 1990; Samuels 2010). The argument from cognitive flexibility is one of the most powerful, because there is no well worked-out alternative explanation of cognitive flexibility, at least for high-level cognitive skills such as problem solving (and according to some, even for lower level cognitive skills, cf. Gallistel and King 2009). Of course, the argument from cognitive flexibility is controversial too, primarily because classical CTC is difficult to reconcile with what is known about neural computation.

An alternative explanation of cognitive flexibility might involve the recruitment of different special-purpose processors, which may or may not be digital, for different tasks. Instead of a general-purpose digital processor, some neurocognitive systems might possess learning and attentional mechanisms that recruit appropriate cortical and subcortical areas as they are needed. When learning and solving arithmetic problems, the system recruits cortical areas involved in representing numbers. When learning and solving geometric problems, it recruits cortical areas involved in the perception and analysis of geometric shapes. When learning and solving puzzles that require various sensorimotor skills, it recruits appropriate cortical (and perhaps subcortical) areas involved in sensorimotor coordination. And so forth.

Language is one area of cognition that requires further explanation. Human language has a recursive grammar that can be used to represent instructions for any program in any programming language whatsoever. In this sense, linguistic cognition approximates computational universality all by itself many times over. Explaining the cognitive flexibility of linguistic cognition is of a piece with explaining linguistic cognition itself. Explaining linguistic cognition is especially difficult, among other reasons, because it has no true analogue in nonhuman species. Because of this, linguistic cognition is not open to the kind of invasive neurophysiological investigations that can be conducted on nonhuman species.

Human beings are wired to acquire their first natural language during a critical period that ends around age nine. How, exactly, humans acquire language is too controversial for me to discuss here. Suffice it to say that language acquisition involves a complex interplay between innate structures that are probably specialized for language acquisition and linguistic stimuli, and that there is no hard evidence that a general-purpose digital processor is involved in language acquisition. Human beings do not acquire their first language by receiving instructions for processing that language, as a general-purpose processor would do, nor do they appear to have such instructions built in.

As a result of acquiring language, human beings do acquire the capacity to process recursive structures and to store and process arbitrary linguistic instructions. In this sense, linguistic cognition alone does approximate a small fragment of the ability of a general-purpose digital processor. I say a small fragment because most human beings can memorize only relatively small bodies of linguistic instructions, and they can execute them proficiently only on relatively short inputs (unless, of course, they rely on external aids such as paper and pencil). At any rate, positing a limited approximation of computational universality within linguistic cognition, as I have just done, falls short of positing general-purpose digital processors as the basis for all cognition, as the argument from cognitive flexibility does.

My speculative explanation of cognitive flexibility is just a promissory note, whose details remain to be worked out. Nevertheless, a more fleshed out version of this alternative hypothesis might explain cognitive flexibility without relying on a general-purpose digital processor. It might even be true.

In conclusion, it is very plausible that neurocognitive processes are computations in the generic sense, but it is difficult to determine which specific kinds of computation—classical, digital but nonclassical, analog, or sui generis—are involved in which cognitive capacities.

9.5 Classicism, Connectionism, and Computational Neuroscience Revisited

The distinctions introduced in the preceding sections allow us to shed light on longstanding debates between classicism, connectionism, and computational neuroscience.

By the 1970s, McCulloch and Pitts were vaguely remembered, but their theory was often ignored or misunderstood. The dominant computational paradigm in cognitive science was classical or "symbolic" AI, aimed at writing computer programs that simulate intelligent behavior without much concern for how brains work (e.g., Newell and Simon 1976). It was commonly assumed that digital CTC is committed to classicism, that is, the idea that cognition is the manipulation of language-like representations. Language-like representations have constituent structure whose computational manipulation, according to classicism, explains productive and systematic psychological processes. On this view, cognition consists of performing computations on sentences with a logico-syntactic structure akin to that of natural languages, but written in the language of thought (Harman 1973; Fodor 1975).

It was also assumed that, given digital CTC, explaining cognition is independent of explaining neural activity, in the sense that figuring out how the brain works tells us little or nothing about how cognition works: neural (or implementational, or mechanistic) explanations and computational explanations are at two different "levels" (e.g., Fodor 1975; Marr 1982).

During the 1980s, connectionism re-emerged as an influential approach to psychology. Most connectionists deny that explaining cognition requires positing a language of thought and affirm that a theory of cognition should be at least "inspired" by the way the brain works (Rumelhart and McClelland 1986).

The resulting debate between classicists and connectionists (e.g., Fodor and Pylyshyn 1988; Smolensky 1988) is somewhat confusing. Different authors employ different notions of computation, which vary in their degree of both precision and inclusiveness. Specifically, some authors use the term "computation" only for classical computation—i.e., at a minimum, algorithmic digital computation over language-like structures—and conclude that (nonclassicist) connectionism falls outside CTC. By contrast, other authors use a broader notion of computation along the lines of what I called computation in the generic sense, thus including connectionism within CTC (see Piccinini 2015; Chapter 13, this volume for more details). But even after we factor out differences in notions of computation, further confusions lie in the wings.

Classicism and connectionism are often described as being at odds with one another, because classicism is committed to classical computation (the idea that the vehicles of digital computation are language-like structures) and—it is assumed—to autonomy from neuroscience, two theses flatly denied by many prominent connectionists. But many connectionists also model and explain cognition using neural networks that perform computations defined over strings of digits, so they should be counted among the digital computationalists (Bechtel and Abrahamsen 2002; Feldman and Ballard 1982; Hopfield 1982; Rumelhart and McClelland 1986; Smolensky and Legendre 2006).

Furthermore, for many years both classicists and connectionists tended to ignore computational neuroscientists, who in turn tended to ignore both classicism and connectionism. Computational neuroscientists often operate with their own mathematical tools without committing themselves to a particular notion of computation (O'Reilly and Munakata 2000; Dayan and Abbott 2001; Eliasmith and Anderson 2003; Izhikevich 2007; Ermentrout and Terman 2010). To make matters worse, some connectionists and computational neuroscientists reject CTC—they maintain that their neural networks, while explaining behavior, do not perform computations at all (Edelman 1992; Freeman 2001; Globus 1992; Horgan and Tienson 1996; Perkel 1990; Spivey 2007).

In addition, the very origin of digital CTC calls into question the commitment to autonomy from neuroscience. McCulloch and Pitts initially introduced digital CTC as a theory of the brain, and some form of CTC or other is now a working assumption of many neuroscientists.

To clarify this debate, we need two separate distinctions. One is the distinction between digital CTC ("cognition is digital computation") and its denial ("cognition is something other than digital computation"). The other is the distinction between classicism ("cognition is algorithmic digital computation over language-like structures") and neural-network approaches ("cognition is computation—digital or not—by means of neural networks").

We then have two versions of digital CTC—the classical one ("cognition is algorithmic digital computation over language-like structures") and the neural network one ("cognition is digital computation by neural networks")—standing opposite to the denial of digital computationalism ("cognition is a kind of neural network computation different from digital computation"). This may be enough to accommodate most views in the debate. But it still doesn't do justice to the relationship between classicism and neural networks.

A further wrinkle derives from the ambiguity of the term "connectionism." In its original sense, connectionism says that behavior is explained by the changing "connections" between stimuli and responses, which are biologically mediated by changing connections between neurons (Thorndike 1932; Hebb 1949). This original connectionism is related to behaviorist associationism, according to which behavior is explained by the association between stimuli and responses. Associationist connectionism adds a biological mechanism to explain the associations: the mechanism of changing connections between neurons.

But contemporary connectionism is a more general thesis than associationist connectionism. In its most general form, contemporary connectionism, like computational neuroscience, simply says that cognition is explained by neural network activity. But this is a truism—or at least it should be. The brain is the organ of cognition, the cells that perform cognitive functions are (mostly) neurons, and neurons perform their cognitive labor by organizing themselves in networks. In Section 9.3 we saw that neurocognitive processes are computations at least in the generic sense. And even digital computers are just one special kind of neural network. So even classicists, whose theory is most closely inspired by digital computers, are committed to connectionism in its most general sense.

The relationship between connectionist and neurocomputational approaches on one hand and associationism on the other is more complex than many suppose. We should distinguish between strong and weak associationism. Strong associationism maintains that association is the only legitimate explanatory construct in a theory of cognition (cf. Fodor 1983, 27). Weak associationism maintains that association is a legitimate explanatory construct along with others such as the innate structure of neural systems.

To be sure, some connectionists do profess strong associationism (e.g., Rosenblatt 1958, 387). But that is beside the point, because connectionism per se is consistent with weak associationism or even the complete rejection of associationism. Some connectionist models do not rely on association at all—a prominent example being the work of McCulloch and Pitts (1943). And weak associationism is consistent with many theories of cognition including classicism. A vivid illustration is Alan Turing's early proposal to train associative neural networks to acquire the architectural structure of a universal computing machine (Turing 1948; Copeland and Proudfood 1996). In Turing's proposal, association may explain how a network acquires the capacity for universal computation (or an approximation thereof), while the capacity for universal computation may explain any number of other cognitive phenomena.

Although many of today's connectionists and computational neuroscientists emphasize the explanatory role of association, many of them also combine association with other explanatory constructs, as per weak associationism (cf. Smolensky and Legendre 2006, 479; Trehub 1991, 243–5; Marcus 2001, xii, 30). What remains to be determined is which neural networks, organized in what way, actually explain cognition and which role association and other explanatory constructs should play in a theory of cognition (cf. Dacey 2016, 2019).

Yet another source of confusion is that classicism, connectionism, and computational neuroscience tend to offer explanations at different mechanistic levels. Specifically, classicists tend to offer explanations in terms of rules and representations, without detailing the neural states that implement the representations and components that process them; connectionists tend to offer explanations in terms of highly abstract neural networks, which do not necessarily represent networks of actual neurons (in fact, a processing element in a connectionist network may represent an entire brain area rather than an actual neuron); finally, computational neuroscientists tend to offer explanations in terms of mathematical models that represent concrete neural networks based on neurophysiological evidence. Explanations at different mechanistic levels are not necessarily in conflict with each other, but they do need to be integrated to describe a multi-level mechanism. Integrating explanations at different levels into a unified multi-level mechanistic picture may require revisions in the original explanations themselves.

Different parties in the dispute between classicism, connectionism, and computational neuroscience may offer different accounts of how the different levels relate to one another. As I pointed out, one traditional view is that computational explanations are not even mechanistic. Instead, computational and mechanistic explanations belong at two independent levels (e.g., Fodor 1975; Marr 1982). This suggests a division of labor: computations are the domain of psychologists, while the implementing neural mechanisms are the business of neuroscientists. According to this picture, the role of connectionists and computational neuroscientists is to discover how neural mechanisms implement the computations postulated by (classicist) psychologists.

This traditional view is untenable for several reasons I have already emphasized in previous chapters: (i) both psychologists and neuroscientists offer computational explanations, (ii) both computational explanations and mechanistic explanations come in many levels of organization, (iii) far from being independent, different mechanistic levels directly constrain one another, (iv) computational explanation is actually a kind of mechanistic explanation.

One alternative to the traditional view is that connectionist or neurocomputational explanations simply replace classicist ones. Perhaps some connectionist computations approximate classical ones (Smolensky 1988), or perhaps not. In any case, some authors maintain that classicist constructs, such as program execution, play no causal role in cognition and will be eliminated from cognitive science (e.g., Churchland 2007).

A better account of the relation between explanations at different mechanistic levels is the framework I have defended here. According to the mechanistic account of computation, computational explanation is just one type of mechanistic explanation. Mechanistic explanations provide components with such properties and organization that they produce the explanandum. Computational explanation, then, is explanation in terms of computing mechanisms and components—mechanisms and components that perform computations. (Computation, in turn, is the manipulation of medium-independent vehicles in accordance with a rule.) Mechanistic explanations come with many levels of organization, where each level is constituted by its components and the way they are organized. If a mechanistic level produces its behavior by the action of computing components, it counts as a computational level. Thus, a mechanism may contain zero, one, or many computational levels depending on what components it has and what they do. Which types of computation are performed at each level is an open empirical question to be answered by studying cognition and the nervous system at all levels of organization.

9.6 A Viable Empirical Hypothesis

The Computational Theory of Cognition is here to stay. We have good reasons to conclude that cognitive capacities have computational explanations, at least in the generic sense that they can be explained in terms of the functional processing of medium-independent vehicles. Moreover, everyone is (or should be) a computational neuroscientist, at least in the general sense of embracing neural computation.

Much work remains to be done to characterize the specific computations on which cognition depends, what role association plays in neurocognitive systems, whether—and to what extent—a satisfactory explanation of cognition requires classical computational mechanisms as opposed to nonclassical digital computation; and whether cognition involves digital computation, analog computation, or computation of a third kind (Figure 9.1). It may turn out that one computational

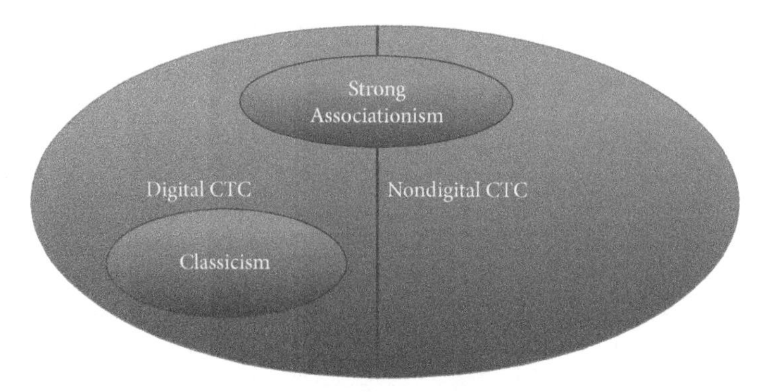

Figure 9.1 Some prominent forms of CTC and their relations.

theory is right about all of cognition or it may be that different cognitive capacities are explained by different kinds of computation.

In Chapter 13, I will argue that in the general case, neural computation is neither digital or nor analog—it is sui generis. We have a few other tasks before we get there. I have mentioned several times that some computational theorists of cognition attempt to derive CTC from the Church–Turing thesis. The next chapter dispels this fallacy.

10
The Church–Turing Fallacy

10.1 The Computational Theory of Cognition and the Church-Turing Thesis

The Computational Theory of Cognition (CTC) is the empirical hypothesis that cognitive capacities are explained by inner computations (Chapter 9). In biological organisms, inner computations are realized by neural processes; to wit, they are neural computations. Typically, CTC theorists also maintain that neural computations are Turing-computable, that is, computable by Turing machines (TMs). This is entailed by the kind of digital CTC initially defended by McCulloch and Pitts (Chapter 5).

The Church–Turing thesis (CT) says that, if a function is computable in the intuitive sense, then it is computable by TMs (or, equivalently, it is Turing-computable) (Church 1936, Turing 1936–7). CT entails that TMs, and any formalism equivalent to TMs, capture the intuitive notion of computation.[1]

Suppose that, as CTC maintains, neural activity—or at least the neural activity that explains cognition—is computation, and suppose that the functions computed by neural mechanisms are computable in the intuitive sense. Then, by CT, for any function computed by a neural mechanism, there are TMs that compute the same function. This is a legitimate argument for McCulloch-Pitts-style digital CTC, according to which neural computations are digital and Turing-computable, from a form of CTC according to which neural processes are computations in the intuitive sense, via CT. It remains to be seen whether neural processes are computations in the intuitive sense and whether we should believe CT.

Should we believe CT? The initial proponents of CT, and most of CT's supporters, appeal to a number of considerations. The main considerations are that there are no known counterexamples, that various attempts at formalizing the notion of computation have yielded computationally equivalent formalisms, and that the notion of TM seems to capture well the intuitive notion of computation (Kleene 1952). Due to these considerations, most logicians and mathematicians believe CT to be true. But some authors have attempted to go beyond these

[1] I am appealing to the standard understanding of CT, according to which CT provides a correct analysis of the intuitive notion of computability employed by mathematicians. Useful references on CT include Kleene 1952; Copeland 2002; Folina 1998; and Sieg 2001.

Neurocognitive Mechanisms: Explaining Biological Cognition. Gualtiero Piccinini, Oxford University Press (2020).

DOI: 10.1093/oso/9780198866282.001.0001

considerations. They have attempted to establish a strong enough version of CTC on independent grounds and use it to support CT.

The first to do so were McCulloch and Pitts (1943). Their CTC entails that all neural activity is digital computation and that every cognitive activity is explained by some neural computation. The neural networks defined by McCulloch and Pitts are computationally less powerful than TMs but, if supplemented with a tape and the means to act on it, then they are computationally equivalent to TMs. Since their neural networks are no more powerful than TMs, McCulloch and Pitts argue that their theory offers a "psychological justification" of CT (McCulloch and Pitts 1943, 35).

McCulloch and Pitts's argument for CT rests on two premises: (i) that the intuitive notion of computable function pertains to what can be computed by neural networks directly, and (ii) that neural networks perform digital computations. Premise (i) is debatable but, more importantly, premise (ii) is a consequence of McCulloch and Pitts's theory, which was not empirically plausible even when it was formulated (cf. Chapters 5 and 9). So, McCulloch and Pitts's justification of CT is dubious. Still, this is no reason to abandon CT, which remains well supported by the usual considerations.

McCulloch and Pitts's result—viz., that the neural networks defined within their theory, when supplemented with tapes, are computationally equivalent to TMs—was given a positive spin by von Neumann. He reversed McCulloch and Pitts's order of justification, appealing to CT in order to support digital CTC:

> The McCulloch-Pitts result . . . proves that anything that can be exhaustively and unambiguously described, anything that can be completely and unambiguously put into words, is ipso facto realizable by a suitable finite neural network. Since the converse statement is obvious, we can therefore say that there is no difference between the possibility of describing a real or imagined mode of behavior completely and unambiguously in words, and the possibility of realizing it by a finite formal neural network. The two concepts are coextensive. A difficulty of principle embodying any mode of behavior in such a network can exist only if we are also unable to describe that behavior completely. (von Neumann 1951, 22–3)

Von Neumann's terminology, "describing behavior completely and unambiguously," is regrettably ambiguous, which makes it hard to interpret von Neumann's assertions. Presumably, by "describing behavior completely and unambiguously," von Neumann means providing a digital computational description of the behavior—what we now call a computational model. Von Neumann might be saying that computational models—including computational models of cognition—can be run on McCulloch-Pitts networks. This would be unobjectionable.

Many interpreted this kind of statement more liberally. There are people who expect that cognition, and perhaps all natural phenomena, can be described completely and unambiguously via digital computation—that is, that a digital computational description exhausts everything there is to say about cognition or other physical processes. In addition, these people assume—reasonably enough—that computational models are Turing-computable. On this more liberal interpretation, then, von Neumann is saying or implying that cognition—and perhaps every physical process—is a (Turing-computable) digital computation, which in turn can be realized in a McCulloch–Pitts neural network. Thus, digital CTC follows from the assumption that cognition can be described computationally combined with CT and McCulloch and Pitts's "result."

Statements like von Neumann's make it sound like digital CTC follows from CT. Because of this, many have been persuaded that digital CTC holds a priori, or almost a priori, and does not need empirical support from neuroscience. Nothing could be further from the truth. A thorough assessment of CTC requires that we disabuse ourselves of this notion.

Interpreted liberally, von Neumann's statement may be the earliest example of what Jack Copeland calls the Church–Turing fallacy. This is the supposition that a version of digital CTC that is committed to cognitive functions being Turing-computable (or, more weakly, the view that cognitive capacities can be simulated by TMs) follows from some reasonable assumptions conjoined with CT or some result established by Church and Turing (Copeland 1998, 2000; cf. also Kearns 1997).[2] CT pertains to functions that are computable in the intuitive sense employed in formal logic. In order to show that something falls under CT, it must first be shown that it is computable in that sense. CT, per se, does nothing to establish that something is computable. Because of this, supposing that CT entails CTC is a fallacy. (The question of what can be *simulated*, in the sense of approximated, by TMs has nothing to do with CT; I address it in Section 10.3.)

Yet, as Copeland shows, the Church–Turing fallacy is widespread in the cognitive science literature. Here is an example:

> [A] standard digital computer, given only the right program, a large enough memory and sufficient time, can compute any rule-governed input-output function. That is, it can display any systematic pattern of responses to the environment whatsoever. (Churchland and Churchland 1990, 26)[3]

There are many problems with this passage; I will only discuss the most obvious one. What the Churchlands *mean* to say appears to be that digital computers are universal in Turing's sense; namely, computers can compute any Turing-computable

[2] The tendency to commit the Church-Turing fallacy is probably promoted, at least in part, by the common conflation between CT and Turing's discovery that there are universal TMs, namely, TMs that compute any function computable by ordinary TMs (e.g., Wolfram 2002, 1125).

[3] Cf. also Guttenplan 1994, 595.

function (until they run out of memory and time). But by using the ambiguous phrases "rule-governed functions" and "systematic patterns of responses," the Churchlands say something much stronger—and radically false.

If we accept CT, we know from Turing's results and from standard computability theory that only countably many functions defined over a denumerable domain, out of uncountably many, are computable by standard digital computers. Yet in computability theory there is a clear sense in which many Turing-*un*computable functions defined over a denumerable domain, such as the halting function, are "rule-governed" and "systematic." For example, the halting function is the function whose value is 1 if and only if a given Turing machine halts on a given input, and whose value is 0 otherwise. The halting function is defined by the rule I just gave, and its pattern of responses is as systematic as any. In this sense (among others), it is far from true that computers can compute all "rule-governed," "systematic" functions. The Churchlands's statement is a straightforward example of the kind of language that leads to the Church–Turing fallacy (cf. Copeland 2000 and 2002).

Copeland contrasts standard digital CTC, which posits that cognitive functions are Turing-computable, with what might be called hyper-digital-CTC. According to hyper-digital-CTC, neurocognitive systems compute Turing-*un*computable functions. I wish to draw a more general contrast. I wish to contrast CTC with the view that nervous systems do *not* compute Turing-computable functions. If neurocognitive systems do not compute Turing-computable functions, it may be for one of three reasons:

(1) Hyper-digital-CTC: neurocognitive systems compute non-Turing-computable functions defined over denumerable domains. This is still a form of digital CTC, albeit of a nonstandard form.[4]

(2) Nondigital CTC: neurocognitive systems compute functions that are not defined over denumerable domains. This is also a form of CTC and I will defend it in Chapter 13. If nondigital CTC is correct, one follow-up question is whether neural computations can be simulated by Turing machines and other classical formalisms, and how efficiently they can be simulated.

(3) Anticomputationalism: neurocognitive systems do not compute anything at all—neurocognitive activity is something other than computation. Anticomputationalism rejects CTC. I address objections to CTC in Chapter 11.

The view that neurocognitive systems do not compute Turing-computable functions is still consistent with mechanistic functionalism (Chapter 4) but is more encompassing than Copeland's hyper-digital-CTC, because it includes Copeland's hyper-digital-CTC, nondigital CTC, and anticomputationalism. Perhaps neural

[4] In addition to Copeland, this appears to be the view defended by Penrose (1989, 1994).

systems are simply not computing mechanisms but some other kind of mechanism. Or perhaps neural systems compute functions that are not defined over denumerable domains. As we shall see in Chapter 13, the latter view is supported by contemporary neurophysiology and fits well with contemporary theoretical and computational neuroscience, where much of the most rigorous and sophisticated work posits computations that are markedly different from digital computations.

In order to properly eradicate the Church–Turing fallacy, however, it is not enough to expose it as fallacious. Several authors have attempted to go beyond von Neumann's quick remark and offer explicit arguments for digital CTC based on CT. The rest of this chapter will evaluate these arguments in more detail and show why they fail.

In order to assess CT's relevance to CTC, it is convenient to formulate a digital version of CTC in terms of Turing-computable functions:

(C) The functions from neural inputs to neural outputs are Turing-computable.

Given this version of CTC, CT is potentially relevant to it, for CT states that functions belonging to a certain class are Turing-computable. If the functions whose values are generated by neural systems belong to the relevant class of functions, then, by CT, (C) follows. There are three main ways in which serious arguments from CT to CTC have been run. Although they are old arguments, they have reappeared in the literature many times. I address each in turn.

10.2 The Physical Church–Turing Thesis

An important and uncontroversial result of philosophical work on CT during the last several decades is the distinction between Mathematical CT, which pertains to functions that are effectively calculable in a sense that is intuitive to formal logicians, and Physical CT, which pertains to functions whose values are generated by physical systems. Whether or not they are aware of the distinction between Mathematical and Physical CT, several authors have used Physical CT to argue for (C).

Physical CT pertains to the computational limitations of physical systems. Since nervous systems are physical systems, it follows from Physical CT that the functions physically computed by brains are Turing-computable. So Physical CT appears to entail (C).[5] In evaluating this argument from Physical CT, we

[5] Cf.:

> Human beings are products of nature. They are finite systems whose behavioral responses to environmental stimuli are produced by the mechanical operation of natural forces. Thus, according to Church's thesis, human behavior ought to be simulable by Turing machine (McGee 1991, 118).
>
> [W]e have good reason to believe that the laws of physics are computable, so that we at least ought to be able to *simulate* human behavior computationally (Chalmers 1996, 313).

should distinguish between two importantly different—though seldom distinguished—versions of Physical CT. Accordingly, we need to examine two versions of the argument from Physical CT.

10.2.1 From Modest Physical CT

A first version of Physical CT pertains to what functions can be *computed* by a physical system that manipulates inputs and outputs defined over a denumerable domain—that is, digital computing mechanisms. Modest Physical CT does not apply to all physical processes, but only to processes that are digital computations. It says that *the functions whose values are generated by digital computing mechanisms are Turing-computable.* In other words, if a mechanism performs digital computations, then Modest Physical CT entails that that mechanism will compute functions that are Turing-computable.

Modest Physical CT has been challenged. It is true if and only if genuine hypercomputers—machines that compute functions that are not Turing-computable—are physically impossible, and whether genuine hypercomputers are physically possible remains a somewhat open question. Hypercomputers are an interesting theoretical possibility, and they are useful in discussions in the foundations of physics. Nevertheless, there is no solid evidence that genuine hypercomputers can be built and used by humans. Insofar as it concerns computer scientists, Modest Physical CT is plausible (Piccinini 2015, chap. 16).

The version of Modest Physical CT that concerns computer scientists is also the one that concerns cognitive neuroscientists. For cognitive neuroscientists are interested in neurocognitive mechanisms, which are relatively medium-sized physical systems confined within medium-sized spatiotemporal regions. There is little reason to believe that neural mechanisms have access to exotic physical resources, such as Malament-Hogarth spacetimes, that are exploited in putative designs for hypercomputers.

If Modest Physical CT applies to brains, the resulting version of CTC is not trivial. For Modest Physical CT applies to physical systems that compute functions over denumerable domains, hence it entails that those systems are genuine digital

The use of the term "simulate" in these quotes may suggest that these authors are arguing for something weaker than digital CTC. Perhaps they are suggesting that digital computations can generate approximations of cognitive capacities, without explaining them. But the term "computational simulation" is used to mean either computational approximation (as when a weather forecasting program simulates the weather) or computational replication (as when a universal TM simulates another TM). Unfortunately, these authors do not state explicitly which sense of "simulation" they are employing. Since they appeal to CT, however, and since CT is only relevant to the stronger notion of simulation (see Section 10.3 for more on computational approximation), they commit themselves to the stronger notion of simulation. At any rate, the context of their remarks makes clear that these authors do intend to argue for digital CTC.

computing mechanisms, whose activities are digital computations—as opposed to noncomputing mechanisms, whose activities are not computations, and nondigital computing mechanisms, whose activities are nondigital computations. In this respect, if we could conclude that Modest Physical CT applies to brains, we would learn something substantive about them.

But Modest Physical CT says nothing about whether any particular physical system is a digital computing mechanism. It leaves open whether the solar system, the weather, or your brain is a digital computing mechanism. Whether nervous systems or any other physical systems are digital computing mechanisms must be established by means other than Modest Physical CT. If we can establish that nervous systems are digital computing mechanisms by other means, then Modest Physical CT applies to them and, if Modest Physical CT is true, then the functions nervous systems compute are Turing-computable. What Modest Physical CT establishes (if true) is only that if brains are digital computing mechanisms, then they are not hypercomputers.

10.2.2 From Bold Physical CT

A second version of Physical CT pertains to all physical systems, whether or not they perform computations. Bold Physical CT says that any physical process is Turing-computable. In other words, *any function whose values are generated (by digital computation or any other means) by physical systems are Turing-computable.* Hence, assuming that brains are physical, Bold Physical CT does entail that the functions whose values are generated by brains are Turing-computable, which establishes (C). But this is a Pyrrhic victory.

To begin with, Bold Physical CT is falsified by any genuinely random process. For, as Turing knew well, genuinely random processes are not Turing-computable (cf. Piccinini 2003b).[6] Even if we restrict it to deterministic systems, Bold Physical CT is difficult to make precise. The main reason is that the mathematical functions that are normally used to describe physical systems are functions of real (continuous) variables, whose domain and range include uncountably many values, whereas Turing-computable functions are functions of discrete variables, whose domain and range include only countably many values. Because of this, functions of real variables cannot be directly mapped onto Turing-computable functions.

[6] My point is simply that we cannot pick a Turing machine (or any other process) and ask it to generate the same outputs as a genuine random process unless we already know all the outputs of the random process. (Almost all sequences generated by genuinely random processes are not Turing-computable.)

There are several ways in which computability theory has been extended to functions of real variables. One such extension defines primitive computational operations that manipulate real-valued quantities instead of the strings of symbols of classical computability theory (Blum et al. 1998). Another proposal maintains the usual computations over strings of symbols but allows computations to rely on the exact values of real-valued constants (Siegelmann 1999). We may call the functions that are computable under these extensions of computability theory *real-computable* functions. Under these extensions, Bold Physical CT may be reformulated as stating that a function is real-computable if and only if it is Turing-computable. Unfortunately, this version of Bold Physical CT is far from true. Under either of the above extensions of computability theory, all functions from strings to strings—including all those that are not Turing-computable—are real-computable. So, this formulation of Bold Physical CT is false for mathematical reasons. The resulting argument from Bold Physical CT to (C) is valid but unsound.[7]

Even if there were a true version of Bold Physical CT that could be used to entail (C) for some significant class of systems, however, this would yield only cold comfort to the digital CTC theorist. The main price of using Bold Physical CT to support digital CTC is that digital CTC is thereby trivialized. The original motivation for CTC is that the notion of computation can be used to *distinguish* mental processes from other processes—to find a mechanistic explanation that is specific to cognitive capacities (e.g., cf. Fodor 1998). Bold Physical CT cannot do this, because it applies indifferently to nervous systems as well as other physical systems by virtue of their being physical.

Any view of neural processes derived from Bold Physical CT is not a genuine form of CTC, according to which cognitive capacities are *explained* by neural computations as opposed to some noncomputational process. In theorizing about cognitive capacities, we are looking for mechanistic explanations that are specific to them. If digital computation is used in a sense that applies to any physical process, then it cannot be the basis for an explanation that applies specifically to cognitive capacities. So, anyone who wishes to claim that brains are digital computing mechanisms in a sense that is specifically suited to explaining cognitive capacities, even if she believes Bold Physical CT, must look for a more stringent version of digital CTC and support it independently of Bold Physical CT.

To summarize, neither version of Physical CT helps the supporter of CTC, digital or otherwise.

[7] Another way to link computation to real-valued processes asks whether, when a deterministic system's initial conditions are defined by a computable real number (i.e., a real number whose expansion can be printed out by a TM), the system's dynamical evolution always leads to states defined by computable real numbers. This question is quite removed from the usual concerns of CTC. At any rate, there are field equations for which the answer is negative (Pour-El and Richards 1989).

10.3 Between the Modest and the Bold Physical Church–Turing Thesis

In the study of physical systems and their mathematical descriptions, computability is relevant in several ways that do not fit comfortably within discussions of Physical CT. I will briefly review some of them, which I think are more relevant to CTC, and more fruitful to investigate, than Bold Physical CT.

10.3.1 Mathematical Tractability

It may be useful to remind ourselves of the main reason why scientists, either in cognitive neuroscience or in any other sciences, resort to computational descriptions. Generally speaking, the need for computational descriptions in science has nothing to do with CTC or the attempt to explain phenomena computationally. It has to do with the analytical intractability of most mathematical dynamical descriptions.

Given a system of equations describing a physical system, an analytic solution is a formula such that, given any initial condition of the system and any subsequent time t, the formula yields the state of the system at time t. The question of which systems of equations can be solved analytically is not directly relevant to computability, but its answer leads to the important issue of computational approximation of physical systems, which is relevant to computability. It is well known that most systems of equations have no analytic solutions. In particular, the majority of nonlinear systems, which make up the majority of systems of equations, are not solvable analytically.[8]

In order to study systems that are not analytically solvable, a geometrical, qualitative approach has been developed by mathematicians.[9] This approach allows mathematicians to identify important qualitative features of a system's state space (e.g., its fixed points) without solving the system analytically. Unfortunately,

[8] Sometimes, the question of whether a system of equations is analytically solvable gets confused with whether a function is computable:

> Most dynamical systems found in nature cannot be characterized by equations that specify a computable function. Even three bodies moving in Newtonian space do not satisfy this assumption. It is very much an open question whether the processes in the brain that subserve cognition can be characterized as the computation of a computable function (Cummins 2000, 130).

This brief argument is a nice counterpoint to the arguments by McGee and Chalmers cited above. Like Chalmers and McGee, Cummins is mistakenly assuming that the notion of computability applies directly to the functions described by ordinary dynamical descriptions, such as differential equations. Unlike them, Cummins is assuming that whether a system of equations is solvable is the same question as whether a function is computable. What Cummins should have said is that most dynamical systems are characterized by equations that are not analytically solvable.

[9] For an introduction, see Strogatz 2015.

the geometrical approach is suitable only for relatively simple systems, in which the number of state variables can be reduced to (at most) three, one per axis of a three-dimensional space. This limitation is mainly due to the fact that (ordinary) human beings are unable to visualize a space with more than three-dimensions, and hence to apply this geometrical approach to systems whose state variables cannot be reduced to less than four. Nevertheless, the development of these geometrical techniques remains a fertile area of mathematical investigation. Overcoming the limitations of current methods for studying complex nonlinear dynamical systems, either by extending existing methods or by inventing new methods, is a research project of many mathematicians.

There is yet another way to tackle dynamical systems, whether solvable or unsolvable analytically, simple or complex. It is the use of computational methods for approximating the dynamical evolution of physical systems. The modern study of dynamical systems has exploded over the last half-century, to a large extent, thanks to the advent of digital computers. This is because computers, by offering larger and larger amounts of processing speed and memory, allow scientists to develop methods of approximation for systems of equations that are not analytically solvable, so as to study their behavior based on those approximations. Computational approximation, which is one of the crucial tools in contemporary science, is what we now turn to.

10.3.2 Computational Approximation

There are at least two importantly different ways to approximate the behavior of a system computationally. One relies on the equations describing the dynamics of the system and on numerical methods for finding successive states of the system from those equations. The other does not rely on equations but represents the dynamics of the physical system itself as discrete. I will now briefly discuss these two methods.

Given a system of equations describing a physical system, whether or not the system is analytically solvable, it may be possible to develop methods for computing approximations of the behavior of the system. Working out specific numerical methods for specific sets of equations and showing that the resulting approximations are accurate within certain error bounds is another fertile area of mathematical investigation. These numerical methods, in turn, are behind the now widespread use of most computational models in science. These models are computer programs that exploit appropriate numerical methods to compute representations of subsequent states of a system on the basis of both the equations representing the system's dynamical evolution and data representing the system's initial conditions. Models of this kind can be constructed for any system whose behavior is described by known systems of equations that can be approximated by known numerical methods. This kind of computational approximation is the most widespread form of contemporary scientific modeling.[10]

A different method of computational approximation relies on a computational formalism called cellular automata. Cellular automata are lattices of cells, each of which can take a finite number of discrete states and changes state in discrete time. At any given time step, the state of each cell is updated based on the state of its neighboring cells at that time. Different updating rules give rise to different cellular automata. In modeling a physical system using cellular automata, the system is spatially discretized in the sense that distinct spatial regions of the system are represented by distinct cells of a cellular automaton, and the dynamics of the system is temporally discretized in the sense that the state changes of the system's spatial regions are represented by the updating of the cells' states. The pattern generated by the cellular automaton can then be compared with the observations of subsequent states of the system, so as to evaluate the accuracy of the approximation.[11]

The popularity and usefulness of computational approximations of physical systems, not only in physics but in many other sciences, may have been a motivating factor behind Bold Physical CT. Some authors state forms of CT according to which every physical system can be "simulated," by which they appear to mean computationally approximated in the present sense, by TMs.[12] But the question of whether every physical system can be computationally approximated is only superficially similar to Bold Physical CT.

The importance of computational approximation is not that it embodies some thesis about physical systems and how to explain their behavior, but that it is the most flexible and powerful tool ever created for scientific modeling. An approximation may be closer or farther from what it approximates. The way in which and the degree to which an approximation accurately represents the system it models vary greatly depending on the nature of the target system as well as the goals and resources of the investigators who are building the model.

[10] For more on computational models in science, see Rohrlich 1990; Humphreys 2004; Winsberg 2010.

[11] For more on cellular automata as a modeling tool, see Rohrlich 1990; Hughes 1999.

[12] This is one possible interpretation of von Neumann's remark, cited above. Here are is another example: "theories of cognition are formulated in terms of processes that could be emulated by programs running on a digital computer" (Scott 1997, 68); see also Baum 2004, 47. The reason given by Scott for his view is that *all* scientific theories "must be the embodiment of a Turing computable function" (Scott 1997, 63), and his reason for that is that due to CT together with the requirement that scientific theories be publicly understandable, "any genuinely scientific theory must embody an effective procedure that will allow its intended audience to determine what it entails about the range of situations to which it applies" (Scott 1997, 66). Scott's argument is a non sequitur. For even if, for the sake of the argument, we accept the otherwise dubious premise that any scientific theory must embody an effective procedure for deriving its logical consequences, it doesn't follow that the processes *described* by the theory either are computational or may be simulated computationally.

If we allow computational approximations to be arbitrarily distant from the dynamical evolution of the system being approximated, then the thesis that every physical system can be computationally approximated becomes trivially true. If we are stricter about which approximations are acceptable, then that same thesis becomes nontrivial but much harder to evaluate. Formulating stricter criteria for acceptable approximations and evaluating which systems can be approximated to what degree of precision is a difficult question, which would be worthy of systematic investigation. Here, I can only make a few obvious points.

First, strictly speaking, unpredictable (e.g., nondeterministic) systems cannot be computationally approximated in any strict way. A computational approximation can only indicate the possible dynamical evolutions of such systems, without indicating which path will be followed in any particular case.

Second, if there are any (deterministic or nondeterministic) physical systems whose state transitions are not Turing-computable—e.g., if genuine hypercomputers are possible—then there is a strict sense in which those systems cannot be computationally approximated (by current computational methods).

Finally, as soon as the state-variables of a system are more than two and they interact nonlinearly in a sufficiently complex way, the system may exhibit chaos (in the mathematical sense). As is well known, chaotic systems are so sensitive to initial conditions that their dynamical evolution can only be computationally approximated for a relatively short time before diverging exponentially from the observed behavior of the system.

In conclusion, the extent to which physical systems can be computationally approximated depends both on the properties of physical systems and their mathematical descriptions, and on the criteria that are adopted for adequate approximation. The same computational model may count as producing adequate approximations for some modeling purposes but not for others. At any rate, on any nontrivial criteria for adequate approximation, it is far from true that every physical system can be computationally approximated. Having thus clarified the relation between computation and physical systems, we can go back to arguments from CT to CTC that do *not* rely on *Physical* CT. From now on, unless otherwise noted, by "CT" I mean Mathematical CT (as opposed to Physical CT).

10.4 Cognitive Processes as Implementing Effective Procedures

A straightforward way of arguing from CT to (C) would be to show that cognitive processes are effective computations in the sense analyzed by Church and Turing. If cognitive processes are effective in this sense, then the functions whose values are generated by brains when they exhibit cognitive capacities fall under CT properly so called. Then, by CT itself, (C) follows.[13] This section evaluates the thesis that cognitive processes implement effective procedures.

Several authors believe that the claim that cognitive processes implement effective procedures is an empirical hypothesis, to be supported on empirical grounds such as the successes of AI, psychology, or linguistics. This view does not concern us here. Here, I only discuss arguments that attempt to establish that cognitive processes implement effective procedures without waiting for the empirical science of cognition to run its course. The most explicit of such arguments is due to Judson Webb (1980, 236ff.; a similar argument is in Baum 2004, 33–47).

Webb introduces his argument as if it were an explication of Turing's argument for CT (Webb 1980, 220ff.). Webb is not alone in reading Turing as offering an argument for CTC,[14] but this is a misunderstanding of Turing. Turing expressed the following view:

> If the untrained infant's mind is to become an intelligent one, it must acquire both discipline and initiative. So far [i.e., by discussing effective procedures] we have been considering only discipline... But discipline is certainly not enough in itself to produce intelligence. That which is required in addition we call initiative... Our task is to discover the nature of this residue as it occurs in man, and try to copy it in machines. (Turing 1948, 21)

As I mentioned in Chapter 5, the consensus among Turing scholars is that in arguing for CT, all that Turing was attempting to establish is that human *computation* processes are computable by TMs—he was not attempting to establish that all cognitive processes implement effective procedures.[15]
Regardless of what Turing thought about this matter, here is what Webb says:

> To show that human beings are not abstract (universal) Turing machines it would be sufficient to show that at least one of the following conditions is false:
> (i) Humans are capable of only a finite number of internal (mental or physical) states $q_i \in Q$.
> (ii) Humans are capable of discriminating only a finite number of external environmental states $s_i \in S$.
> (iii) Human memory is described by a function f from pairs $<q_i, s_i>$ to Q.
> (iv) There is a finite set B of atomic human behaviors, including some which may be identified with states s_i of S, which they effect (as in printing a symbol), and each $<q_i, s_i>$ determines a unique element of B. A human

[13] Cf: "Human cognitive processes are effective; by the [Church–Turing] thesis it follows they are recursive relations" (Nelson 1987, 581).

[14] Cleland 1993, 284; Shanker 1995, 55; Fodor 1998; Baum 2004, 33.

[15] See Sieg 1994; Copeland 2000. I have argued at length against this misinterpretation of Turing in Piccinini 2003b.

> (molar) behavior is comprised of a finite sequence of atomic behaviors of *B*, some of which have neural dimensions, while others may be molar behaviors in their own right. (Webb 1980, 236; I have altered Webb's language to make it gender neutral)

Webb seems to believe conditions (i)–(iv) obtain. We should add that, for his argument to go through, at least two further implicit conditions must obtain. First, the function from pairs $<q_i, s_i>$ to *B* implicit in (iv) must be Turing-computable. Second, humans must be capable of going through at most finitely many memory states and atomic behaviors in a finite time. It is well known that systems that do not satisfy these further conditions can compute Turing-uncomputable functions (Giunti 1997; Pitowsky and Shagrir 2003). If Webb's explicit and implicit conditions are satisfied by human brains, then their behaviors are Turing-computable—that is, (C) holds.

As to (i), Webb appeals to Turing's argument that a human who is performing calculations is capable of only finitely many internal states, otherwise some of them would be arbitrarily close and would be confused (Webb 1980, 221). This is justified in an analysis of human calculation, where the internal states must in principle be unambiguously identified by the computing humans on pain of confusion in performing the calculation. In his argument, Turing makes it clear that his "internal states" should be replaceable by explicit instructions.[16] Since the instructions have finite length, they can only distinguish between finitely many states.

But this says nothing about the number of internal states humans are capable of outside the context of calculation. Ordinary mathematical descriptions of physical systems, such as systems of differential equations, employ continuous variables. Because of this, ordinary mathematical descriptions ascribe physical systems state space trajectories that include uncountably many states. There is no a priori reason to suppose that humans are different from other physical systems in this respect. In fact, theoretical neuroscientists make extensive use of ordinary mathematical descriptions, which ascribe to neural mechanisms uncountably many states (Dayan and Abbott 2001; Ermentrout and Terman 2010; Eliasmith 2013).[17]

As to (ii), Webb also attributes it to Turing. But again, Turing was only concerned with effective calculability by humans, an activity that must be describable by an effective procedure. Since effective procedures are finite, it seems plausible that they can only be used to discriminate between finitely many "environmental states"

[16] This point has been emphasized by Sieg (e.g., 2001).

[17] There may be denumerably many equivalence classes of an uncountable set that are enough to explain cognitive capacities, but even a denumerable infinity violates (i).

(i.e., symbols). Again, this says nothing about how many environmental states humans can discriminate outside the context of calculation.

Webb adds that (ii) "would follow from a finiteness condition in physics itself to the effect that there *were* only finitely many states of the environment there to be discriminated" (Webb 1980, 236). Webb, however, gives no reason to believe that such a physical condition obtains.

As to (iii), Webb makes it clear that the function f from pairs $\langle q_i, s_i \rangle$ to Q should be Turing-computable. He admits the possibility that f be nondeterministic, but submits that that could be taken care of by a nondeterministic TM (that is, a TM whose state transitions are not deterministic). But nondeterministic TMs can take care of this situation only if there are at most finitely many q_i and s_i, i.e., if (i) and (ii) obtain, and we've already seen that (i) and (ii) are unjustified. Webb is also skeptical that it could be "shown effectively" that there is no such Turing-computable f. If by "showing something effectively" Webb means proving something rigorously, Webb's statement is surprising. For one of Turing's greatest achievements was precisely to prove rigorously that there is no Turing-computable f for solving uncountably many problems, including the halting problem for TMs. Given CT, there is no principled difficulty in showing that a function is not Turing-computable.

The main difficulty with (iii), however, is not that f may not be Turing-computable, which there is no evidence for. The main difficulty is that Turing-computability may be simply irrelevant to "describing" human memory in an explanatorily relevant way. The relationship between human memory and Turing-computability can be divided into two clusters of issues. One cluster belongs with a general analysis of the relationship between computability and physical systems. This set of issues has nothing to do with whether cognitive processes implement effective procedures, and I already covered its relevance to CTC in Sections 10.2 and 10.3. The other cluster belongs with assessing the empirical hypothesis that neurocognitive mechanisms are computational. That empirical hypothesis cannot be settled a priori.

As to (iv), Webb says, "it is . . . hard to imagine anything but Descartes' mechanical organisms existing at the dawn of evolution" (Webb 1980, 236), where the context makes clear that "Descartes' mechanical organisms" satisfy condition (iv). That something is "hard to imagine" is hardly conclusive evidence for (iv).

As to the further conditions implicit in Webb's argument, they seem no more a priori true than those explicitly stated by Webb. In the end, Webb has offered little support for his view that cognitive processes implement effective procedures. This, of course, is not to say that cognitive processes implement non-effective procedures, or Turing-uncomputable procedures. It may be that procedures are just irrelevant to scientific theories of cognition. To determine the relevance of procedures, whether effective or not, to cognition, we should develop and test empirical theories of cognition rather than argue a priori about these matters.

10.5 Effective Procedures as a Methodological Constraint on Psychological Theories

A final argument from CT to (C) invokes a methodological constraint on psychological theories, to the effect that psychological theories should only be formulated in terms of effective procedures. If this is the case, then, by CT, (C) follows. This argument is originally due to Daniel Dennett, and it has found many followers.[18] Here is Dennett's original:

> [C]larity is ensured for anything expressible in a programming language of some level. Anything thus expressible is clear; what about the converse? Is anything clear thus expressible? The AI programmer believes it, but it is not something subject to proof; *it is, or it boils down to, some version of Church's Thesis* (e.g., anything computable is Turing-machine computable). But now we can see that the supposition that there might be a nonquestion-begging nonmechanistic psychology gets you nothing, unless accompanied by the supposition that Church's Thesis is false. For a nonquestion-begging psychology will be a psychology that makes no ultimate appeals to unexplained intelligence, and that condition can be reformulated as the condition that whatever functional parts a psychology breaks its subjects into, the smallest, or most fundamental, or least sophisticated parts *must not be supposed to perform tasks or follow procedures requiring intelligence*. That condition in turn is surely strong enough to ensure that *any procedure admissible as an "ultimate" procedure in a psychological theory falls well within the intuitive boundaries of the "computable" or "effective" as these terms are presumed to be used in Church's Thesis*. The intuitively computable functions mentioned in Church's Thesis are those that "any fool can do," while the admissible atomic functions of a psychological theory are those that "presuppose *no* intelligence." If Church's Thesis is correct, then the constraints on mechanism are no more severe than the constraints against begging the question in psychology, for any psychology that stipulated atomic tasks that were "too difficult" to fall under Church's Thesis would be a theory with undischarged homunculi [fn. omitted]. So our first premise, that AI is the study of all possible modes of intelligence, is supported as much as it could be, which is *not quite* total support, in two regards. The first premise depends on two unprovable but very reasonable assumptions: that Church's Thesis is true, and that there can be, in principle, an adequate and complete psychology.
>
> (Dennett 1978a, 83; emphasis added)

[18] Webb 1980, 220; Haugeland 1981, 2; Fodor 1981, 13–15; Pylyshyn 1984, 52, 109; Boden 1988, 259. A previous discussion of this argument can be found in Tamburrini 1997.

Dennett asserts that CT yields a methodological constraint on the content of psychological theories. This is because, he says, any theory that posits operations or procedures that are "too difficult" to fall under CT is positing an "undischarged homunculus," that is, an unexplained intelligent process. And any psychological theory that posits undischarged homunculi should be rejected because it begs the question of explaining intelligence.[19]

Dennett's reference to "procedures admissible as ultimate procedures in a psychological theory" implies that psychological theories are formulated in terms of procedures. Since what fall under CT are *effective* procedures, Dennett's argument entails that the only ingredients of psychological theories are effective procedures. As a matter of fact, with the rise of cognitive psychology, some psychologists did propose that psychological theories be formulated as effective procedures, or computer programs, for executing the behavioral tasks explained by the theories (Miller, Galanter, and Pribram 1960). This view was elaborated philosophically by Fodor (1968b), which is one of the works referred to by Dennett (1978a).

To the extent that psychological theories are or should be formulated in terms of effective procedures, Dennett's appeal to CT is well motivated. For suppose that a psychologist offered an explanation of a behavior that appeals to a nonmechanical effective procedure of the kind imagined by Gödel (1965), or to arbitrary correct means of proof like those hypothesized by Kálmar (1959). Suppose that this psychologist refused to give effective instructions for these putative procedures. Dennett would be justified in concluding that these putative procedures are undischarged homunculi. Such a psychological theory purports to explain a behavior by positing an unexplained intelligent process, which begins an infinite regress of homunculi within homunculi. Gödel and Kálmar's proposals may have a legitimate role to play in the philosophy of mathematics, but not in a naturalistic explanation of behavior, as a nonquestion-begging psychological explanation should be. In other words, any psychologist who wants to posit effective procedures that do not fall under CT should give a mechanistic explanation of how they can be followed by people. By so doing, this psychologist would falsify CT. So far, Dennett's argument is sound.[20]

It remains to be seen the extent to which psychological theories are or should be formulated in terms of effective procedures. In Chapter 6, I pointed out that explaining a behavior by positing an effective procedure (or computer program)

[19] For more by Dennett on homunculi and question begging in psychology, cf. Dennett 1978, 57–9, and Dennett 1978b, 119ff. For more by Dennett on AI as study of all mechanisms (as opposed to computing mechanisms) for intelligence, see Dennett 1978b, esp. 112.

[20] Except for his implicit claim that a psychology that falsifies CT would be a "nonmechanistic psychology." Unlike me, Dennett is implicitly operating with a narrow notion of mechanism, according to which only computational processes that are Turing-computable count as mechanistic. As I discussed in Chapter 6, Dennett's implicit restriction on what counts as mechanistic is a relic of an obsolete notion of mechanism.

is only one species of computational explanation. Computational explanation, in turn, is only one species of mechanistic explanation, which consists of positing a set of components and ascribing functions and organization to those components. I also pointed out that in the philosophy of psychology tradition that goes from Fodor to Dennett and beyond, explanation by appeal to effective procedure and mechanistic explanation were often conflated. One of the effects of this conflation is Dennett's conclusion that CT constitutes a methodological restriction on all psychological explanation rather than a restriction on explanations that appeal to effective procedures (or computer programs).

Mechanistic explanations in cognitive neuroscience face the same constraint against begging the question of explaining intelligence that Dennett exploits in his argument. That is to say, the components posited by a neurocognitive explanation should not contain undischarged homunculi. If a component is ascribed intelligence (or other high-level cognitive abilities), this intelligence should be discharged by the lesser intelligence (or cognitive abilities) of its components, until a level of organization whose components have no intelligence (or other cognitive abilities) is reached. This methodological restriction on mechanistic explanations of cognition was already formulated by Attneave (1961), who did not mention either effective procedures or CT. He did not mention them because, to the extent that psychological theories are formulated without positing effective procedures, CT is irrelevant to them.

CT is only relevant to mechanistic explanations that posit effective procedures, not to other kinds of mechanistic explanation, including computational explanation that do not posit effective procedures. And vice versa: a psychological theory that explains behavior without positing effective procedures does nothing by itself to falsify CT. Modest Physical CT, however, is relevant to any mechanistic explanation that posits a process of digital computation. If a theory posits a genuine hypercomputation as a neurocognitive process, then it falsifies Modest Physical CT. If it posits nondigital computations or no computations at all, then it is irrelevant to Modest Physical CT too. Finally, the issues of computational approximation I discussed in Section 10.3 (Between Modest and Bold Physical CT) are relevant to any mechanistic explanation. They are relevant because they are relevant to any dynamical description—they have nothing in particular to do with psychological or neurocognitive theories. In conclusion, CT poses a methodological constraint on theories of cognition that posit effective procedures—it poses no general constraint on mechanistic theories of cognition.

10.6 Assessing the Computational Theory of Cognition Independently of the Church–Turing Thesis

This chapter addressed three attempts to support CTC on the grounds of CT. I argued that, given a proper understanding of CT, all those arguments are unsound. CT does entail that *if* the brain follows an effective procedure, *then* that procedure is Turing-computable. And Modest Physical CT does entail that *if* the brain performs digital computations, *then* those computations are Turing-computable. But neither CT nor Modest Physical CT is of any use in determining whether neurocognitive systems follow effective procedures or perform digital computations.

There is another way that computability is relevant to the explanation of mental capacities. Neural mechanisms are complex nonlinear dynamical systems par excellence, which lie at the frontier of what is mathematically analyzable by dynamical systems theory (cf. Barabási 2002; Strogatz 2003; Dayan and Abbott 2001; Ermentrout and Terman 2010; Eliasmith 2013). Because of this, methods of computational modeling are crucial to their scientific study. This is independent of whether neurocognitive mechanisms are computing mechanisms in any non-trivial sense.

Where does this leave CTC? Even though CT does not help to establish CTC, CTC remains the most plausible theory of cognition. The science of cognition belongs with physiology and engineering, which explain the behavior of systems by finding mechanisms. Some mechanisms perform computations (e.g., digital computers) and some don't (e.g., stomachs). Of those mechanisms that perform computations, some are digital and some aren't.

CTC is true if and only if neurocognitive mechanisms perform computations and those computations explain cognitive capacities. Digital CTC is true if and only if neural computations are digital. In Chapter 9, I argued on largely empirical grounds that the neural processes that explain cognition are computational, so CTC does hold. In Chapter 13, I will argue that those same neural computations are not digital but sui generis. Before we get there, the next chapter will take a close look at the main objections against CTC.

11
The Resilience of the Computational Theory of Cognition

11.1 Objections to the Computational Theory of Cognition

The Computational Theory of Cognition (CTC) says that cognition is (a kind of) computation, or that cognitive capacities are explained by the agent's computations. Cognitive processes and behavior are the explanandum. The computations performed by cognitive systems are the proposed explanans. Since the cognitive systems of biological organisms are more or less their nervous systems, we may say that according to CTC, the cognitive processes and behavior of organisms are explained by neural computations (Chapter 9).

CTC has met a wide range of objections ever since Warren McCulloch and Walter Pitts (1943) first proposed it. There are two types of objection: (1) insufficiency objections maintain that computation is insufficient for some cognitive phenomenon, and (2) objections from neural realization maintain that cognitive processes are realized by neural processes, but neural processes have at least one feature that is incompatible with being (or realizing) computations. We will look at each type in turn.

In this chapter, I explain why CTC survives these objections. Insufficiency objections are at best partial: for all they establish, computation may be sufficient for most cognitive processes, may be part of the explanation for all cognitive processes, or both. Objections from neural realization are based on a false contrast between neural processes and computation. Our guiding framework is the mechanistic account of physical computation: a physical computation is the processing of medium-independent vehicles by a functional mechanism according to a rule defined over the vehicles (Chapter 6).

11.2 Insufficiency Objections

There is a long tradition of arguments to the effect that computation is insufficient for genuine cognition. These are the most discussed objections to CTC. Their general form is as follows: cognitive phenomena include *X*, but computation is insufficient for *X*. Therefore, cognition is not computation. Candidates for *X* include but are not limited to mathematical insight (Lucas 1961; Penrose 1994),

Neurocognitive Mechanisms: Explaining Biological Cognition. Gualtiero Piccinini, Oxford University Press (2020).

DOI: 10.1093/oso/9780198866282.001.0001

qualia (Block 1978), intentionality or understanding, abduction (Fodor 2000), embodiment (Thompson 2007), and embeddedness (van Gelder 1998).

Insufficiency objections suffer from a fundamental weakness: even if computation is insufficient for some cognitive phenomenon *X*, CTC can be retained in its most plausible form. In its strongest formulation, CTC says that all cognitive phenomena can be explained by computation alone. Virtually no one believes this. The reason is precisely that some aspects of cognition, such as at least some of the candidate *X*'s listed above, appear to involve more than computation.

A weaker and more plausible formulation of CTC is that computation is an important proper part of the explanation of all or most cognitive processes and capacities. This formulation is entirely adequate for the role CTC plays in cognitive neuroscience. Cognitive neuroscientists offer computational explanations of cognitive capacities. Such explanations are usually formulated in terms of computations over representations, presupposing some account of how the representations get their representational content. In some cases, such explanations are formulated in terms of conscious states, presupposing some account of how the states get to be conscious. Finally, such explanations are often formulated in terms of states and processes that are coupled with the organism's body and environment. Again, this presupposes that neural computations are coupled with the body and environment in the relevant way.

Thus, a complete explanation of cognition may require, in addition to an appeal to computation, an account of consciousness, representational content, embodiment, and embeddedness. Computation may be insufficient for some aspect of cognition, but it may still be an important part of the explanation of most—or even all—cognitive capacities.

This concessive strategy only goes so far. If insufficiency objections turned out to be not only correct, but correct to the point that computation only plays a minor role in the explanation of a few cognitive processes, then CTC would lose most of its luster.[1] In other words, if computation contributed little to explaining cognition, then CTC would no longer be a very interesting and substantive view. But, as they stand, insufficiency objections do not undermine the most interesting version of CTC: that neural computations are a necessary and important part of the explanation of most cognitive processes and capacities of organisms.

In conclusion, insufficiency objections—to the effect that computation is insufficient for *X*, for some feature *X* that cognition has—show at best that a complete theory of cognition involves more than an appeal to computation—something that virtually no one denies. This being said, it will be instructive to take a closer look at some aspects of cognition that computation is said to be insufficient for, to see how computation may fit within a more complete explanation of X.

[1] I owe this observation to Mark Sprevak.

CTC, I have said, is the view that cognition is explained at least in part by neural computation. In light of Chapter 9, this may mean a number of different things. The original form of modern CTC is *digital CTC*—cognition involves (a kind of) digital computation. A weaker thesis is *generic CTC*—cognition involves (a kind of) computation in the generic sense, which may not be digital. In evaluating objections to CTC, we should keep in mind whether they impugn only digital CTC or also generic CTC, and whether they give rise to an alternative way of explaining cognition.

11.2.1 Consciousness

Let's stipulate that the mind includes both cognition and phenomenal consciousness. According to mainstream cognitive neuroscience, many cognitive processes are completely unconscious and much of conscious cognition can be explained in terms of neurocognitive mechanisms without worrying much about phenomenal consciousness. Still, some cognitive processes are conscious. When cognition is conscious, cognition and consciousness might be intimately tied in important ways. If cognition involves phenomenal consciousness in some way or another at least some of the time, the nature of consciousness may well be relevant to a complete explanation of cognition.

Some argue that computation is insufficient for phenomenal consciousness (e.g., Block 1978). The main reason is that consciousness is an eminently qualitative phenomenon, whereas computation appears to be an eminently functional phenomenon (Chapters 4 and 6). If this is correct and performing functions is insufficient for possessing certain qualities, then computation is insufficient for phenomenal consciousness.

Whether performing functions is sufficient for possessing qualities is a difficult question in the metaphysics of properties. As I have pointed out in Chapter 1, there are many views about the relationship between qualities and functions (i.e., causal powers). I will not enter that debate. Suffice it to say that according to some prominent views, performing functions is insufficient for possessing certain qualities. If this is right and computation is a matter of performing functions while consciousness is a matter of possessing certain qualities, then computation is insufficient for consciousness. If computation is insufficient for phenomenal consciousness and cognition involves phenomenal consciousness, then, at the very least, computation is insufficient for cognition.

Whether and in what way cognition involves phenomenal consciousness is controversial. Some philosophers maintain that consciousness is epiphenomenal and that cognition can be explained without involving consciousness. If so, then the objection from consciousness is defused. Even if cognition does involve consciousness, some cognitive scientists maintain that consciousness is reducible to computation and information processing (cf. Baars, Banks, and Newman 2003), and several philosophers attempt to explain consciousness in broadly computational and informational terms. I will discuss this possibility in more detail in Chapter 14.

Finally, even if cognition involves consciousness and consciousness requires something beyond computation, this is no reason to doubt that computation is an important aspect of the explanation of cognition. In any case, the objection from consciousness does not give rise to a noncomputational approach to cognition; at best, it says that there might be more to cognition than computation alone.

11.2.2 Intentionality and Understanding

Some argue that cognition involves intentionality, but computation is insufficient for intentionality. Therefore, cognition is not computation. A well-known variant of this objection is that computation is insufficient for understanding.

I will begin with a brief note about understanding versus intentionality. Understanding can be divided into a conscious feeling—the conscious feeling that we understand a linguistic stimulus—and a cognitive capacity—the capacity to respond to the linguistic stimulus. The problem posed by our capacity to respond to linguistic stimuli is that the stimuli have intentionality and this intentionality needs to be explained. In this respect, the objection from our ability to understand reduces to the objection from intentionality. The conscious feeling of understanding poses whatever problem is posed by phenomenal consciousness. Therefore, the objection from understanding reduces to a combination of the objection from phenomenal consciousness and the objection from intentionality. Since we've already dealt with the former, let's address the latter.

One possible response is that intentionality is an illusion. According to this response, explaining behavior does not require mental representations or similarly intentional notions. Another response is that computation does suffice for intentionality. The semantic account of computation may seem to help here. For if computation presupposes representation in a robust enough sense ("original intentionality"), then computation is automatically intentional in the right way. The problem with this second response is that computation per se hardly presupposes the right notion of representation ("original intentionality"). At any rate, I reject the semantic account of computation (Chapter 6).

Most realists about intentionality prefer a third response: computation is insufficient for intentionality but computation still explains cognitive capacities. According to this response, the kind of cognition that is involved in explaining intentionality is *intentional* computation. Where does the intentionality come from? Some take intentionality as a primitive or give it a nonreductive account, while others offer naturalistic accounts of intentionality. Most of today's defenders of CTC accept that a complete explanation of cognitive phenomena requires not only a computational explanation of behavior, but also an account of intentionality.

I favor this last response—with a twist. In Chapter 12, I will discuss neural representation in detail, arguing that it's an observable and well-established feature of the nervous system and outlining an Informational Teleosemantic theory of its semantic content. This is nowhere near enough for full-blown linguistic intentionality, but it's a crucial ingredient in such an account. A full-blown account of intentionality will require a lot of additional work to articulate the specific kinds of neural representations and neural mechanisms involved (for a start, see Morgan and Piccinini 2017 and Piccinini 2011, forthcoming).

11.2.3 Embodied, Embedded, Enacted, and Extended Cognition

Obviously, ordinary cognition is embodied—that is, dynamically coupled with a body. Equally obviously, ordinary cognition is embedded—that is, dynamically coupled with an environment. Perhaps ordinary cognition is not just embodied and embedded but extends into the environment as well—that is, parts of an agent's environment are also part of that agent's cognitive processes (e.g., Wilson 1994; Clark and Chalmers 1998). Perhaps ordinary cognition is enacted—that is, it's self-directed and not representational (Thompson 2007). At any rate, some people assert that, unlike cognition, computation is disembodied, unembedded, not enacted, and wholly within the boundaries of the agent. If so, then cognition is not computation (cf. Varela, Thompson, Rosch 2017). This objection is based on a false premise.

Computation *can* be disembodied, unembedded, not enacted, and wholly within the boundaries of the agent. But it *need not* be. Let's look at embodiment and embeddedness first. Computing systems can be dynamically coupled with a body, an environment, or both. There is a whole field of computer science devoted to embedded computing. In recent years, neuroscientists have begun to pay more attention to the tight interconnection between neural systems and their milieu. And although ordinary Turing machines are a poor model for embodied and embedded cognition, even Turing machines can be modified to include a memory tape separate from the input and output tape. A separate memory tape can preserve the outcomes of prior computations, thereby modeling the long-term coupling between the neurocognitive system, its body, and its environment (Maley and Piccinini 2016). In sum, for any substantive version of the thesis that cognition is embodied or embedded—any version that does not build the rejection of CTC into its definition—computation remains the primary process driving cognition.

With respect to extended cognition, Robert Wilson (1994) proposes a version of CTC according to which the computations themselves extend into the environment. In fact, many of us use computers, smartphones, or other computing artifacts every day. When we do so, some of the computations that explain our behavior occur outside our bodies—they occur within the computer, smartphone, or other artifact we are using. I don't know whether this is a case of extended cognition and I don't know whether any extended computations occur when we are not using computing artifacts. Regardless, extended cognition is no obstacle to CTC.

As to enactivism, it seems to boil down to two main claims: cognition is autonomous, or self-directed, and cognition does not involve representations. This is supposed to be in conflict with CTC because computation is often assumed to presuppose representation (Fodor 1981) and to be heteronomous, i.e., controlled from outside the system. These are misconceptions. As to autonomy, computation *can* be heteronomous, and many of the computations that we are familiar with are completely programmed from outside the system. But nothing prevents computational systems from being autonomous. Computational systems can select their inputs, store memories of their interactions with the environment, and even modify their own computational structure. As to representation, I've been at pains to emphasize that computation does not presuppose representation (Chapter 6), which means that CTC is entirely compatible with rejecting representation (cf. Dewhurst and Villalobos 2017, 2018). That said, I will show in Chapter 12 that cognition does involve representation. Insofar as enactivism rejects representations, it is false.

In conclusion, CTC is compatible with embodied, embedded, extended, and even enacted cognition (cf. Milkowski 2017; Milkowski et al. 2018; Isaac 2018).

11.2.4 Dynamical Systems

Some argue that cognition is dynamical as opposed to computational (cf. Port and van Gelder 1995, 4; van Gelder 1998; Spivey 2007). They propose explaining cognitive capacities not as computations but rather as the behavior of dynamical systems.

This argument presupposes that there is a contrast between dynamical systems and computational ones. There is no such contrast! Dynamical systems are systems that change over time as a function of their current state. Dynamical systems are usefully modeled using systems of differential equations or difference equations, and the study of such equations is called "dynamical systems theory." Cognitive systems, of course, are dynamical systems—they change over time as a function of their current state. Thus, theoretical and computational cognitive neuroscientists fruitfully employ differential equations to study cognitive systems at various levels of organization. By the same token, computational systems are dynamical systems too, and they are often studied using differential or difference equations. In fact, mathematical tools derived from dynamical systems theory figure prominently in most types of neurocomputational explanations. Thus, while there may be opposition between a particular type of dynamics (say, an irreducibly continuous dynamics) and a particular type of computation (say, algorithmic digital computation), in general there is no genuine opposition between being dynamical and being computational.

To get around this rather pedestrian fact, some proponents of the dynamical objection attempt to define a special sense in which cognition is dynamical but computation is not. A dynamical system in this specialized sense must have an independent metric—namely, a metric over states, related to behavior, but specifiable independently of behavior (van Gelder 1998). Requiring such an independent metric is ad hoc and, in any case, it does not help the dynamical objection. It is a requirement that the cognitive system be *described* a certain way; it does not affect the nature of the system. If a system is computational, all that this proposal does is require that there be an independent metric defined for this system. There is no reason why physical computing systems cannot be described this way—any more or less than any other dynamical system.

A related objection is that the proper *explanation* of cognition is dynamical as opposed to computational or mechanistic. A so-called dynamical explanation shows that a phenomenon can be *predicted* given a set of equations that describe the system plus local conditions (Stepp, Chemero, and Turvey 2011, 432). According to this view, dynamical explanation is not mechanistic: "Dynamical explanations do not propose a causal mechanism" (Stepp, Chemero, and Turvey 2011; cf. Chemero and Silberstein 2008). This is relevant because, as I have argued, computational explanation is mechanistic (Chapter 6). If computational explanation is mechanistic but cognitive explanation is dynamical and hence nonmechanistic, then cognitive explanation is noncomputational. If this objection works, CTC offers the wrong kind of explanation.

This objection from dynamical explanation does not work for two reasons. On one hand, prediction is insufficient for explanation, as decades of debate have made abundantly clear (Salmon 1984; Kaplan and Craver 2011). I can predict the weather using a barometer, but the barometer does not explain the weather. So dynamical "explanation," as defined above, is not really a kind of explanation. On the other hand, as I've already said, dynamical models are perfectly compatible with mechanistic explanation. Some dynamical models are not explanatory, but surely many dynamical models—including many dynamical models employed in computational cognitive neuroscience—are explanatory. They are explanatory *because* they describe mechanisms. More precisely, they describe causal and dynamical relations between aspects of mechanisms (Craver 2006; Bechtel and Abrahamsen 2010; Zednik 2011). Therefore, mechanistic explanation and dynamical models are two aspects of the same explanatory enterprise (Bechtel 2013; Bechtel and Abrahamsen 2013; Kaplan 2015; Lyre 2017; Chemero and Faries 2018; see also Beer and Williams 2015; Vernazzani 2017).

In sum, there is no conflict between computation and dynamics. They are two aspects of neurocognitive activity.

11.2.5 The Mathematical Objection

As Alan Turing showed, any digital computing system that follows a fixed set of procedures—e.g., any Turing machine—is equivalent to a formal system for proving mathematical theorems. In addition, there is a precise limit to the range of theorems that any digital computing system with a fixed set of procedures can prove. But, as Turing also pointed out, mathematicians are capable of inventing new methods of proof. Because of this, mathematicians can prove more theorems than any fixed computing system. Therefore, the objection concludes, the cognitive systems of human mathematicians are not just (fixed) computing systems; some ingredient beyond computation is required to account for human mathematical ability (Turing 1950).

This mathematical objection can also be couched in terms of Gödel incompleteness. As Kurt Gödel showed, any consistent formal system with sufficient expressive power is incomplete, in the sense that there are true arithmetic statements that the system cannot prove. But, as Gödel showed and others point out, mathematicians can give novel arguments, not relying on any given formal system, to the effect that a given system is consistent or that certain arithmetical statements that cannot be proved within the system are true. Therefore, the objection concludes, the cognitive systems of human mathematicians are not just (fixed) computing systems; some ingredient beyond computation is required to account for human mathematical ability (Gödel 1951; Wang 1974; Lucas 1961, 1996; Penrose 1989, 1994).

Some people respond to the mathematical objection by positing that cognition involves hypercomputation—that is, computation more powerful than computation by Turing machines (Copeland 2000; Siegelmann 2003; Bringsjord and Arkoudas 2007). Based on our taxonomy (Chapter 9), this is still a version of digital CTC—it just says that the computations performed by cognitive systems are more powerful than those of Turing machines. But there is no persuasive evidence that human beings can perform hypercomputations: no one has shown how to solve a genuinely Turing-uncomputable problem, such as the halting problem,[2] by using human cognitive faculties. Therefore, there is no evidence that cognition involves hypercomputation.[3] Still, we do need an answer to the mathematical objection.

[2] The halting problem is the problem of determining whether any given Turing machine will ever halt while computing on any given input. There is no Turing machine that solves the halting problem.

[3] Cubitt et al. (2015) define a physical problem (spectral gap) that turns out to be undecidable. It remains speculative whether this physical problem corresponds to actual physical systems in such a

Paraphrasing Turing, intelligent cognition can be divided into two aspects: discipline and initiative. Discipline consists in applying known methods of investigation to a problem—searching for a solution patiently, step by step, applying whatever operations or rules we already know. Initiative, by contrast, consists in asking a new question, developing a new idea or method, or finding a novel way of solving a problem.

As Turing put it, computation that employs a fixed set of methods is pure discipline. It is based on making known rules explicit and following them patiently, step by step, until they yield a solution to our problem. What about discovering new solutions, which are not generated by existing rules? What about discovering new rules? Or what about discovering entirely new problems, for that matter? How can we account for that?

One way to account for initiative is by positing computation procedures for discovering new solutions, new rules, and new problems. To some extent, this can be done. Another way to account for initiative is by inserting what Turing called "a random element" in a computational process. Although our computations will generally follow existing methods, we can also set them up to, once in a while, do something random and see if that helps.

As Turing pointed out, computing systems that change their procedures over time and are allowed to make mistakes are not limited in the way that fixed computing systems are. They can add to their methods and the theorems they can prove. Therefore, the cognitive systems of human mathematicians might still be computational, so long as they change over time (in a noncomputable way) and can make mistakes. Since this last claim is independently plausible, CTC might escape the mathematical objection (Piccinini 2003b).

One way to introduce randomness in our computation is via genetic algorithms. These are algorithms that change existing computation procedures in more or less random ways to see if they can be improved. Lo and behold, genetic algorithms often improve computation procedures beyond anyone's expectations.

So, some aspects of initiative may be accounted for either in terms of computation alone or in terms of computation plus randomness. Yet it remains to be seen whether all aspects of initiative—more generally, all aspects of intelligent cognition—can be accounted for by computation plus randomness.

It's important to note that there are no serious alternative research programs for studying what intelligent cognition might be if it goes beyond computation plus randomness. The closest thing I can think of to an alternative conception of cognition is a not-entirely-random, medium-dependent functional process that yields solutions to cognitive problems, including mathematical problems. Such a process might rely on consciousness if consciousness is noncomputational

way that the problem can be solved by physical means. There is no positive evidence that neural systems can solve undecidable problems by physical means.

(Chapter 14). Or, without relying on consciousness, such a process might involve medium-dependent inferences between neuronal representations plus testing such connections for adequacy in solving cognitive problems. If such a process were possible, cognition would go beyond computation plus randomness. The question of whether intelligent cognition goes beyond computation plus randomness remains one of the hardest questions pertaining to CTC.

11.3 Objections from Neural Realization

In addition to insufficiency objections, there are objections to CTC based on alleged differences between neural processes and computations. Compared to insufficiency objections, these objections from neural realization have been neglected by philosophers. I will now discuss some important objections from neural realization and why they fail.

11.3.1 Nonelectrical Processes

Like signals in electronic computers, neural signals include the propagation of electric charges. But unlike signals in electronic computers, neural signals also include the diffusion of chemicals. More specifically, there are at least two classes of neural signals that have no analogue in electronic computers: neurotransmitters and hormones. Therefore, according to this objection, neural processes are not computations (cf. Perkel 1990).

As it stands, this objection is inconclusive at best. Computations are realizable by an indefinite number of physical substrates, so computing mechanisms can be built out of an indefinite number of technologies.[4] Today's main computing technology is electronic, but computers used to be built out of mechanical or electromechanical components, and there is active research on so-called "unconventional computing" such as optical, DNA, and quantum computing. There is no principled reason why computations, including digital computations, cannot be realized by processes of chemical diffusion.

Even if the chemical signals in question were essentially noncomputational, pointing out that they occur in the brain would not show that neural processes are noncomputational. Here, different considerations apply to different cases. Many hormones are released and absorbed at the periphery of the nervous system. So, the release and uptake of such hormones might be treated simply as part of the input–output interface of the neural computations, in the same way that sensory and motor transducers are.

[4] This is not to suggest that every process is a computation—far from it (Chapter 6).

With respect to neurotransmitters and neuropeptides, we need to remember that computational explanations apply at some mechanistic levels and not others (Chapters 7–8). For example, at one mechanistic level, the activities of ordinary electronic computers are explained by the programs they execute; at another level, they are explained by electronic activity in their circuits; at yet another level, by quantum mechanical processes. We should also remember that not all processes that occur in a computing system are computations. Consider temperature regulation: it occurs in most artificial computers, but it is not part of their computations. So, anyone who wishes to appeal to chemical signals as an objection to CTC needs to show both that those chemical signals are not computational and that they occur at the mechanistic levels at which neural systems allegedly perform computations.

As far as we can now tell, chemical signals have one of two functions: transmit information between neurons, or transmit information between neurons and non-neuronal cells. Information transmission is an essential aspect of computing in systems with many parts. Chemical signals do not appear to serve as computational vehicles—vehicles manipulated in the process of performing computations—but rather information transmission functions within a complex computing system and between the computing system and the rest of the organism. In any case, nothing prevents chemical signals from being computational vehicles. In conclusion, the presence of chemical signals in nervous systems does not threaten CTC.

11.3.2 Temporal Constraints

Obviously, neural processes are temporally constrained in real time. Some argue that computations are not so constrained; therefore, neural processes are not computations (cf. Globus 1992; van Gelder 1998).

This objection trades on an ambiguity between mathematical representation of time and real time. True, computations are temporally unconstrained in the sense that they *can* be defined and individuated in terms of computational operations, independently of how much time it takes to complete an operation. This point applies most directly to algorithmic (i.e., step-by-step) computations. Many neural networks that perform computations do so without their computations being decomposable into algorithmic steps (Section 9.1; see also Piccinini 2015, sect. 13.6), so the premise of this objection does not apply to them in the same direct way.

But the deeper flaw in this objection is that abstraction from temporal constraints is a feature not of physical computations themselves, but of (some of) their descriptions. The same abstraction can be performed on any dynamical process, whether computational or not. Consider differential equations—a type of description favored by some anticomputationalists. Differential equations contain time variables, but these do not correspond to real time any more than the discrete time intervals during which a Turing machine performs a step correspond to any particular real time interval. In order for the time variables of differential equations to correspond to real time, a temporal scale must be specified (e.g., whether time is being measured in seconds, nanoseconds, years, etc.). By the same token, the time steps of a computer can be made to correspond to real time by specifying an appropriate time scale. When this is done, computations are no less temporally constrained than any other process. As any computer designer knows, you can't build a well-functioning computer without taking the speed of components and the timing of their operations into account—in real time. In fact, even many defenders of classical CTC were concerned with temporal constraints on the computations that, according to their theory, explain cognitive phenomena (Pylyshyn 1984; Newell 1990; Vera and Simon 1993). And temporal constraints are a main concern, of course, in the neurocomputational literature—especially on motor control (VanRullen et al. 2005).[5]

11.3.3 Analog vs. Digital

Some have argued that neural processes are analog whereas computations are digital; therefore, neural processes are not computations (cf. Dreyfus 1979; Perkel 1990). This is one of the oldest and most often repeated objections to CTC. Unfortunately, it is formulated using an ambiguous terminology that comes with conceptual traps.

As we have seen, computations may be analog as well as digital. Analog computers were invented before digital computers and were used widely in the 1950s and 60s. If neural processes are analog, they might be analog computations. If so, then CTC remains in place in an analog version. In fact, the original proponents of the analog-versus-digital objection did not offer it as a refutation of CTC, but only of McCulloch and Pitts's digital version of CTC (Gerard 1951). This reply employs the ambiguous terms "analog" and "computation." Depending on what is meant by these terms, an analog process may or may not be an analog computation.

In a loose sense, "analog" refers to the processes of any system that, at some level of organization, can be characterized as the temporal evolution of real (i.e., continuous, or analog) variables. Some proponents of the analog-vs.-digital objection seem to employ some variant of this broad notion (cf. van Gelder 1998). It is easy to see that neural processes fall in this class, but this does not impugn

[5] Another rebuttal of the objection from temporal constraints—developed independently of the present one—is given by Weiskopf (2004). Some of his considerations go in the same direction as mine. See also Milkowski 2018 for an independently developed assessment of objections to CTC, with similar conclusions.

CTC. On one hand, most processes that are analog in this sense, such as the weather, planetary motion, and digestion, are paradigmatic examples of processes that are not computations in any interesting sense. Neural processes may fall into this class. On the other hand, it is also easy to see that virtually all systems, including computing systems such as computers and other neural networks, fall within this class. Continuous variables have more expressive power than discrete ones, so continuous variables can be used to express the same information and more. More importantly, most of our physics and engineering of midsize objects, including much of the physics and engineering of digital computers, is couched in terms of differential equations, which employ continuous variables. Therefore, even digital computers are analog in this sense. So, the trivial fact that neural mechanisms are analog in this loose sense does nothing to either refute CTC or establish an analog version of it.[6]

In a more restricted sense, "analog" refers to analog computers—a class of machines that used to be primarily employed for solving certain differential equations (Chapter 6). Analog computers are either explicitly or implicitly invoked by many of those who formulate the analog-vs.-digital objection. Claiming that neural systems are analog in the sense that they are analog computers is a strong empirical hypothesis. Since this hypothesis claims that neural mechanisms are analog computing mechanisms, it is a form of CTC (Rubel 1985; Maley 2018). I will discuss it in Chapter 13.

11.3.4 Spontaneous Activity

Neural processes are not the effect of inputs alone because they also include a large amount of spontaneous activity; hence, neural processes are not computations (cf. Perkel 1990). The obvious problem with this objection is that computations need not be the effect of inputs alone either. Computations may be the effect of inputs together with internal states. In fact, only relatively simple computations are the direct effect of inputs; in the more interesting cases, internal states also play a role.

11.4 How to Test the Computational Theory of Cognition

As we have seen, objections to CTC fall into two main classes: insufficiency objections and objections from neural realization. According to insufficiency

[6] It has been argued that at the fundamental physical level, everything is discrete (e.g., Wolfram 2002). This remains a speculative hypothesis for which there is no direct evidence. Moreover, it would still be true that our best science of midsize objects, including digital computers at certain levels of organization, is carried out in terms of continuous variables.

objections, computation is insufficient for some cognitive phenomenon *X*. According to objections from neural realization, neural processes have feature *Y* and having *Y* is incompatible with being a computation. In this chapter, I explained why CTC survives these objections. Insufficiency objections are at best partial: for all they establish, computation may be sufficient for cognitive phenomena other than *X*, may be part of the explanation for *X*, or both. Objections from neural realization are based on a false contrast between feature *Y* and computation.

The way to test CTC is to identify the most general properties of computation and see if neurocognitive processes have those properties. This is the method I follow. Computation is a functional process whose most general properties are that its vehicles are medium-independent and they are processed according to a rule (Chapter 6). Neurocognitive processes are computations because they have these properties (Chapter 9). In addition, neural computations carry information and are dynamical, temporally constrained, embodied, and embedded. Computation is not all there is to cognition but it's an important aspect of its explanation.

Having clarified that CTC is a viable and well-supported empirical hypothesis, we are finally in a position to address one of the most controversial issues surrounding it. Are neurocomputational vehicles representations? This is the topic of the next chapter.

12

Neural Representation

12.1 From Theoretical Posits to Observables

Chapters 9–11 defended the Computational Theory of Cognition (CTC) as an empirical hypothesis, according to which neurocognitive processes are computations. Chapter 5 pointed out that computation does not require representation. Therefore, CTC does not entail that neurocomputational vehicles are representations. Still, one reason for CTC is that neural processes integrate multiple sources of information in the service of controlling the organism. Might this be enough to conclude that neurocomputational vehicles are representations?[1]

To a first approximation, representations are internal states that "stand in" for X and can guide behavior with respect to X. For example, a representation of yogurt in your refrigerator "stands in" for the yogurt and can guide your behavior with respect to the yogurt. In other words, representations are entities that have both a semantic content—e.g., "there is yogurt in the refrigerator"—and an appropriate functional role (W. Ramsey 2016), and their specific functional role depends on their semantic content.

Representations are traditionally construed as unobservable entities posited by a theory or model to explain and predict behavior. The theory may be either folk psychology (e.g., Sellars 1956) or scientific psychology (e.g., Fodor 1975, 1981). The traditional debate centers on whether the representations posited by some theory or model are real, whether they have semantic content, and whether their semantic content plays a functional role.

Several grades of antirealism about representation have been proposed. Some argue that representations are real entities that play a functional role but their semantic content is merely a matter of interpretation—it is not a real property and plays no functional role (e.g., Dennett 1987; Cummins 1983, 1989). A closely related view is that representations are real in the sense that there are internal states individuated by what *we* take them to represent, but the internal states themselves have no semantic content (Chomsky 1995). Finally, there is full-blown representational eliminativism: representations have no place in a mature science of cognition, which will explain and predict behavior without appealing to representations (Stich 1983; Brooks 1991; van Gelder 1995; Keijzer 1998; Garzon 2008;

[1] This chapter is a substantially revised descendant of Thomson and Piccinini 2018, so Eric Thomson deserves partial credit for most of what is correct here.

Neurocognitive Mechanisms: Explaining Biological Cognition. Gualtiero Piccinini, Oxford University Press (2020).

DOI: 10.1093/oso/9780198866282.001.0001

Chemero 2009; Hutto and Myin 2013, 2017; Downey 2017; Varela, Thompson, Rosch 2017).

In recent years, the debate about representation has come to include cognitive and computational *neuro*science (P.S. Churchland 1986; P.M. Churchland 1989, 2012; P.S. Churchland and Sejnowski 1992; Bechtel 2008; Milkowski 2013). In spite of this shift, the debate continues to center on representations qua theoretical posits. Skeptics argue that neuroscientific models dispense with representations, or at least with representations properly so called (W. Ramsey 2007; Burge 2010; Hutto and Myin 2014; Raja 2017). Other skeptics argue that, even if neuroscientific models do not dispense with representations, they dispense with semantic content in the sense that content plays no causal or explanatory role (Egan 2014). Supporters reply that representations, complete with semantic content, do play an explanatory role within at least some neuroscientific models (Shagrir 2012, 2017; Sprevak 2013; Colombo 2014; Clark 2016; Gładziejewski 2016; Williams 2018; Kiefer and Hohwy 2018; Maley 2018).

Participants in this debate have mostly neglected the role of representation in *experimental* neuroscience. Some philosophers of neuroscience do discuss experimental practices (Bickle 2003; Craver 2007; Craver and Darden 2013; Kästner 2017; Silva, Landreth, and Bickle 2014; Sullivan 2009) but they do not focus on neural representations. On the rare occasions when philosophers discuss neural representations within experimental neuroscience, typically they mention experimental neuroscience either as inspiring information-based theories of content (e.g., W. Ramsey 2016, 7) or as supporting some particular theory, model, or explanation (e.g., Grush 2004; Sullivan 2010). Two exceptions are Churchland and Sejnowsky (1992, chap. 4) and Bechtel (2008, chap. 5, 2016), who suggest that experimental evidence supports the existence of neural representations. This chapter builds on their insight.

Since Ian Hacking's (1983) groundbreaking work on experimental practices, philosophers of science have pointed out that experimental science often has a life of its own: through observation and manipulation, experimentalists can establish that something is real. This conclusion is especially robust when it's supported by multiple distinct methods that use different instruments and techniques (Salmon 1984, 2005; see also Galison 1987, 1997; Staley 1999; Franklin 2002, 2013; Chang 2004; Weber 2005, 2014; Franklin and Perovic 2016). To be sure, theories and experiments coevolve, and theories are often used in designing experimental procedures, calibrating instruments, and processing and interpreting data. Nevertheless, experimentalists use theories and methods that are often established independently of current points of dispute. On that basis, experimentalists can establish that a type of entity exists. In Hacking's slogan, "if you can spray them, then they are real" (1983, 23).

For example, in the nineteenth and early twentieth century there was a dispute about whether nerve fibers form a continuous network or are made of distinct yet interconnected cells. Resolving it in favor of the neuron doctrine required the development of reliable staining techniques, electron microscopy, and other experimental methods. Later on, recording the precise shape of action potentials required the development, with the help of Ohm's law, of sensitive means of measuring millivolt deflections across individual nerve fibers (Hodgkin and Huxley 1939). Yet Ohm's law, the reliability of staining techniques and microscopes, and other background conditions were established independently of debates about neuronal connectivity and action potential generation. In addition, neurons and their action potentials can be recorded and manipulated in a variety of ways using a variety of techniques. By now, it is a well-established experimental observation both that neurons are distinct cells and that they fire action potentials.

In this chapter, I reframe the debate on representation by focusing on standard results from experimental neuroscience. Experimental neuroscientists began talking about representations in the nervous system *almost a century before the beginning of the cognitive revolution*, which is so often associated with the contemporary dispute (Chirimuuta 2019; Kraemer unpublished). At first, neuroscientists posited what are now called motor representations (Hughlings Jackson 1867, 1868); later they added what are now called sensory representations (Horsley 1907, 1909). While neural representations began as theoretical posits, neuroscientists soon found ways to observe and manipulate representations using multiple methods and techniques in multiple model systems, just as they observe and manipulate neurons and action potentials. The techniques and procedures they use are validated independently of any debates about neural representation. As a result, there are at least three kinds of empirically well-established neural representations: sensory representations, representations uncoupled from current sensory stimulation, and motor representations.

I will sample research at two levels of resolution. On one hand, I give a panoramic overview of the neurobiology of representational systems, without going into much detail about the evidence. This is meant to give a sense of the ubiquity of neural representational systems, including some that have not received much attention from philosophers (e.g., the birdsong system). On the other hand, in each section I zoom in on one example to give a more detailed sense of the types of experiment and reasoning that are involved in discovering and manipulating neural representations. For sensory representations, I focus on representation of motion in area MT; for uncoupled representations, I highlight working memory; and for motor representations, I highlight efference copy in the eye movement system. (For a couple of more detailed case studies, see also Bechtel 2016; Burnston 2016a, b.)

In the next section, I will articulate the notion of representation at play, allaying concerns that neuroscientists use an overly permissive notion of representation (W. Ramsey 2007). In the following three sections, I will review some of the many experimental observations and manipulations of sensory, uncoupled, and motor representations. With that, I rest my case.

12.2 Structural Representation

For something to count as a representation, it must have a semantic content (e.g., "there is yogurt in the fridge") and an appropriate functional role (e.g., to guide behavior with respect to the yogurt in the fridge). Let's look more closely at these two features of representation.

Let's begin with functional role. A representation's role is to serve as a "stand in" for X so as to guide behavior with respect to X. Following Frank Ramsey (1931), Fred Dretske (1981, 197) describes beliefs as internal maps by which we steer (cf. Armstrong 1973; Dretske 1988, 78). Similarly, Kenneth Craik (1943) describes representations as mental models that we use to plan and guide our behavior. This perspective is sometimes cashed out more formally in terms of functioning homomorphisms (Gallistel 1990, 2008; Gallistel and King 2009) or, equivalently, exploitable similarity (Godfrey-Smith 1996; Shea 2007, 2018; Gładziejewski and Miłkowski 2017; Williams and Colling 2018). The resulting notion of representation is often called *structural representation* (W. Ramsey 2007).[2]

For an internal state to *stand in* for a target in a strong sense, it should be possible for the internal state to be present within a system and play its role in guiding behavior even when its target is not immediately present—it is not in the system's immediate, directly observable environment. That is, the functional role of representation includes the possibility of being uncoupled from its target under at least some circumstancecs. Accordingly, structural representation includes four elements: (i) a homomorphism between a system of internal states and their targets, (ii) a causal connection from at least some targets to the internal states that target them, (iii) the possibility for some of the internal states to be decoupled from their target at least under certain circumstances, and (iv) a role in action control.

Along with a fitting functional role, a representation has semantic *content* (Anscombe 1957), which is either indicative or imperative (Millikan 1984). Representations with *indicative* content represent how things *are*; they are satisfied to the degree that they track the actual state of the world. By contrast, representations with *imperative* content represent how things *will* be made to

[2] Similar notions of representation are defended by Shepard and Chipman 1970; Swoyer 1991; Cummins 1996; Grush 2004; O'Brien and Opie 2004; Ryder 2004, unpublished; Bartels 2006; Waskan 2006; Bechtel 2008; Churchland 2012; Shagrir 2012; Isaac 2013; Hohwy 2013; Clark 2013; Morgan 2014; Neander 2017, chap. 8.

be; they are satisfied to the degree that the world comes to track them (cf. Mandik 2003).

Indicative representations include *sensory* representations. For example, a sensory representation of a small dark spot moving to the left is satisfied to the degree that there is indeed a small dark spot moving to the left. If there is no such object, or if it's *not* moving left, the system has *misrepresented* the environment.

Imperative representations include *motor* representations, whose function is to *bring about a new state of affairs* by generating behaviors. They are satisfied to the degree that the body appropriately carries out the commands. For example, the command to pick up an orange is satisfied just in case the agent picks up the orange. Since we are focusing on neural representations, from now on I will focus on sensory and motor representations, leaving other kinds of indicative and imperative representations aside.

To specify the content of representations more precisely, I will adopt a version of the best-developed and most plausible theory of representational content in biological systems: Teleosemantics (Stampe 1977; Dretske 1988; Fodor 1987, 1990, 2008; Millikan 1984, 1993; Ryder 2004, forthcoming; Neander 2017; this literature is surveyed in Adams and Aizawa 2010; Neander 2012; and Neander 2017, chap. 4). Different versions of Teleosemantics differ in their details; this is not the place to discuss their differences (cf. Neander 2017). Here I adopt a specific version of Teleosemantics that incorporates both functional role and semantic content.

In the philosophical literature, most of the attention has gone to sensory representations (exceptions include Millikan 1984, 1993; Papineau 1984, 1993; Mandik 2003; Butterfill and Sinigaglia 2014; Ferretti 2016; Mylopoulos and Pacherie 2017). In experimental neuroscience, motor and sensory representations are both extremely important, so I discuss both. I will also discuss indicative representations that are not directly coupled to current sensory stimulation.

When it comes to sensory representation, Teleosemantics takes an *informational* form (Neander 2017). That is, Teleosemantics assigns semantic content to sensory representations based on the natural semantic information they carry. A state (or signal) S carries natural semantic information that P just in case it raises the probability that P (Chapter 6). This can be quantified in different ways, including in terms of mutual information (Thomson and Kristan 2005).

Carrying information about something is not enough to represent it in the relevant sense. For example, fingerprints carry information about who touched what, but they do not represent it. According to my version of Informational Teleosemantics, what is needed in addition to information is (teleological) function—specifically, the function to carry information in the service of action guidance. Fingerprints, by themselves, lack functions, including the function of carrying information in the service of action guidance. A detective, though, may use fingerprints to guide her actions. The representations in the detective's mind, including representations of fingerprints, have the function of carrying the relevant information. Accordingly, the function of a sensory system is to carry natural semantic information about events in the local environment (including the body), so that the neuronal signals that transmit this information can subsequently guide how the agent responds to such events.

When most teleosemanticists talk about teleological functions, they typically give a selectionist account of functions, according to which functions are selected effects. In contrast, I rely on the goal-contribution account I defended in Chapter 3. When I talk about the functions of representational systems, I mean regular contributions to the biological goals of organisms.

My version of Informational Teleosemantics for sensory representations is this:

> (INDR) A state (or signal) S within an agent's representational system R *indicatively represents* that P $=_{\text{def}}$ A function of R is to produce S, so that S carries natural semantic information that P, S is one among a range of similar states that map onto a range of similar external states, and S can guide the agent's behavior with respect to the fact that P.

In other words, a state S indicatively represents that P just in case S tracks that P, where tracking that P consists in carrying the natural information that P, and the system R that produces S does so in order to guide the organism's behavior with respect to P. That is the system's *indicative function*. The content of sensory representations will track the state of the world to the degree that the system fulfills its indicative function. If S represents that P but it is not the case that P, then S misrepresents the state of the world.

Motor representations have a different function, and hence a different type of semantic content, than sensory representations. They do not serve to carry information about the world into the system, but to *generate new states of the world*. That is, they carry imperative, not indicative content.

A motor representation is satisfied if the world *becomes* the way it is represented to be. Here is my version of Teleosemantics for motor representations:

> (IMPR) A state (or signal) S within an agent's representational system R *imperatively represents* that P $=_{\text{def}}$ A function of R is to produce S in coordination with R's indicative representations, S can cause that P, and S is one among a range of similar internal states that map onto a range of similar future states of the environment.

In other words, a state S imperatively represents that P just in case the system that produces S has the function of producing S in order to bring it about that P. That is the system's *imperative representational function*. The world will track the content of motor representations to the degree that the system fulfills its imperative representational function. The environmental state that P is the *goal* of the internal state, which can be used to assign an error measure to the behavior. If S imperatively represents that P, nothing outside the nervous system prevents P from happening, but P doesn't come about, then there was a mistake in execution.

As I hinted in Chapter 9, there are powerful theoretical reasons to conclude that nervous systems process semantic information and construct internal models that guide the organism. We can now strengthen those considerations into an argument that nervous systems process structural representations.

The Argument from Complex Control

1. Nervous systems perform complex control functions in a computationally tractable way.
2. Performing complex control functions in a computationally tractable way requires processing structural representations.

Therefore, nervous systems process structural representations.

Needless to say, some nervous systems are more complex than others, and so are their control functions. All that matters here is that some nervous systems' control functions are complex enough that performing them in a computationally tractable way requires structural representations. That nervous systems perform their control functions in a computationally tractable way is demonstrated by their survival: any organism stuck with a computationally intractable control system would have gone extinct long ago.

Typical nervous systems perform very complex control functions. They can't just follow a light source or a chemical gradient, like many bacteria and plants can. They can't simply respond to aspects of the environment that are directly detectable using one or two physical variables, like centrifugal governors do (cf. van Gelder 1995, 1998). Nervous systems must select very complex responses to very complex stimuli, many aspects of which are not directly detectable, and they must do so in real time while their noisy environment keeps changing. Nervous systems have to distinguish between preys, predators, family, potential mates, and so forth, all of which may produce very similar sensory inputs. To distinguish between stimuli in the relevant ways, nervous systems need to integrate information about shapes, colors, distances, sounds, pressures, chemicals in the air, and so forth, many of which belong to the same physical sources. Many of these properties are environmental invariants that cannot be detected directly; they must be extracted from the proximal stimuli by building internal states that correlate with them.

To integrate these physically disparate inputs, nervous systems need to use some sort of medium-independent variables that can carry information about all the different sources and process such variables in a way that does not depend directly on the particular physical properties of any of the specific physical sources, because what works for processing sounds as such need not work for processing chemicals diffused in air, etc. Thus, nervous systems need internal states that are medium independent.

Furthermore, nervous systems control up to millions of muscle fibers at a time. Coordinating so many muscle fibers to respond to distal stimuli appropriately requires internal states that correlate with appropriate responses, which can be transformed into specific motor commands. Finally, nervous systems need to keep track of aspects of the environment that are not perceivable (because they are occluded or distant), keep track of the past, as well as anticipate aspects of the future. This requires internal states that can be decoupled from their targets.

A related factor is that there are many orders of magnitude more possible proximal inputs of the sort complex animals receive (patterns of retinal stimulation, patterns of cochlear stimulation) than possible kinds of distal stimuli (preys, predators, family, mates, etc.). The same is true on the output side: there are many orders of magnitude more possible proximal outputs (patterns of muscle fiber contractions) than possible ecologically meaningful actions (chasing a prey, fleeing a predator, courting a possible mate, etc.). Trying to figure out how to respond appropriately to every possible proximal input is a computational problem vastly less tractable than extracting information about invariants from the proximal inputs, using information about invariants to select a representation of an appropriate response, and then turning such motor representations into motor commands. This is true whether the organism is trying to learn how to respond by trial and error or trying to find an appropriate response by searching through a state space of possible responses.

Therefore, nervous systems need to first extract the structure of themselves and their environment out of their inputs by building (medium-independent) structural representations. Only by identifying the relevant properties of stimuli via structural representations can nervous systems learn, and later select, motor representations of appropriate responses in a computationally tractable way. The bottom line is that, at least for sufficiently complex organisms, selecting behavior in a computationally tractable way requires building highly selective structural representations of themselves within their environment and using representations to guide action (cf. Buckner 2018, Schulz 2018, and Poldrack 2020). Not only that: complex enough cognition requires structural representations that are medium-independent and represent invariants in their environment.

The argument from complex control already shows that neural representations are not merely theoretical posits to be tested by building computational models—their existence is supported by compelling theoretical considerations. In the rest of this chapter, I will add that neural representations are routinely observed and manipulated by experimental neuroscientists. Observing and manipulating neural representations means establishing, via experimental observations and interventions, that a system of neural states fits the criteria for representation. For sensory representations, the criteria are that (i) the neural states carry information about states external to the system, (ii) there is a systematic mapping between a range of similar neural states and a range of similar external states, and (iii) the system can use these neural states to guide behavior. For motor representations, the criteria are that (iv) the neural signals correlate with future states of the environment (where the environment includes the body), (v) there is a systematic mapping between a range of similar neural states and a range of similar future states of the environment, and (vi) such neural states can *cause* movements that bring about the future states of the environment. An additional criterion, which applies to the representational system as a whole, is that under appropriate circumstances it should be possible for some representations with indicative content to occur without any information coming directly from their targets—in other words, (vii) some representations are decouplable from their targets.

As I shall point out, (i)–(vii) have been established again and again, beyond reasonable doubt, in many different neural systems using many independent techniques. Therefore, even if experimental neuroscientists had *not* already reached consensus that neural states that satisfy the above accounts are representations—*which they have*—that's what they *should* conclude. Experimental neuroscientists *in fact* discover, observe, and manipulate neuronal representations in the sense discussed above.

Before getting into the details, let's briefly consider some of the methods used in physiology and anatomy. Physiologists *measure* neuronal activity at scales from the very small (patch clamp to record currents and voltages generated in single neurons or even patches of neuronal membrane, extracellular recordings, calcium- and voltage-sensitive dye imaging) to the very large (EEG or fMRI) (P. S. Churchland and Sejnowski 1992; Kandel 2013). In sensory systems, such neural activity is observed to correlate with some specific features of an external stimulus; this establishes what the activity carries semantic information about and that there is a broader mapping between similar neural signals and similar external states. To examine the causal role of individual neurons or neuronal areas and show that it affects downstream processing and ultimately behavior is more complicated: it typically involves ablation studies, or manipulating currents and voltages in individual neurons (current and voltage clamp) or clusters of cells via extracellular microstimulation (Kandel 2013). Recent years have also seen the explosion of the powerful technique of optogenetics, in which genetically targeted classes of neurons are excited or inhibited via light, in real-time, using light-activated ion channels (Yizhar, Fenno, Davidson, Mogri, and Deisseroth 2011). Transcranial magnetic stimulation is sometimes used to noninvasively perturb circuits on much larger spatial scales (Hallett 2000).

Neuroanatomy lets us probe the closely related *structural* features of neurons and circuits at multiple scales, which can be useful for determining the trajectory of information flow and causal relationships in a system, including its connections within the brain and to states of the external environment. At the lowest scale is scanning electron microscopy (EM), the gold standard. Not only does it reveal the fine-grained structural features of individual neurons (such as dendritic tree organization), but it has recently given birth to *connectomics* (the large-scale study of connections among all neurons in a system, or part of a system such as a piece of the retina). Connectomics was inaugurated by the recent development of serial block-face scanning EM (Denk and Horstmann 2004), but it is also being pursued with the help of diffusion-tensor imaging in MRI (Mori and Zhang 2006). Tract tracing is also pursued with more traditional methods such as anterograde and retrograde tracing that allows you to find the neurons presynaptic to, or postsynaptic to, a given neuron, respectively (Purves 2018; Wickersham et al. 2007). In addition, there are traditional histological techniques such as filling individual neurons with dye to reveal their morphology, and staining sliced sections of the brain, which helps reveal the brain's gross cytoarchitectural features. Obviously, anatomy and physiology are closely related: if two areas are revealed, via anterograde tracing techniques, to be tightly connected, then they will be a good candidate for dual intracellular recordings to find functional chemical or electrical synapses.

Before getting to the empirical evidence, a few caveats are in order.

First, a common objection to Teleosemantics is that it's too liberal—it includes too many systems as representational (e.g., Burge 2010). For example, plants and bacteria contain circadian clocks one of whose functions is to carry semantic information about the time of day; circadian clocks continue to function in the absence of stimuli and can control behavior (Bechtel 2009; Morgan 2014). According to Informational Teleosemantics, circadian clocks of plants and bacteria indicatively represent the time of day. Is this too inclusive?

No. There is nothing wrong with plants and bacteria containing representations so long as Teleosemantics does not turn everything into a representation, and it does not. Furthermore, there is a more stringent and relevant notion of representation than the one that applies to plants. That is, there are representational systems that integrate physically different sources of information by transducing them into a medium-independent system of vehicles and perform computations over such vehicles. This is the kind of representational system that nervous systems typically contain and use to guide behavior. This is what I focus on.

Second, the present account is limited to sensory, uncoupled, and motor neural representations. I am not giving an account of what is distinctive about *mental* representations. I am not attempting to explain full-blown mental or linguistic intentionality, including the ability to represent nonexistent objects like unicorns or abstractions like numbers, the ability to attach different senses to the same referent (Frege 1892), or the possession of nonnatural meaning in Grice's (1957) sense. Nevertheless, neural representations are building blocks that will be part of a complete theory of mental representation and intentionality, given suitable extensions (Neander 2017; Morgan and Piccinini 2017, Piccinini forthcoming).

Third, I am not endorsing any specific psychological theory of representation, such as the language of thought (Fodor 2008, Schneider 2011) or connectionism (P. M. Churchland 1989; P. S. Churchland and Sejnowski 1992; Clark 1993; Horgan and Tienson 1996). My account is focused on the best available evidence from experimental neuroscience. The framework I defend is limited to the notion of representation articulated above in combination with some basic ideas from information theory (Cover and Thomas 2006) and control theory (Wolpert and Miall 1996).

Fourth, accepting neural representations does not negate that cognition is dynamical, embodied, and embedded (cf. Chapter 11). Neural representations are states of dynamical neural systems, which are often tightly coupled to the organism's body and environment. As we shall see, representations acquire semantic content thanks to the coupling between nervous systems, bodies, and their environment. Perhaps biological minds also include aspects of the body and environment beyond the neurocognitive system, but this does not affect my conclusions.

12.3 Sensory Representations

When neuroscientists discuss sensory representations, they mean activation patterns in the nervous system that carry information about the current environment, are part of a broader mapping between neural states and states of the world, and help guide behavior. These are observable properties. For instance, activity in the topographic map of the visual environment contained in primary visual cortex may carry the information that the traffic light has just turned green, and this activity is used by the brain to decide to do things like pull the foot off the brake pedal. This conforms with (INDR), my teleosemantic account of indicative content.

Experimental neuroscientists have discovered sensory representations at multiple processing stages in the nervous system, from peripheral sensory representations in the retina to representations many steps removed from the periphery. Even the leech has tactile representations that guide behavior with respect to touch (Lewis and Kristan 1998b). Sensory representations provide the basic informational intersection between the local environment and the brain, giving the brain indirect access to the environment and providing organisms fallible but reliable information channels upon which to base decisions. In the next few subsections, let's examine sensory representations using examples from the visual system.[3]

[3] For a fuller treatment, see Hubel and Wiesel 2005; Rodieck 1998; Wandell 1995.

12.3.1 Image Compression in the Retina

The eye projects the visual scene onto the photoreceptor mosaic. The photoreceptors in the retina convert a specific type of physical stimulus—light—into signals that can be used by the rest of the nervous system to guide behavior.

Photoreceptors (Figure 12.1A) are not evenly distributed across the retinas, but highly concentrated in one region, the *fovea* (Figure 12.1B). Photoreceptor density drops off precipitously as you move away from the fovea into the periphery of the visual field. The optic nerve exits the eye from the back the retina. In that region of the retina, there are *no* photoreceptors, so an object projected to this region of the retina will disappear from view. This region is called the *blind spot*.

But the retina is much more than a sheet of photoreceptors. It is a complex information processing system in its own right, where a good deal of convergence and integration of information occurs (Figure 12.1A; for a review see Dowling 2012). While each human eye contains about 100 million photoreceptors, only one million axons make their way through the optic nerve. Thus, information from multiple photoreceptors is pooled and transformed, in complex ways across multiple processing stages, until the signals reach the retinal ganglion cells (RGCs) (Masland 2012). The RGCs are the output neurons of the retina—their axons snake their way to the brain via the optic nerve (Figure 12.1A). Therefore, the brain receives a kind of compact summary of the original information stream transduced by the photoreceptor mosaic.

Recording from RGCs shows that they are not merely light detectors but *feature detectors* (Sanes and Masland 2015) that respond to specific features present in the visual field in a circumscribed region of space, a region of space known as that neuron's *receptive field* (Spillmann 2014). RGCs tend to respond to *spatial contrast* in light intensity, or *differences* in light levels. One classical type of RGC has what is known as an "on-center/off-surround" receptive field, and it responds with a brisk burst of activity if you present a bright spot of light surrounded by a dark annulus (Figure 12.1C) (Dowling 2012). There is an entire subfield in visual neuroscience devoted to the study of how such receptive fields are assembled in retinal circuits (Briggman, Helmstaedter, and Denk 2011; Ding, Smith, Poleg-Polsky, Diamond, and Briggman 2016; Mangel 1991).

While neuroscientists rarely say that individual photoreceptors *represent* the visual world in primary research publications (Baylor 1987; Korenbrot 2012), once we reach the RGCs, such attributions are pervasive (P. H. Li et al. 2014; Roska and Werblin 2001; Soo, Schwartz, Sadeghi, and Berry 2011; Wandell 1995). That is, once we reach neurons with receptive fields that compactly encode information about a relatively high-capacity stimulus space, the language of representation is used frequently. This conforms to the notion of indicative representation (INDR) discussed above in Section 12.2, in which neuronal

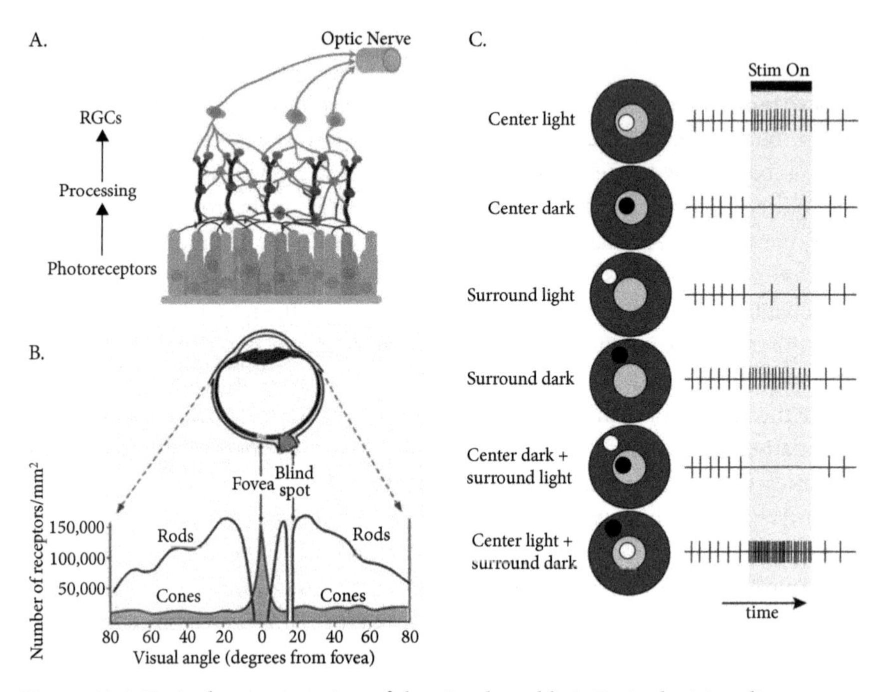

Figure 12.1 Retinal representation of the visual world. A. Retinal wiring diagram. Light activates photoreceptors, and information flows through multiple neuronal processing stages before reaching the retinal ganglion cells (RGCs), whose axons leave the eye via the optic nerve. B. Rod and cone density as a function of distance from the fovea. The blind spot is where photoreceptor density is zero, as that is where the optic nerve exits the eye. (Figure based on Wandell 1995, Fig. 3.1.) C. Cartoon showing response properties of an on-center, off-surround RGC. The entire receptive field is about the size of a dime held at arm's length from the eye. Each neuron has some background firing rate before a stimulus is presented. When lights is incremented in the receptive field center, or light is dimmed in the surround, the firing rate increases. Conversely, when brightness is incremented in the surround, or dimmed in its center, the firing rate *decreases*: Center and surround responses combine, as shown in the bottom rows. For instance the cell responds maximally when light is incremented in the center, and dimmed in the surround (bottom row).

processes carry information into the brain, forming an internal map of an aspect of the environment, and this map is used to control behavior.

12.3.2 Visual Maps in V1 and the Representation of the Blind Spot

Once leaving the eye, visual information flows through the optic nerve into the CNS, where it makes its way to the primary visual cortex (V1) (Figure 12.2A). The map of the visual world in V1 is retinotopically organized, such that two points close to one another on the retina are also represented by nearby positions in V1 (Figure 12.2B).

The representation of visual space in V1 is not an exact mirror of the visual world, but highly biased. For instance, there is much more cortical real estate devoted to the foveal region compared to the periphery. Moreover, the neurons representing the fovea have smaller receptive fields (Daniel and Whitteridge 1961; Dow, Snyder, Vautin, and Bauer 1981; Tootell, Switkes, Silverman, and Hamilton 1988). That is, V1 yields a distorted picture of visual space, with the fovea receiving disproportionately large amount of cortical territory, with more spatial acuity, compared to the periphery (Tootell et al. 1988).

In general, V1 receptive fields are more complex than those found in RGCs, with neurons tuned to more complex properties such as movement, oriented

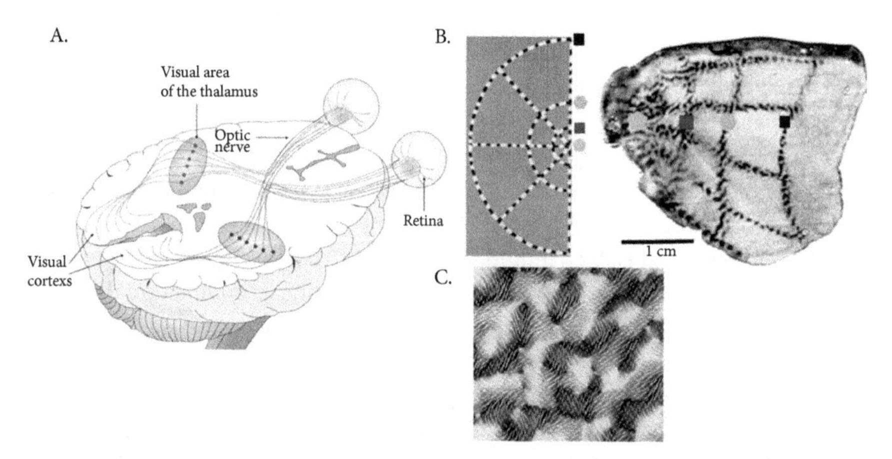

Figure 12.2 Main visual pathway. A. The visual pathway carries information from the retina to the thalamus via the optic nerve. From the thalamus, visual information makes its way to primary visual cortex (V1). (Reproduced with permission from Stangor 2011.) B. Topographic organization of V1. On the left is the visual stimulus that was presented on the left half of a video screen. On the right is the corresponding metabolic activity profile in the right-hand side of V1. The polygons on the left are shown for reference: for instance, the pentagon is next to the foveal region of the visual field, and is also shown in the foveal region of V1 on the right. The topographic organization is maintained, but with more territory devoted to regions closer to the fovea. Even though the pentagon and square are very close together in visual space, they are relatively far from one another on the surface of the cortex. (Reproduced with permission from Tootell et al. 1988.) C. Orientation map in V1. A flattened view of V1, as in 2B, but here each shade corresponds to a region of V1 with neurons tuned to bars of a particular orientation. For instance, some regions contain neurons that tend to prefer vertical edges, while other regions contain neurons that prefer horizontally oriented bars. (Reproduced with permission from Blasdel 1992.)

bars, and depth (Hubel and Wiesel 1968). V1 contains multiple fine-grained topographically organized *feature maps* of such properties embedded in the larger-scale retinotopic representation of space. For instance, those neurons selective for horizontally oriented bars tend to cluster together in *cortical columns* in V1, and nearby columns contain neurons that are tuned to similar orientations (Figure 12.2C).

What is the fate of the blind spot within V1? Dennett argued that the blind spot is not explicitly represented in the brain, as much as *ignored* (Dennett 1991). In fact, experiments using electrophysiology and fMRI have shown that the blind spot is actively represented in V1 (Awater, Kerlin, Evans, and Tong 2005; Azzi, Gattass, Lima, Soares, and Fiorani 2015; Komatsu, Kinoshita, and Murakami 2000; Matsumoto and Komatsu 2005). For instance, one study performed single-unit recordings in the region of V1 corresponding to the blind spot in capuchin monkeys (Matsumoto and Komatsu 2005). After finding the "hole" that corresponds to the blind spot in the V1 topographic map—namely, neurons that do not respond to stimuli that fall completely within the blind spot—they meticulously mapped these neurons' receptive fields using standard techniques: they presented a range of stimuli (e.g., randomly flickering lights, or oriented moving bars) to the monkey. Interestingly, they found a *topographic map* of visual space *within* the blind spot that serves to preserve the general topography found in the rest of V1 (Matsumoto and Komatsu 2005).

This representation is, by definition, not a direct response to retinal activation in the relevant region of visual space. It seems to be due to completion mechanisms that "fill in" the blind spot, interpolating based on the cues at its edges. This amounts to the system's best estimate of what is happening in the world and provides a compelling example of how even perpiheral sensory representations can be uncoupled from direct sensory stimulation. While the activity corresponding to the blind spot doesn't directly respond to sensory cues in isolation, it is still a sensory representation in the sense discussed above, as it still serves to carry natural semantic information about what is happening in that region of the world, interpolative and defeasible as it is.

12.3.3 Motion Representation in Area MT

Unless we suffer from akinetotopsia, we don't see a series of disconnected snapshots of the world, but events unfolding smoothly in time (Schenk and Zihl 1997). Time is an ineliminable factor in motion perception: turn down the frame rate of a movie of water dripping, and the drips will appear to move more slowly, because the brain takes into account the temporal delays between frames. There is a certain window into the past within which the brain integrates sensory inputs to construct its estimate of what we are presently seeing.

In the primate brain, the motion-sensitive neurons in V1 project to a region known as area MT (Movshon and Newsome 1996). Single unit recordings have revealed that almost all of the neurons in MT are tuned to movement, with individual neurons showing strong preferences for stimuli that move in a certain direction (Albright 1984; Born and Bradley 2005). Just like the feature maps in V1, these neurons are topographically organized: neurons with similar motion preferences are clustered together into cortical columns (Figure 12.3).

Activity in MT neurons is strongly correlated with behavioral performance on motion discrimination tasks (Newsome, Britten, and Movshon 1989). Indeed, when MT activity is enhanced via cortical microstimulation at a single location, perceptual discrimination is altered in predictable ways: if the stimulated region contains neurons that prefer a certain direction of motion, monkeys are biased to respond as if stimuli are moving in that direction (Salzman, Britten, and Newsome 1990; Salzman, Murasugi, Britten, and Newsome 1992). Further, if MT is chemically inactivated, performance on motion discrimination tasks is severely

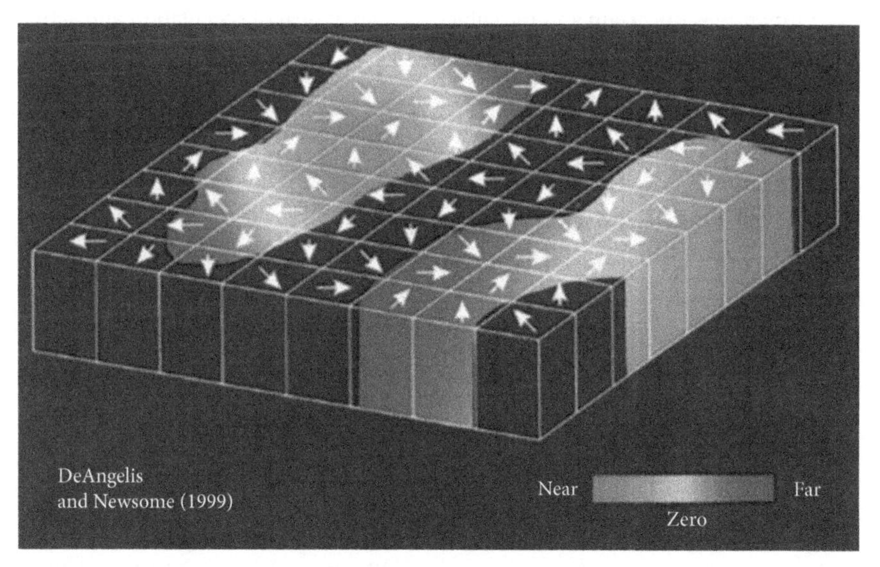

Figure 12.3 Topographic representation of motion in monkey area MT. Each block indicates a cortical column that contains neurons tuned to motion in a particular direction. The regions with shades that fade into one another represent neurons tuned to stereoscopic depth. (Reproduced with permission from DeAngelis and Newsome 1999.)

compromised, while performance on other visual discrimination tasks remains unharmed (Newsome and Pare 1988).

Cumulatively, such data suggest that MT is not accidentally correlated with visual movement, but it's used by downstream networks that adaptively control behavior with respect to visual motion in the environment. That is, area MT contains a representation of visual motion, in the sense discussed in Section 12.2 (INDR). I am not claiming that area MT *only* represents motion, or that motion representations are not learned: just as in V1, there seems to be multiplexed representation of many stimulus features (for instance, as shown in Figure 12.3, it also represents depth; for recent work on the plasticity of MT representations, see Liu and Pack 2017). Area MT is just one of many visual regions: there are dozens in the primate brain, each specialized for processing different types of information (Felleman and Van Essen 1991).

Despite the relatively simple-sounding picture painted so far, the visual system does not process information in a strict feedforward hierarchical fashion. Dense feedback connections are the norm, such as connections from MT to V1 (Maunsell and van Essen 1983; Rockland and Knutson 2000; Ungerleider and Desimone 1986). While such feedback is surely important (Gilbert and Li 2013; Hupe et al. 1998; Lamme, Super, and Spekreijse 1998; Muckli and Petro 2013), its exact function is uncertain and the subject of a great deal of speculation and active research.

12.3.4 Illusions and Other Anomalous Perceptual Phenomena

Anomalous perceptual phenomena (hallucinations, dreams, illusions) are extremely difficult to explain if perception is just skilled engagement with the world, without any need for representational intermediaries (Varela, Thompson, and Rosch 2017). This is most obvious when behavior is held fixed while perceptual content varies. Consider bistable perceptual phenomena such as the Necker cube or binocular rivalry, in which percepts alternate even if you do not change your response toward the stimulus (Figure 12.4).

Why do such perceptual alternations happen? They occur because perceptual states toggle back and forth between two "interpretations" of the stimuli. We find that peripheral sensory representations, even in V1, track the perceptual states during the presentation of rivalrous stimuli like those in Figure 12.4, and their representational contents match the perceptual contents: this has been observed using both single-unit recordings in monkeys and fMRI in humans (Leopold and Logothetis 1996; Polonsky, Blake, Braun, and Heeger 2000). For instance, when you perceive a red house, neurons tuned to the color red tend to be more active than neurons tuned to the color green.

Figure 12.4 Binocular rivalry demonstration. Project each image to a different eye (e.g., by placing a piece of paper perpendicular to the images), and fuse the checkered circles. Most people do not see a sum of the house and face, but rather the patterns alternate, seeing the house for a few seconds, and then the face for a while, and so on. This phenomenon is known as *binocular rivalry*. During transitions between percepts, the new percept will spread relatively quickly across the old as a kind of traveling wave. (Reproduced with permission from Tong et al. 1998.)

To sum up, nervous systems contain myriad sensory representational systems. Sensory representations have indicative content in the sense defined in INDR, which means that they also guide behavior. For instance, if the visual map in V1 were suddenly inverted, we would expect an animal to become completely disoriented. Such disorientation is exactly what happens in sensory prosthetic systems when the spatial mapping from sensor to brain is scrambled after an animal has learned to use the prosthesis (Hartmann et al. 2016).

12.4 Uncoupled Representations

While sensory representational systems are prototypical, representations that are uncoupled from current sensory stimulation allow us to think about things that are not present to the senses, which underwrites much of our planning, counterfactual reasoning, and abstract thought (Gardenfors 2005).

Let's consider a few simple examples of uncoupled indicative representations. Like their coupled sensory analogues, these are activation patterns in the nervous system that carry information about the (possibly past) state of the environment as part of a broader mapping between internal and external states that guides action. Thus, these uncoupled representations fit (INDR), my informational teleosemantic account of indicative representations.

12.4.1 Working Memory: From Receptive Fields to Memory Fields

Present a red square on a computer monitor, and then let the screen go blank for ten seconds. After this *delay period*, present a red square and a blue circle. A monkey or a person can reliably select the red square, indicating that they held information about the stimulus in memory during the delay period (Quintana, Yajeya, and Fuster 1988). Monkeys are excellent at working memory tasks, in some cases significantly better than humans (Inoue and Matsuzawa 2007). While neuroscientists are still actively investigating the mechanisms underlying working memory tasks, a few key observations of the underlying mnemonic representations have emerged. Let's focus on results from the prefrontal cortex (PFC) (for reviews, see Leavitt, Mendoza-Halliday, and Martinez-Trujillo 2017; Riley and Constantinidis 2016).

As demonstrated by multiple studies recording from single units during working memory tasks, a large percentage of neurons in the prefrontal cortex maintain a stimulus-specific representation of the target during the delay period (Fuster and Alexander 1971). Consider the classic *oculomotor delayed response* (ODR) task, which explores memory-guided saccades (Figure 12.5A) (Funahashi, Bruce, and Goldman-Rakic 1989; Takeda and Funahashi 2002). In the ODR task, the subject fixates on a central cue (such as a plus sign), and then a visual cue will briefly appear at a random location on the screen. After a delay period (usually between 1 and 40 seconds) the fixation point will disappear and the subject is rewarded for making a saccade to the location where the visual cue was presented.

A majority of individual neurons (recorded using single-unit extracellular recordings) in the PFC show sustained, stimulus-specific activity during the delay period. These mnemonic units have *memory fields*, an extension of the receptive field idea from sensory neuroscience: a memory field is the set of sensory cues that evoke a sustained response during the delay period (see Figure 12.5B, from Funahashi et al. 1989). PFC neurons tend to be broadly spatially tuned, responding to stimuli in the contralateral visual field, just like neurons in peripheral sensory cortical areas like V1 (Takeda and Funahashi 2002).

One problem with the basic ODR task structure is that it cannot differentiate sustained activity that is mnemonic in nature from activity devoted to motor planning and execution. Does sustained activity in response to a cue represent the sensory cue or is it just used to generate the motor output, the saccade? To overcome such problems, which emerge when sensory and behavioral variables are so tightly coupled, researchers have come up with many clever variants of working memory tasks that tease apart sensory and motor components (reviewed in Riley and Constantinidis 2016). The majority of the neurons in PFC carry sensory signals.

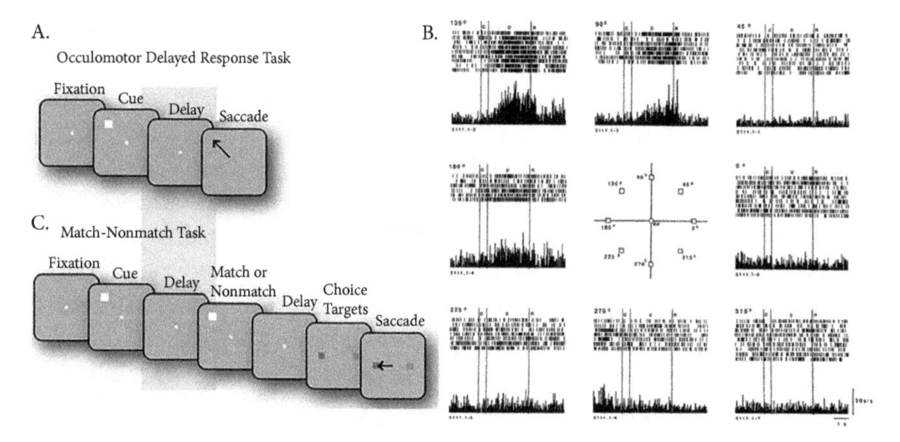

Figure 12.5 Working memory: behavioral and neuronal perspectives. A. Basic structure of ODR task described in the text. After fixation on a central spot, a visual cue appears at a random location on the screen. After a delay period of a few seconds, the fixation spot disappears, and the subject must saccade to the location of the original sensory cue to receive a reward. (Reproduced with permission from Riley and Constantinidis 2016.) B. Memory field in PFC neuron: the central graph shows the location of each possible saccade cue in the ADR, and the raster plots and poststimulus time histograms in the outer edge show the responses during the fixation, cue, delay, and saccade periods that were shown in panel A. (Reproduced with permission from Funahashi et al. 1989.) C. Schematic showing the match/nonmatch task structure. After fixation, a visual cue appears. After an initial delay period, a second visual cue appears, either at the same location (match) or a different location (nonmatch). Then, after a *second* delay period, two cues appear that contain information about where to saccade: toward the left square if the visual cues matched, and the right square if they did not match. (Reproduced with permission from Riley and Constantinidis 2016.)

For instance, when the intensity of the initial sensory cue is modulated—dim on some trials and bright on others—the sustained activity during the delay period is also modulated, with higher sustained responses to brighter stimuli (Funahashi et al. 1989). This suggests a sensory memory function, rather than a motor function, because the saccade amplitudes and trajectories are the same for the dim versus bright stimuli. More directly, in the delayed *anti-saccade* task, subjects saccade in the *opposite* direction of the initial visual cue on some trials, and in the same direction on other trials. This way, researchers can factor out which responses correlate with sensory cues, and which correlate with motor variables. The majority (59%) of individual neurons showed stimulus-selective memory effects, with 25% tuned to movement parameters (Funahashi, Chafee, and Goldman-Rakic 1993). Similar results are seen in the ingenious *delayed choice* version of the task, called the *match/nonmatch* task (Figure 12.5C). In this case, the subject initially doesn't know the appropriate saccade direction—information is only provided *after* the initial delay period (Qi et al. 2010).

Importantly, disrupting activity in the frontal cortex drastically impairs performance in delayed response tasks (Bauer and Fuster 1976; Funahashi, Bruce, and Goldman-Rakic 1993; Fuster and Alexander 1970; Mishkin and Manning 1978). For instance, localized unilateral lesions to the prefrontal cortex produces significant deficits in the ODR task when the subject is required to saccade to locations contralateral to the site of ablation (Funahashi, Bruce, et al. 1993). Monkeys with such ablations can still perform *sensory-guided* saccades, in which the visual cue stays present during the delay period (Funahashi, Bruce, et al. 1993). This suggests the deleterious effects on the ODR task are based on disruptions of sensory cue processing and maintenance, rather than motor deficits per se.

I'm not suggesting that PFC is a working memory *module*. The active maintenance of recent sensory events is not localized solely in the PFC, and the PFC has many other functions. On the first point, patterns of sustained activity during working memory tasks are distributed across multiple cortical and subcortical areas (Bolkan et al. 2017; Watanabe and Funahashi 2004), and the network mechanisms for generating this sustained activity are extremely active research topics (Murray, Jaramillo, and Wang 2017; Wimmer, Nykamp, Constantinidis, and Compte 2014; Zylberberg and Strowbridge 2017).

For instance, the posterior parietal cortex, which is involved in spatial attention and a key locus of visuospatial processing (Goodale and Milner 1992), has strong reciprocal connections to PFC. It also shows significant sustained activity during the delay period (Qi et al. 2010). However, there are asymmetries in their response properties, which suggests that PFC neurons tend to be more *directly* involved in maintaining task-specific information. For instance, the sustained activity in PFC neurons is relatively resistant to visual distractors presented during the delay period, while the sustained activity in parietal cortex is disrupted by such distractors (Qi et al. 2010). This has been replicated in other central visual areas (E. K. Miller, Erickson, and Desimone 1996).

While the general story has held up quite well for some time, we will need additional behavioral experiments and simultaneous neuronal recordings from multiple areas to nail down the full story about the working memory system, in particular its relationship to its closely related cousin, attention (Gazzaley and Nobre 2012).

Note that well-crafted behavioral experiments, coupled with the general fact that maintenance of stimulus-specific information is required to solve working memory tasks, lets us infer that working memory is representational. In particular, the *data processing inequality* from information theory (Cover and Thomas 2006) is frequently used for "big picture," constraint-based thinking about how neuronal systems operate in a behavioral context, implicitly guiding a lot of back-of-the-napkin thinking in neuroscience. Roughly, this inequality states that if an animal's behavior carries information about a stimulus (e.g., which stimulus was presented a few seconds ago in a working memory tasks), then there must be internal states in the animal that carry *at least* as much information about the stimulus in the meantime.[4] It is no coincidence that among working memory researchers there is no significant debate about whether working memory involves representations; rather, debates and experiments are guided by the desire to discover the nature and distribution of the representations. *Inferring* representations from behavior and generic informational considerations in this way is a useful, and relatively ubiquitous, first move toward representation *observed*.

12.4.2 Birdsong Learning: Memory-guided Error Correction

Birdsong learning has long gone underappreciated among philosophers as a window into neural representation. It has proved to be an extremely fruitful model system for socially learned vocal communication, with many interesting parallels with human language evolution and acquisition (Bolhuis and Everaert 2013; Doupe and Kuhl 1999; Pfenning et al. 2014; Sereno 2014).

Birdsong learning happens in two main stages (Brainard and Doupe 2002; Mooney 2009). In the first, *sensory learning* stage, young songbirds listen to a conspecific tutor sing a song and acquire a memory of the song appropriate for its locale. In the next, *sensorimotor learning* stage, they finally start to sing and ultimately come to reproduce the tutor song that they heard during the sensory learning stage. Some songbird species go months between the sensory learning and sensorimotor learning stages, with no rehearsal during the intermission (Marler and Peters 1981). Where was the song between the two phases? Not in behavior. A *memory of the song* was stored in long-term memory.

While sensory learning happens quickly, sensorimotor learning is slow and requires vocal experimentation. At the beginning of the sensorimotor stage, birds emit rambling vocalizations similar to the babbling of human infants. Eventually they begin producing sections of song resembling those of the original tutor, and then slowly shape their vocalizations until they match the template stored in memory (Mooney 2009).

Sensory feedback is crucial during this sensorimotor learning stage. If a songbird is deafened before this phase, the animal never converges on the tutor song, but it ends up producing distorted vocalizations that bear little resemblance to the tutor song (Konishi 1965). That is, songbirds undergo *feedback-guided error*

[4] Technically, the inequality states that if $X \rightarrow Y \rightarrow Z$ is a Markov Chain, then $I(X;Y) \geq I(X;Z)$, where I () is mutual information. Using this to theorize about internal states of the animal assumes that the behavior of the animal in the working memory task depends on some internal state of the animal after the stimulus was presented. This is easy enough to demonstrate by removing the brain of the animal.

correction during the sensorimotor learning phase. The error signal that guides learning is the difference between the song they produce and the memory of the tutor song (Mooney 2009).

A good deal of effort has been put into tracking down the representation of the tutor song (Hahnloser and Kotowicz 2010). There were breakthroughs on this in two recent studies of the swamp sparrow, a species that learns many different songs as juveniles in addition to the tutor song taught by its main tutor. Using extracellular recordings in a nucleus (HVC) known from ablation studies to be important for song learning and production, Richard Mooney's group found a population of neurons that selectively responds to songs in the adult bird's repertoire but exhibits stronger responses to its unique tutor song (Prather, Peters, Nowicki, and Mooney 2010). In a follow-up study (Roberts, Gobes, Murugan, Olveczky, and Mooney 2012), they discovered that disrupting activity in HVC (via optogenetic or electrical stimulation) prevents the acquisition of the tutor song.

12.4.3 Fetch! Recognition and Memory-guided Search

Consider a dog looking for a ball, or waiting patiently for its owner as it sits on the front porch. This behavior suggests there exists an internal state guiding the behavior of the dog that allows it to recognize when the relevant condition is satisfied. While there are many technical arguments in the literature about object permanence in dogs (Miller, Rayburn-Reeves, and Zentall 2009), such arguments don't block the general point that dogs seek targets and, when they reach their target, their search ends in ways that are easy to identify. Dogs are like heat-seeking missiles when it comes to games of fetch. They can fetch tens, hundreds (Kaminski, Call, and Fischer 2004), and sometimes on the order of *a thousand* toys in a home *by name* (Pilley and Reid 2011).[5]

It is hard to come up with a plausible story about such goal-directed behaviors that does not involve internal representations of the target used by the dog, a *memory of the object* that explains the animal's ability to recognize when the target has been reached. This is much like the sensorimotor learning stage of birdsong, but on a much shorter time scale: the dog is comparing current sensory cues to an internal memory of the object for which they are searching. While an animal's behavior often provides the best evidence that an animal is searching (Ryle 1949), our *explanation* of such behavior will ultimately draw on facts about internal representations, in particular *memories of specific events* and *objects*.

[5] Humans beat dogs by a good order of magnitude. One study presented 10,000 pictures to passive observers in one sitting, and they were later able to recognize them with 90% accuracy (Standing 1973).

Long-term memory stores are harder to observe and measure than occurrent sensory representations and working memories: long-term memories are not stored directly in ongoing electrical activity patterns in the brain, but in some latent variable such as long-term modifications of synaptic weights between neurons (Feldman 2012). Luckily for researchers, the study of long-term memory formation has been greatly helped by clues provided by neuropsychological patients. The most famous is HM who, after surgery for intractable epilepsy, began to suffer from severe anterograde amnesia. That is, he was unable to form new memories even though he retained memories from before the surgery. For instance, he could easily recognize faces of people he knew before the surgery, but not those he met after the surgery (Squire 2009). Interestingly, his disability was fairly specific to long-term memory: he had ordinary perceptual abilities, only slightly impaired language abilities, and in many other ways he seemed normal (Skotko, Andrews, and Einstein 2005). He could even acquire new motor skills like learning to ride a bicycle or using a new tool, even though he could not consciously remember learning them (Shadmehr, Brandt, and Corkin 1998).

The discovery of such syndromes was a watershed in the study of memory formation across the animal kingdom (Kandel 2006; Squire and Wixted 2015; Squire, Stark, and Clark 2004). It suggested that, in mammals, the hippocampal formation is involved in forming and consolidating long-term memories, but that extra-hippocampal regions (such as the cerebral cortex) are the long-term stores for such memories. This view has held up to more rigorous tests in model systems (Kitamura et al. 2017; Morris, Garrud, Rawlins, and O'Keefe 1982; Nabavi et al. 2014; Whitlock, Heynen, Shuler, and Bear 2006). For instance, activity-dependent cell labeling is a recently developed technique that allows researchers to selectively stimulate neurons that are active in a given context (e.g., a room with specific visual cues) (Reijmers, Perkins, Matsuo, and Mayford 2007). In a recent "false memory implant" paradigm, researchers used this technology to genetically tag a group of cells active in a certain chamber; and then they used optogenetic stimulation on these same neurons, but paired this stimulation with a foot shock, which evokes a behavioral "freezing" response. Later, when introduced back to the original chamber, the animals exhibit a freezing response, even though they had never actually been shocked in that room (Ramirez et al. 2013).

12.5 Motor Representations

While sensory representations carry information from the world into the brain, motor representations have a different role. They function as commands: instructions sent from the brain telling the body where to move in the world.

Whereas the experimental study of sensory representations typically involves observing neuronal responses to different stimuli, the study of motor processes is often quite different. Since the motor system is sending *commands* to the body to move, one mainstay in the study of motor control is the artificial generation of movements via the electrical stimulation of the motor regions of the nervous system. Neuroscientists then assign content to motor representations by observing the resultant patterns of muscle contractions, movements, and adjustments of the external environment. This is in line with (IMPR), my teleosemantic account of the semantic content of imperative representations.

While the content of some sensory representations is potentially phenomenologically available, the same is not obviously true of motor representations. When gross *disorders* of the motor system emerge, however, the results are usually obvious to everyone. Errors in motor systems involve a breakdown in either the production or readout of commands sent to the body to move. In extreme cases, such as Tourette's syndrome or hemiballismus, the body engages in large-scale involuntary behaviors such as flailing or vocalizations. In cases of peripheral nerve deficits, such as spinal paralysis or Lou Gehrig's disease, intentions to move are still produced centrally, but not extracted appropriately by the downstream circuits that normally control behavior (Truccolo, Friehs, Donoghue, and Hochberg 2008).

Let's consider some examples of motor representations, from motor maps to efference copies, and finally end with a brief discussion of *mixed* representations, which have both indicative and imperative content.

12.5.1 Motor Maps: From Homunculi to Ethological Action Maps

In the primate brain, the primary motor cortex (M1) is part of a large collection of regions that act together to control movement. If you briefly stimulate a small region of tissue in M1, you will typically produce a localized muscle twitch in response. Based on such studies, a somatotopic map of the motor representation in M1 can be built, a *motor homunculus* (Figure 12.6A) (Penfield and Boldrey 1937). These somatotopic maps are analogous to the retinotopic maps we met on the sensory side (Figure 12.2). When a particular region in the homunculus is damaged, animals, including humans, typically experience immediate deleterious side effects in the form of paresis—an impairment in voluntary control that includes partial paralysis and loss of fine motor-control—in the corresponding body part (Darling, Pizzimenti, and Morecraft 2011).[6]

[6] Note that ablating M1 does not always lead to *permanent* paresis, but more short-lived and subtle motor deficits (Schwartzman 1978). Sometimes such ablations show no notable motor deficits, but instead deficits in motor *learning* (Kawai et al. 2015 though see Castro 1972, Makino et al. 2017). Such results undermine simple stories in which M1 is the final common output driving all movement. Some of the most recalcitrant movement deficits such as Parkinson's disease result from damage to

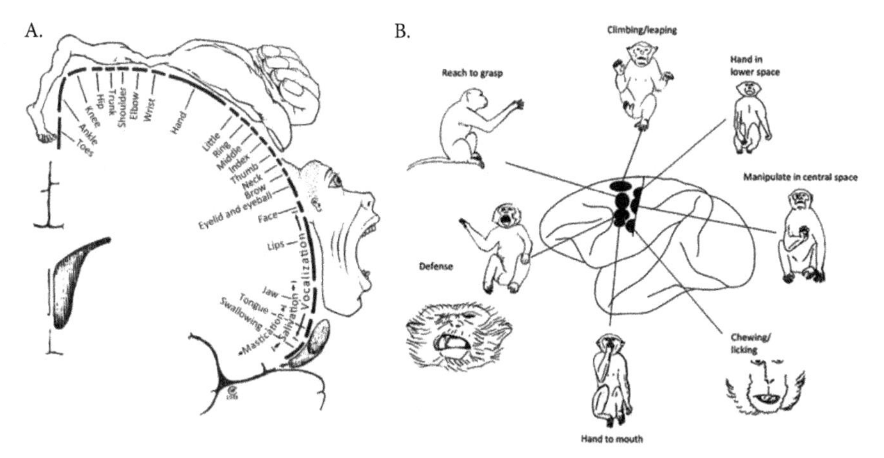

Figure 12.6 From muscles to action maps. A. Traditional motor homunculus, showing where muscles twitch when (human) M1 is briefly electrically stimulated. From (Penfield and Boldrey, 1937). B. Ethological action maps show the clusters of action types evoked by longer trains of electrical stimulation delivered to M1 in unrestrained monkeys. (Reproduced with permission from Graziano 2016.)

When researchers provide stimulation trains on ethologically relevant time scales (around half a second) and do not restrict animal movement, the motor cortex exhibits an additional organizational principle. Instead of individual muscle twitches, ethologically meaningful behavioral patterns involving coordinated activity among multiple muscle groups are observed. These actions include taking defensive postures, moving the hand toward the mouth, or reaching the hand forward as if to grasp an object (Figure 12.6B) (Graziano, Taylor, Moore, and Cooke 2002). Such *ethological action maps* have now been observed in microstimulation studies in multiple primate and rodent species (Graziano 2016).

While stimulation studies are very informative, they are somewhat artificial. The brain doesn't work by indiscriminately shocking localized voxels in the cortex, or individual neurons (Brecht, Schneider, Sakmann, and Margrie 2004). Real motor control is subtler, with commands broadly distributed across a diverse collection of cortical areas, subcortical nuclei, and the cerebellum (Hewitt, Popa, Pasalar, Hendrix, and Ebner 2011; Houk and Wise 1995; Shadmehr and Krakauer 2008). To examine the representations that the *brain* (as opposed to experimenters) generates, we must observe how the brain's activity naturally unfolds as animals engage with the world in real time.

subcortical structures like the basal ganglia. As discussed briefly at the end of this section, motor control is distributed across multiple cortical and subcortical areas, and the focus on M1 here is a convenience meant to keep the discussion contained, not an endorsement of strict localizationist theories of M1 motor control.

12.5.2 Receptive Field Envy: Movement Fields, Force Fields, Goal Fields?

If you were to observe the brain of a monkey moving about in the world, you would notice that neuronal activity in M1 precedes movement of its body in highly reproducible ways. For instance, before the monkey moves its hand, there is elevated activity in its hand representation in M1 about 100 ms before the movement actually starts (Kakei, Hoffman, and Strick 1999). Indeed, the cortex is such a reliable indicator of future movement that some of the most promising clinical work lies in extracting intended movements from activity in the motor cortex of paralyzed patients, and then using such signals to control prosthetic limbs (Alexander and Crutcher 1990; Ganguly and Carmena 2009; Hochberg et al. 2012; Truccolo et al. 2008).

What shape do motor representations take in M1? This is a very active area of research, but there are a few results with which everyone should be familiar. In one classic study, researchers trained monkeys to move their hands in one of eight directions while recording from individual neurons. Similar to the movement-sensitive sensory neurons in area MT, neurons in M1 showed pronounced *direction tuning*. That is, they fired more action potentials before the monkeys moved their hands in a particular direction (Figure 12.7: Georgopoulos, Kalaska, Caminiti, and Massey 1982). This led to the concept of a *movement field* for M1 neurons.

It turns out that movement direction is just one of many features to which individual neurons in M1 are tuned. There have been vigorous debates in the literature about whether M1 neurons are better described as tuned to velocity, force, activity in individual muscles, muscle synergies, or goal-directed acts such as grasping small objects (H. T. Chang, Ruch, and Ward 1947; Georgopoulos and Ashe 2000; Griffin, Hoffman, and Strick 2015; Holdefer and Miller 2002; Kakei et al. 1999; Moran and Schwartz 2000; Scott 2000; Todorov 2000a, 2000b; Umilta et al. 2008). Populations of neurons in M1 appear to represent all of these different (and often highly correlated) variables in flexible ways that can change rapidly with task context, posture, and learning (for a review, see Kalaska 2009).

It seems that the same neuron does not always encode a single parameter, but can be recruited to help produce different behaviors in different contexts, such as when an external force is applied to the arm (C. S. Li, Padoa-Schioppa, and Bizzi 2001), and the same neuron will produce different activity patterns when the animal is preparing to move versus actually moving (Elsayed, Lara, Kaufman, Churchland, and Cunningham 2016).

There have been many recent debates about M1 representations, with two main axes along which opinions have tended to diverge: the kinetic/kinematic axis, and the implicit/explicit axis (Figure 12.8). I have already implicitly discussed the first,

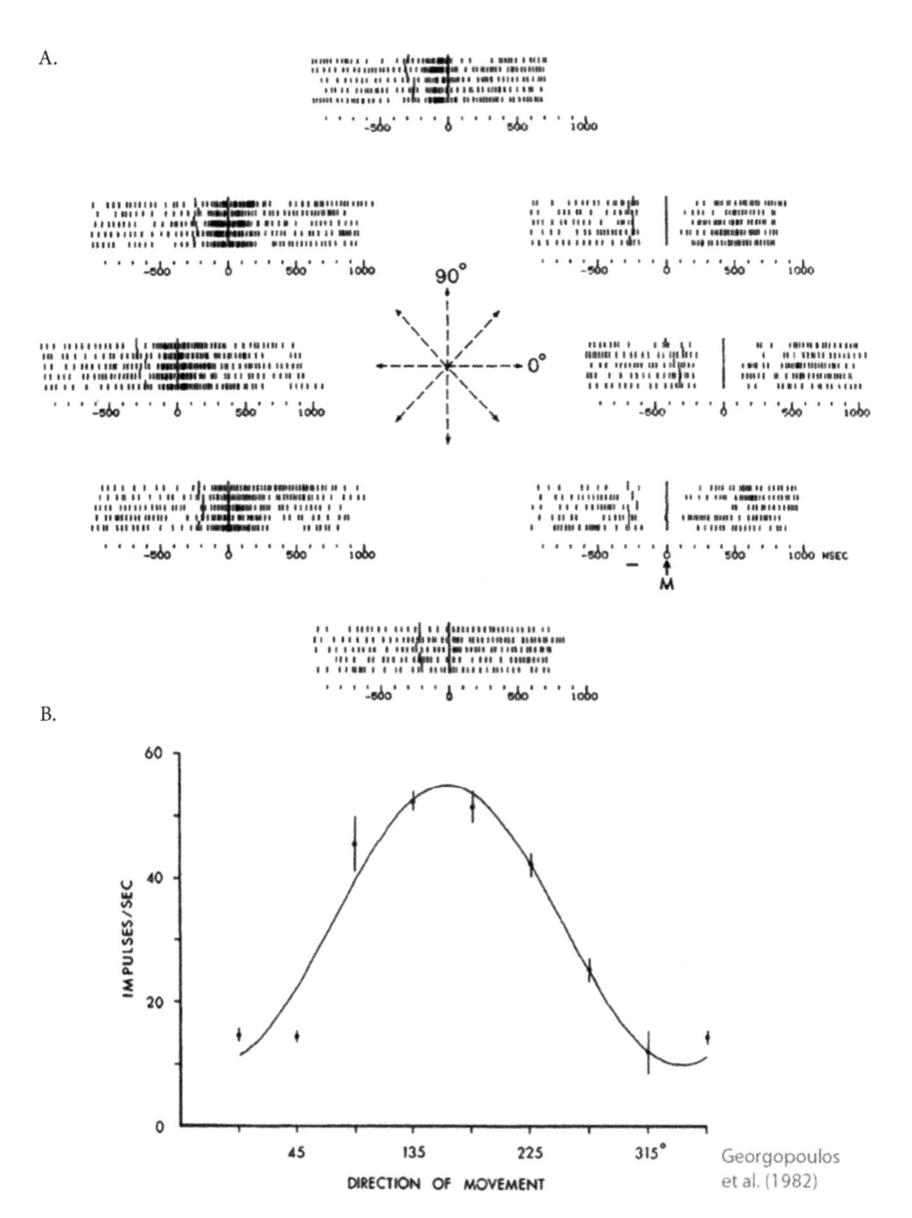

Figure 12.7 Movement field in M1. A. Response of a neuron in M1 to hand movements in eight different directions. Response is displayed as a raster plot for each direction. A raster plot shows a different trial on each row, with individual action potentials as tic marks. (Reproduced with permission from Georgopoulos et al. 1982.) B. Movement field shows the mean number of spikes as a function of the direction in which the monkey moved its hand. This neuron preferred leftward movement (Georgopoulos et al. 1982).

kinetics/kinematics, axis. Briefly, this is about whether M1 neurons encode kinematic features of movement, such as velocity and position (Georgopoulos et al. 1982; Georgopoulos, Schwartz, and Kettner 1986; Wessberg et al. 2000), or kinetic factors that cause movements, such as muscle forces or electrical activity in muscle fibers (Holdefer and Miller 2002; Sussillo, Churchland, Kaufman, and Shenoy 2015).

On the implicit/explicit front, many researchers have argued that there are no fixed parameters encoded by individual M1 neurons (Fetz 1992; Pruszynski, Omrani, and Scott 2014; Sussillo et al. 2015). Rather, it is entire *populations* of M1 neurons that encode such parameters, in such a way that the experimenter (or downstream consumer networks that actually control behavior) must extract the relevant parameters from such activity.

Implicit coding is a quantitative notion, such that a neuronal population *implicitly* represents some feature proportional to the computational cost required to extract information about that feature (Kirsh 2006). "Computational cost" is typically cashed out in terms of the types of computations that neural networks can easily perform, such as taking weighted sums (Koch 2004), or linear classification (Rust 2014).[7] This analysis makes intuitive sense: if someone merely *implies*

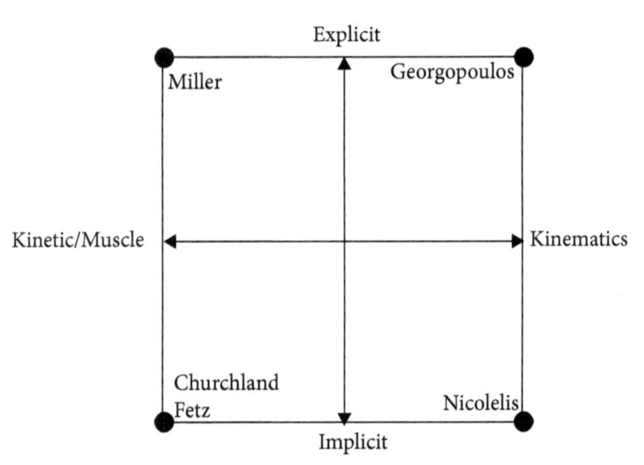

Figure 12.8 Space of M1 representational theories. Debates about M1 representational structure have centered around where in this space M1 representations best fit. See text for details.

[7] Discussions of explicit/implicit coding have always taken place in the sensory system, so it is not actually clear if these are the correct standards to use for M1, which tends to send its outputs to muscles and central pattern generators (Kalaska 2009).

something, this suggests there is a cognitive cost incurred in *extracting* the implied content. The idea is the same here. For instance, if there is a face in your visual field, there is an implicit representation of a face in your retinal ganglion cells because it would require a great number of computational steps to extract this information. Compare this to visual area IT in monkeys, which has individual neurons that fire like crazy when faces are present (Chang and Tsao 2017; Perrett, Rolls, and Caan 1982). These provide explicit representations of faces, because the computational cost required to extract the information that there is a face present in the visual field is relatively low.

For the past few decades, this distinction played out in arguments about how M1 represents movement (Fetz 1992). For instance, in Georgopoulos' framework, velocity is explicitly coded in the firing rate in M1 neurons (recall Figure 12.7).[8] On the other hand, other researchers have suggested that information about movement is not reliably coded in individual M1 neurons, and to extract the intended movement based on M1 activity requires relatively sophisticated decoding algorithms applied to large populations of M1 neurons, often taking into account their dynamical (time-dependent) nature. This can be found in the work of Nicolelis (Wessberg et al. 2000) as well as more recent work from Mark Churchland (Sussillo et al. 2015). In practice most people are not extreme advocates of either pole, falling somewhere in the middle in Figure 12.8 (e.g., you can find elements of both types of codes discussed in Wessberg et al. 2000).

Presumably, those who think M1 only encodes movement implicitly would say that the downstream spinal networks that control behavior must "extract" the information implicitly contained in the M1 commands and generate the appropriate movement sequences, such as walking, based on such commands. This is not at all far-fetched. In fact, it is sort of the standard model of motor control: it has been known for over a century that spinal networks contain central pattern generators and sensory-feedback mechanisms that allow mammals to carry out extremely complex motor trajectories like walking even when the cerebral cortex has been completely severed from the spinal cord (Rossignol and Bouyer 2004; Sherrington 1910).

Recently, Mark Churchland has defended an allegedly nonrepresentationalist account of M1 function. Specifically, he argues against the existence of traditional velocity-tuned neurons, and he focuses on the heterogeneous and dynamical nature of populations of M1 neurons in the direct control of movement (M. M. Churchland, Cunningham, Kaufman, Ryu, and Shenoy 2010; Sussillo et al. 2015). When you dig deeper, however, his work seems to fall cleanly into

[8] Note that "implicit" is not the same as "distributed" or "population" code. "Implicit" implies "population" but not vice-versa. Even in Georgopoulos' work, perhaps the *locus classicus* of explicit motor representations, to know the velocity of the animal's arm you must know the firing rate of the population of M1 neurons. That is, everyone accepts that motor control involves a distributed code.

the implicit/kinetic corner of the representational space in Figure 12.8. That is, his research seems to be pushing against *a species of* representational theory, rather than representations *tout court*. Indeed, in a recent modeling paper, they say, "[A]lthough *the model certainly contains an implicit representation of the upcoming EMG*, individual-neuron responses rarely match the patterns of EMG" (Sussillo et al. 2015, 1032, emphasis added). EMG is a measure of electrical activity in muscles, so this falls into the kinetic side of the representational landscape.

If M1 representations are indeed implicit, one implication is that researchers will need to record activity from populations of neurons in order to faithfully reconstruct the representational content of the activity. It is likely no coincidence that as multielectrode recording techniques become more prevalent, the importance of implicit coding in M1 and other areas has become clearer.

It is important to note that M1 does not represent movement simply because its activity occurs *before*, or *causes* movement. If that were the case, then even electrical activity in muscle fibers would represent movement. Interestingly, just as individual photoreceptors are not described in (sensory) representational terms, muscles are *never* described in (motor) representational terms. This may seem a trivial point, but the transform from electrical activity in muscle fibers to bodily motion is quite complicated—electrical activity in muscles causes muscle shortening via complex biochemical cascades; the forces generated are filtered through the springs and masses of the musculoskeletal system, and are strongly influenced by what the other muscles in the body are doing (Buchanan, Lloyd, Manal, and Besier 2004; Lloyd and Besier 2003).

That is, *despite* the time delays and inherent complexity involved in the transform from muscle activity to movement, muscles are not representational. Instead, motor representations take the shape of an ongoing superposition of more centrally generated behavioral commands. The particular superposition produced at a given moment depends on the current goals of the organism, sensory representations, and recent movements. It is *this ongoing, goal-directed command, and its construction*, that fits (IMPR), my account of the semantics of imperative representations—not electrical activity in muscle fibers. Muscles and bones "consume" motor representations—they are *part of the body* and the ultimate recipients of motor commands. After all, as discussed above, you can have active intentions to move without overt movement, as in cases of paralysis or other disorders of motor control.

In general, in both sensory and motor representational systems, it is standard to quantify the *accuracy* of the system (or, inversely, its error level). In motor systems, this accuracy is quantified in terms of the closeness to the goal (e.g., did the eye saccade to the target, did the hand reach the apple). In sensory systems, it is in terms of how accurately you can reconstruct the stimulus, given the neuronal response (Bialek and Rieke 1992; Thomson and Kristan 2005). These two error measures conform to the different representational functions of motor and sensory systems, respectively.

It is in virtue of the goal of a motor command that we can assign an error measure to the command, and track this error to multiple sources. It could be representational (e.g., in Tourette's syndrome, which seems to be partly due to unintended disinhibition of basic motor commands in the basal ganglia), have to do with faulty readout of the command, which can be quite drastic (e.g., spinal cord damage), or due to musculoskeletal failure (e.g., you fail to lift that barbell a tenth time because of muscle fatigue). In the case of muscle fatigue, the failure is due to the readout mechanisms of your commands, much like you could see a shape incorrectly when seen through a distorting lens (the sensory analog of Tourette's would be a centrally generated visual hallucination).

12.5.3 Efference Copy and Sensory Cancellation

When the brain sends commands to the body to move, *the brain sometimes signals to sensory parts of the brain that it has sent this command.* In the literature, these signals are called *corollary discharge*, or *efference copies*.[9] Evidence is growing that efference copy is important in both perception and motor control (Crapse and Sommer 2008; Wolpert and Miall 1996). Let's consider its role in each process. One reason that efference copy bears emphasis is that some theorists have expressed skepticism about its existence (Clark 2016).

The most direct evidence that corollary discharge is important in perception comes from the eye movement system (Collins 2010). When you voluntarily move your eyes via the rapid, ballistic eye movements known as *saccades*, the world doesn't seem to jump about erratically. However, if you poke your eye (gently) with your finger, the world *does* appear to jump. When you perform saccades, the retinal motion is actually much larger than when you poke your eye with your finger, but somehow the world looks relatively stable. These phenomena suggest that the visual system uses efference copies generated by your eye-movement system to "subtract out" the sensory consequences of eye movement, a phenomenon known as *saccadic suppression* (McFarland, Bondy, Saunders, Cumming, and Butts 2015).

[9] Sometimes these terms are used differently. For instance, "corollary discharge" is sometimes taken to be the output of a forward model (Section 12.5.4). However, the two terms are typically used as synonyms in the literature. For instance, "A ubiquitous strategy is to route copies of movement commands to sensory structures. These signals, which are referred to as corollary discharge (CD), influence sensory processing in myriad ways" (Crapse and Sommer 2008). It would be a mistake to conclude that a paper doesn't support the existence of efference copy just because it uses the phrase "corollary discharge."

In general, it is often important for animals across the phylogenetic spectrum to keep track of which sensory responses are self-generated and which have other causes. There is evidence of efference copy from multiple sensory systems in multiple phyla (for a review, see Crapse and Sommer 2008). Let's consider a few examples, starting with the superior colliculus in the eye movement system.

The superior colliculus (SC) is a complex multi-layered brainstem nucleus that is involved in visual selective attention and orienting responses, including saccades (Krauzlis, Lovejoy, and Zenon 2013). Neurons in its superficial layers are visually responsive, displaying keen motion sensitivity. Unlike neurons in area MT (Section 12.3.3), most SC neurons are "pandirectional," preferring no particular direction: they simply fire when something is moving at a particular location in space (Goldberg and Wurtz 1972). Neurons in the deeper layers contain a (motor) representation of saccade direction and velocity, and exhibit classical movement fields (as in Section 12.5.2). If you stimulate deep-layer SC you will generate saccades or even whole-body orienting toward stimuli; ablation of deep SC layers disrupts saccades (Gandhi and Katnani 2011; Sparks, Lee, and Rohrer 1990). It has been suggested that the superficial visual layers of SC form a kind of "salience map" of the visual field, and that there is a sensorimotor transformation from superficial to deep SC, with the latter generating orienting responses to the salient stimuli represented in the superficial layers (Kustov and Robinson 1996; White et al. 2017).

The problem animals face is that it would be maladaptive to constantly sense salient motion that they have generated *by moving their own eyes*: this would be a constant source of distraction if the goal was to use the SC to orient toward stimuli moving in the environment. The solution to this problem is powerful saccadic suppression in the superficial layers of the SC (Robinson and Wurtz 1976). That is, the same moving stimulus that would generate a large response in superficial SC generates no response to identical visual inputs when such inputs are due to self-generated saccades.

There are a few reasons to think this extra-retinal SC suppression is generated by efference copy rather than sensory inputs from eye movements (such as proprioception). One, even in complete darkness, superficial SC baseline activity is suppressed during saccades (Robinson and Wurtz 1976), which suggests the suppression is not generated by visual stimuli. Two, saccadic suppression persists in animals that are sending out saccade motor commands even when eye movements are prevented by paralyzing the eye muscles (Richmond and Wurtz 1980). This suggests such saccadic suppression is not generated by proprioception or actual movement, but *intended* movement. Third, recent anatomical analysis of SC circuits, single-unit recordings in brain slices revealed that deep SC neurons send out an axonal arbor that loops back to excite superficial SC inhibitory interneurons, providing a channel for the efference copy along with a mechanistic account of saccadic suppression (Phongphanphanee et al. 2011).

Efference copy is not confined to vision, or mammals. In the cricket, there is a single interneuron, aptly named the "corollary discharge interneuron" (Poulet and Hedwig 2006) that responds to the central pattern generator that produces the wing movements responsible for their chirps. This interneuron exerts enough inhibitory control on auditory neurons to filter out its own auditory signals, but the cricket can still respond to externally generated auditory cues (Poulet and Hedwig 2003).

Note that efference copy and its resultant sensory compensation do not always produce generic suppressive responses (Confais, Kim, Tomatsu, Takei, and Seki 2017). One interesting case comes from weakly electric fish of the family Mormyridae (Bullock 1982). Such fish contain an electric organ that periodically generates an electromagnetic (EM) field, an *electric organ discharge (EOD)*: electric fish use the resulting distortion of the local EM field to sense objects in their environment (Heiligenberg and Bastian 1984; Krahe and Maler 2014). How do the mormyrids differentiate self- and environmentally-generated changes in EM fields? The efference copy mechanism has been worked out relatively well (Bell, Libouban, and Szabo 1983; Carlson 2002). Interestingly, in the brain region that processes the response to the EOD, efference copy doesn't produce some generic suppressive effect, but a *negative image* of the sensory consequences of the EOD, a negative image that cancels out the response usually produced by the animal's EOD (Bell 1981). Figure 12.9 is a schematic depiction of this sensory cancellation process.

What if you artificially *distort* the EM field in the animal's environment after every EOD? Amazingly, within minutes, they acquire the ability to cancel such artificially generated EM fields. In other words, the electrosensory system rapidly learns to cancel out novel sensory consequences of its EOD, seeming to interpret the predictable EM field as a self-generated environmental perturbation whose consequences need to be filtered out during its search for externally generated

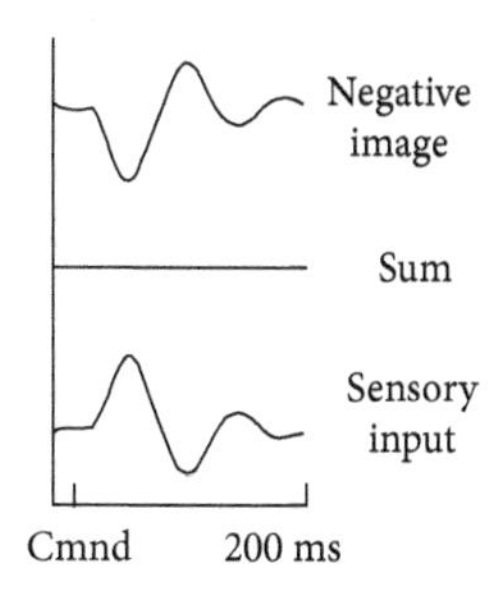

Figure 12.9 Sensory cancellation in weak electric fish. Cartoon representation of input to sensory neuron in the PLL in mormyrid. The sensory input comes in from the ampullary electrosensory system, the negative image is generated via processes downstream from the command nucleus that triggers the EOD. The two signals converge in cells in the PLL, cancelling each other. (Reproduced with permission from Kennedy et al. 2014.)

sources of change (Bastian 1996; Bell 1982). This is now one of the better-studied instances of efference copy and sensory suppression. It has been studied in depth at the cellular level (Kennedy et al. 2014).[10]

Efference copy does not directly control behavior, so it does not fit IMPR. Therefore, efference copy is not an imperative representation. But efference copy does carry natural semantic information about impending movements and their effects. More precisely, a function of the system is to produce efference copy so that it carries natural semantic information about P, efference copy is among a range of similar states that map onto a range of external states similar to P, and efference copy can guide the agent's behavior with respect to P. Thus, efference copy fits INDR: it is an indicative representation.

12.5.4 Efference Copy in Motor Control: Forward Models

Efference copy is not only important for sensory processing. It is integral to what has become one of the main models of voluntary motor control. This involves the construction of *forward models that predict the sensory consequences of particular behaviors*. The following summary will necessarily be brief (for more detailed reviews, see Shadmehr, Smith, and Krakauer 2010; Wolpert and Miall 1996).

When moving around in the world, we do not passively respond to incoming sensory inputs such as the weight of a heavy box we are lifting. Our brain builds up expectations and predictions about what is going to happen. It uses those predictions, coupled with our goals, to shape our behavior *before* those consequences happen. We brace ourselves *before* picking up heavy objects. We prepare our bodies to catch balls well before the ball arrives (Lacquaniti, Borghese, and Carrozzo 1992). There is growing evidence that such predictive anticipatory processes are knitted into basic motor control mechanisms.

Let's consider a concrete example. When you are shaking a saltshaker, you do not grip it with the exact same force the whole time. Rather, you typically unconsciously change your grip force in ways that *anticipate* the changes in torque and load force that will be exerted against your hand, so that the saltshaker will not fall out of your hand. Your grip increases to compensate for changes in the frictional forces at key times, such as when you jerk the shaker at the bottom of its trajectory and is most likely to slip out of your hand (Johansson and Cole 1992).

[10] Why would such plasticity be useful in the electric fish? The local EM fields produced by the same EOD can change depending on changes in water resistivity, or if the animal is swimming, or spending considerable time next to a nonconducting surface such as a rock or air at the water's surface (Bell 1982), so the sensory consequences of the EOD are likely variable enough that it is helpful to learn them (Bell 1981).

Obviously, the motor commands coming from the motor system are sent *before* muscle contraction. Hence, the increase in grip forces that are synchronized to load force changes cannot be happening in response to sensory feedback: our motor control machinery is somehow *anticipating* what is going to happen. For that matter, people will often change their grip *before* the relevant environmental events even occur: for instance, people will increase grip force just *prior* to lifting an object (Forssberg et al. 1992). Thus, adjusting motor commands to anticipated changes requires memory of previous interactions with objects (Johansson and Cole 1992).

Evidence has started to converge that the brain accomplishes this feat using internal feedback loops that predict sensory feedback, also known as *forward models*. In the simplified model in Figure 12.10, the motor controller generates a motor command to directly control behavior (e.g., from M1 to the spinal cord). But this motor command also branches off into an efference copy that is delivered to a system that contains a model of the controlled domain and predicts the sensory consequences of the behavior that will result from the command. This forward model receives efference copies and has the function to *predict the sensory consequences of that motor command*. That is, what sensory feedback will this action create? In the case of gripping an object, what tactile responses will the

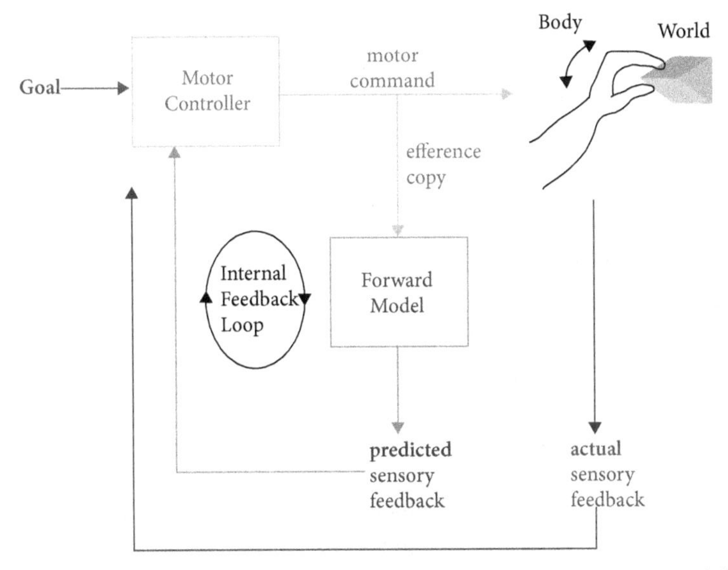

Figure 12.10 Forward models and efference copy. Cartoon representation of the use of forward models for anticipatory motor control. See text for details. Forward models use efference copies of the motor command to predict the sensory consequences of that motor command, generating internal predictions that can be used by the motor control systems to generate appropriate behaviors before those sensory consequences actually happen.

saltshaker produce: in particular, will it slip? If so, go ahead and create the appropriate changes in grip strength *before* that deleterious consequence actually happens.

I have already implicitly discussed one forward model in the weakly electric fish. There the motor command was the same every time: "Produce an EOD." When shaking a salt shaker, the motor command will continuously vary over time. However, if the internal model is accurate, it will predict the sensory feedback that will be generated by your behavior with the particular object you are using. Calculating this sensory prediction will take some time, but the point of the forward model is that it produces internal feedback significantly faster than *actual* sensory feedback from the body, so the motor controller can beat it to the punch (Wolpert and Miall 1996).

I have left out many details. One, how is the forward model supposed to be acquired and updated? As you might guess, this would involve tuning model parameters based on sensory prediction error: the difference between actual feedback and predicted feedback (Shadmehr et al. 2010). I also left out discussions of how predicted sensory feedback is combined with *actual* sensory feedback during online motor control—after all, if you are shaking a salt shaker and feel it starting to slip out of your hand, you will not just keep on going as if nothing is happening (Scott 2016). Third, how detailed is the internal model: does it contain explicit representations of the body and environment or does it construct simple lookup tables, something just simple enough to help an animal get by, like in the electric fish? There are reasons to think that the models are more generalizable than that and show nuanced sensitivity to physical and dynamical features of the body and manipulanda (Lisberger 2009).

Stepping back from its theoretical virtues, is there *evidence* that forward models actually exist, that they are instantiated in brains? Early on, much of the evidence for these models was behavioral, or based on engineering or mathematical considerations (Volkinshtein and Meir 2011). But in neuroscience the armchair isn't good enough: experiments are needed. Initially, because of its intrinsic anatomical structure and connectivity patterns with the rest of the brain, and the fact that damage to this region impaired feedback-guided motor control, the *cerebellum* was the prime suspect to be involved in many of the operations attributed to the forward model (Ito 1970; Wolpert, Miall, and Kawato 1998).

Decades have passed since Ito first posited forward models in the cerebellum, and there has been an ongoing confluence of theory and experiment, with many studies confirming that the cerebellum is indeed a locus of many expected features of forward models:

- The cerebellum directly receives efference copies. For instance, researchers showed that a population of spinal neurons crucial for motor control also sends collateral axonal branches to a cerebellar input nucleus, effectively sending an efference copy from the spinal cord up to the cerebellum. Then, using optogenetics, they silenced this efferent branch, and the animal's reaching behavior was severely compromised, thereby showing the importance of this efference copy in active motor control (Azim, Jiang, Alstermark, and Jessell 2014).

- The cerebellum generates predictions of sensory consequences, which are the expected output of forward models. Researchers showed that one lobe of the cerebellum contains sensory predictions about visual stimuli: namely, visually responsive neurons respond not just to actual visual stimuli but to *expected* visual responses during a visual tracking task in cats, even when visual stimuli are temporarily shut off during the task (Cerminara, Apps, and Marple-Horvat 2009). This suggests that the cerebellum is not just involved in active sensory representation, but in sensory *prediction.*
- The cerebellum contains sensory prediction errors, which are required for sculpting forward models. A recent study, using extracellular recordings of single units in monkeys, showed that neurons in an output nucleus of the cerebellum track sensory prediction error, or the discrepancy between predicted and actual sensory feedback (Brooks, Carriot, and Cullen 2015).

While neuroscientists have come a long way in our study of forward models, there is still a long way to go. The cerebellum is an extremely complex, variegated structure, containing at least three times as many neurons as the cerebral cortex (Herculano-Houzel 2010). So far, the handshake between theory and experiment has been promising.

12.5.5 Mixed Representations

Another type of sensorimotor processing involves *sensorimotor transformations,* in which sensory representations are quickly and directly converted into motor commands (Lewis and Kristan 1998a; Salinas and Abbott 1995). This is what happens when the frog flicks its tongue toward the fly, or you saccade to where a bright light appears in your visual field. Such sensorimotor transforms often involve individual neurons, and neuronal populations, that construct simultaneously sensory and motor representations. More generally, despite all the distinctions above, I am not suggesting that there are sharply segregated sensory and motor modules in the nervous system (Cisek and Kalaska 2010; Matyas et al. 2010).

12.6 Neural Representations Observed

I have briefly reviewed a small but representative sample of empirical findings about neural representations. Sensory and uncoupled representations are maps of aspects of the environment that are dynamically updated thanks to sensory information (plus, possibly, internal feedback from efference copies and forward models) and guide behavior based on the information they carry. Motor representations are action maps that operate in coordination with indicative representations and guide behavior to accomplish what they represent. Thus, the functional role and explanatory power of neural representations depends on their semantic content.

While much empirical and conceptual work is still needed to fully understand neural representations, one conclusion is safe. A number of theoretical considerations predict that, given the complex control functions of nervous systems, structrural representations must be involved. Using a variety of methods and tools, neuroscientists have empirically confirmed this prediction. Some of the complex neural states interleaved between environment and behavior are structural representations. Neural representations are observable, quantifiable, manipulable, and have received multiple independent lines of empirical support. Therefore, neural representations are real—as real as neurons, action potentials, and other entities routinely observed and manipulated in the laboratory. The long-standing debate over representations is finally settled.

Having established that neurocomputational vehicles are representations, I will examine the computations that construct and manipulate such neural representations. Are neural computations digital, analog, or a third kind? This is what we now turn to.

13
Neural Computation

13.1 The Many Versions of the Computational Theory of Cognition

McCulloch and Pitts (1943) were the first to argue that neural activity is computation in the sense defined by Alan Turing (1936–7) and other logicians and that neural computation explains cognition (Chapter 5). McCulloch and Pitts argued that neural computation is digital; Lashley and others countered that it is analog (Gerard 1951; Lashley 1958, 539). Their view is the origin of the Computational Theory of Cognition (CTC). I explained what that amounts to in Chapters 6-9.[1]

Some authors (Fodor 1975; Newell and Simon 1976; Pylyshyn 1984) formulate CTC as the thesis that cognition is computation or that computation explains cognition—without reference to the nervous system. Since cognition in biological systems is a function of the nervous system, the computations that putatively explain biological cognition are carried out by the nervous system. Following the mainstream literature, I refer to them as *neural computations*. The topic of this chapter is the nature of neural computation. Consequently, for present purposes CTC is the thesis that neural computation (simpliciter) explains cognition. There may be artificial forms of cognition explained by other—nonneural—types of computation.

Over the last seven decades, CTC—in its various classicist, connectionist, and neurocomputational incarnations—has been the mainstream theory of cognition. Roughly speaking, classicism is the view that the type of computation that explains cognition is digital computation over language-like vehicles, like the kind of digital computation that occurs in digital computers (Fodor and Pylyshyn 1988). Both connectionism and computational neuroscience maintain that cognition is explained by (nonclassical) neural network activity, but they differ in whether they constrain their models primarily using behavioral evidence (connectionism) or also emphasizing neuroanatomical and neurophysiological evidence (computational neuroscience). What all versions of CTC agree on is that neurocognitive processes are computations.

I have argued that arguments from the Church–Turing thesis to CTC are fallacious (Chapter 10) and yet CTC holds (Chapter 9) in spite of many objections

[1] This chapter is a revised and expanded descendant of much of Piccinini and Bahar 2013, so Sonya Bahar deserves partial credit for most of what is correct here.

Neurocognitive Mechanisms: Explaining Biological Cognition. Gualtiero Piccinini, Oxford University Press (2020).

DOI: 10.1093/oso/9780198866282.001.0001

(Chapter 11). But which type of computation do neurocognitive systems perform? In this chapter, I reject the common assimilation of neural computation to either analog or digital computation, concluding that neural computation is sui generis. Analog computation requires continuous signals; digital computation requires strings of digits. But typical neural signals, such as spike trains, are graded like continuous signals as well as constituted by discrete functional elements (spikes); thus, typical neural signals are neither continuous signals nor strings of digits. It follows that neural computation is sui generis.

After defending that conclusion, I draw three important consequences of a proper understanding of neural computation for the theory of cognition. First, understanding neural computation requires a specially designed mathematical theory (or theories) rather than the mathematical theories of analog or digital computation. Second, several popular views about neural computation turn out to be incorrect. Third, computational theories of cognition that rely on nonneural notions of computation ought to be replaced or reinterpreted in terms of neural computation.

13.2 Why Neural Processes Are Not Analog Computations

The claim that neural *computations* are analog should not be confused with the claim that neural *representations* are "analog," in the sense that they bear some positive analogy to what they represent. Nervous systems build and manipulate analog models—that is, structural representations—of their environment. I have explicated and defended this claim in Chapter 12. Examples include visual (Hubel and Wiesel 1962) and auditory (Schreiner and Winer 2007) maps, as well as the somatosensory representation of the world touched by a rodent's whiskers in the "barrel cortex" (Inan and Crair 2007). But analog *models* need not be represented and manipulated by analog *computers*.[2]

Claiming that neural systems are analog computers in the strict sense of Pour-El (1974; Rubel 1993; Mills 2008) is a strong empirical hypothesis. Analog CTC is committed to the view that the functionally significant signals manipulated by the nervous system include continuous variables whose values over a time interval must be integrated by the system.

To be sure, both brains and analog computers appear to have an essentially continuous dynamic, that is, their vehicles must be assumed to vary over real time in order to construct successful scientific theories of them (for the case of brains

[2] Unless, of course, we *define* analog computers as computers that manipulate analog representations, as Maley (2018) does. That leads back to the semantic account of computation, which I rejected in Chapter 6, and it obliterates important distinctions between different types of mechanisms that manipulate analog representations. In any case, neural systems do manipulate analog representations, so they *are* analog computers in Maley's sense.

see, e.g., Dayan and Abbott 2001; Ermentrout and Terman 2010). Some neuronal inputs—i.e., neurotransmitters and hormones—are often most usefully modeled as continuous variables; in addition, their release and uptake are modulated by chemical receptors that operate continuously in real time. Also, dendrites and *some* axons transmit graded potentials—that is, more or less continuously varying voltage changes. In these respects, brains appear to be more similar to analog computers than digital ones. Finally, as Rubel (1985) points out, neural integrators do play an important role in at least some processes, such as oculomotor control. (Integrators are the most important components of general purpose analog computers.)

But there are crucial disanalogies between brains and analog computers (Figure 13.1). The principal difference is that the vehicles of analog computation include continuous variables, namely, variables that vary continuously over time, in which the specific values taken by the variable over a time interval—the shape of the signal over that time interval—are functionally significant. By contrast, the main vehicles of most neural processes are neuronal spike trains, which are sequences of all-or-none signals. With respect to spike trains, the specific values of the signal—the exact shape of the signal—is functionally irrelevant or nearly so. What matters most is the *rate* at which neurons fire, and sometimes the *time* at which they fire. Firing rates can increase or decrease in a graded way within certain limits and, for this reason, firing rates are sometimes called "analog". But even firing rates are not continuous variables that can take any real value within

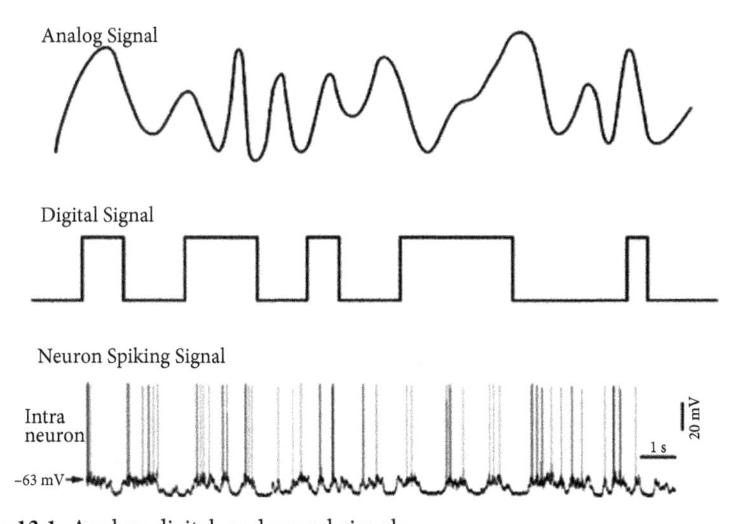

Figure 13.1 Analog, digital, and neural signals.

(credit: http://www.extremetech.com/extreme/144248-neuristors-the-future-of-brain-like-computer-chips)

the relevant interval and whose continuous variation over time needs to be integrated by the computing system.

While the firing threshold is subject to modulation by continuous variables (e.g., ion concentrations), graded potentials vary continuously, and spike timing (in real time) may be functionally significant, none of these aspects of neural signals are similar enough to the processes used by analog computers for the mathematics of analog computers to be directly applicable to brain processes. New mathematical theory—specific to neural processes—had to be invented for this purpose.

Most importantly, spikes are the most functionally significant signals transmitted by most neurons, and what is functionally most significant is not the absolute value of the voltage or the exact shape of the voltage curve but simply whether a spike is present or absent. In most cases, if the total input signal received by a neuron reaches a certain threshold, the neuron fires a spike. If the total signal received by a neuron is below the threshold, no spike is fired. Sometimes neuroscientists say that neurons integrate their dendritic input. In this limited respect, neurons resemble analog integrators. But the result is still very different from that of an analog integrator. Whereas an analog integrator outputs the integral of its input, a typical neuron simply either fires or doesn't fire, depending on whether its input threshold is reached. As a consequence of this all-or-none character of the spike, computational neuroscience focuses on firing rates and spike timing as the most significant functional variables at the level of circuit and network. There is nothing similar to firing rates or spike timing in analog computers (in the present strict sense). They are peculiarly neural phenomena.

Finally, there are specialized neural components that integrate spike trains and are sometimes called neural integrators. They operate on a different type of vehicle—spike trains—than the integrators of analog computers, which operate on continuous variables (Koulakov et al. 2002). Thus, in spite of some similarities, neural systems are not analog computers in the most interesting sense.

13.3 How Spike Trains *Might Have Been* Strings of Digits

The mathematical modeling of neural processes can be traced back to the mathematical biophysics pioneered by Nicolas Rashevsky and his associates (Rashevsky 1938; Householder and Landahl 1945). They employed integral and differential equations to describe and analyze processes that might explain neural and cognitive phenomena. In their explanations, they used neither the notion of computation nor the mathematical tools of computability theory and computer science. Those tools were introduced into theoretical neuroscience by the youngest member of Rashevsky's group, Walter Pitts, in collaboration with Warren McCulloch.

Like Rashevsky's mathematical biophysicists, McCulloch and Pitts (1943) attempted to explain cognitive phenomena in terms of putative neural processes. But McCulloch and Pitts introduced concepts and techniques from logic and computability theory to model what is functionally relevant to the activity of neurons and neural nets, in such a way that—in our terminology—a neural spike can be treated mathematically as a digit, and multiple spikes can be treated as strings of digits.

McCulloch and Pitts's empirical justification for their theory was the similarity between digits and spikes. Spikes *appear* to be discrete or digital—to a first approximation, their occurrence is an unambiguous event relative to the functioning of the system. This was the main motivation behind McCulloch's identification between spikes and atomic mental events, leading to his formulation of digital CTC (Chapter 5). I will set aside whether spikes correspond to atomic mental events and limit our discussion to whether spikes are digits and whether pluralities of spikes are strings of digits.

The assumption that pluralities of spikes are strings of digits requires that spikes be concatenated. In the case of spike trains from a single neuron, the temporal ordering of spikes is a natural candidate for the concatenation relation. In the case of pluralities of spikes from different neurons occurring within well-defined time intervals, a concatenation relation might be defined by first identifying a relevant plurality of neurons and then taking all spikes occurring within the same time interval as belonging to the same string. The plurality of neurons might be defined purely anatomically or by a combination of anatomical and functional criteria. For example, one could try to identify, electrophysiologically or histologically, which neurons are presynaptic and postsynaptic to which.

The suggestive analogy between spikes and digits—based on the all-or-none character of spikes—is far from sufficient to treat a spike as a digit, and even less sufficient to treat a plurality of spikes as a string of digits. In order to build their theory, according to which neural activity is digital computation, McCulloch and Pitts made additional assumptions.

A crucial assumption was that a fixed number of neural stimuli are always necessary and sufficient to generate a neuron's pulse. They knew this is false: they explicitly mentioned that the "excitability" of a neuron varies over time (McCulloch and Pitts 1943, 20–1). In fact, neurons exhibit both absolute and relative refractory periods as well as varying degrees of plasticity ranging from short to long term potentiation and depression. To this, it should be added that spiking is not a deterministic but a probabilistic function of neuronal input. Indeed, the flow of ions through transmembrane channels has a stochastic component (White et al. 1998; Hille 2001), as does neurotransmitter release (Bennet and Kearns 2000; Hille 2001). Even when neurons fire with almost pacemaker-like regularity, their natural frequencies retain some variability (Le Bon-Jego and Raichle 2007; Yassin et al. 2010) due to natural fluctuations in physiological parameters which, in turn, affect the time at which the transmembrane potential reaches the firing threshold.

Another crucial assumption was that all neurons within a network are synchronized so that all relevant events—conduction of the impulses along nerve fibers, refractory periods, and synaptic delays—occur within temporal intervals of fixed and uniform length, which are equal to the time of synaptic delay. In addition, according to McCulloch and Pitts, all events within one temporal interval only affect the relevant events within the following temporal interval. This assumption makes the *dynamics* of the net discrete: it allows the use of logical functions of discrete inputs or states to fully describe the transition between neural events.

As McCulloch and Pitts pointed out, these assumptions are not based on empirical evidence. The reason for these assumptions was, presumably, that they allowed the mathematics of digital computation to describe, at least to a first approximation, what they thought to be the most functionally relevant aspects of neural activity.

The similarity between digits and spikes can hardly be doubted. But there are significant dissimilarities as well, and there are further dissimilarities between spike pluralities—either spike trains from the same neuron or more or less synchronous spikes from different neurons—and strings of digits. In 1943, relatively little was empirically known about the functional properties of neural spikes and spike trains. Since then, neurophysiology has made great progress, and it is past time to bring what is currently known about neural processes to bear on the question of whether they are digital computations.

Although McCulloch and Pitts's theory influenced computer design and computability theory (Chapter 5), as a theory of neural activity it was soon abandoned in favor of more sophisticated and empirically adequate mathematical models. The following are a representative sample: conductance-based models (Hodgkin and Huxley 1952); integrate-and-fire neurons (Lapicque 1907/2007; Caianiello 1961; Stein 1965; Knight 1972); and firing rate models (Wilson and Cowan 1972).

From a mathematical point of view, the models employed by theoretical neuroscientists after McCulloch and Pitts resemble those of Rashevsky's original mathematical biophysics, in that they do not employ concepts and techniques from computability theory. Recall that by "computability theory" I mean the theory of recursive functions and their computation (Davis et al. 1994). In the models currently employed by theoretical neuroscientists, spikes are not treated as digits, and spike pluralities are not treated as strings of digits. The following section identifies principled empirical reasons for why this is so.

The point of the exercise is not to beat a long-dead horse—McCulloch and Pitts's theory. The point is that if we reject McCulloch and Pitts's assumptions, which were the original motivation for calling neural activity digital computation, it remains to be seen whether there is any way to replace McCulloch and Pitts's assumptions with assumptions that are both empirically plausible and allow us to retain the view that neural activity is digital computation in the relevant sense. If no such assumptions are forthcoming, then digital CTC needs to be either abandoned or grounded in (nondigital) neural computation.

13.4 Why Spikes Are Not Digits and Spike Pluralities Are Not Strings

Current neuroscientific evidence suggests that typical neural processes are not digital computations. The structure of the argument is this:

(1) Digital computation requires the manipulation of strings of digits (Chapter 6);
(2) Neural spikes are not digits, and even if they were digits, spike pluralities would not be strings;
(3) Therefore, the manipulation of spike pluralities is not digital computation.

This argument challenges not only classical CTC, which is explicitly committed to digital computation, but also any form of connectionist or neurocomputational theory that is either explicitly or implicitly committed to the thesis that neural activity is digital computation.[3]

Caveat 1: A defender of classicism might argue that while classicism is committed to digital computation at the "cognitive" level, it is not committed to digital computation at the "implementation" level I am considering here. It may be that nondigital neural computations give rise to digital computations at the "cognitive" level. This speculative proposal faces two problems. First, biological cognitive systems—i.e., neurocognitive systems—cannot be analyzed into two levels, one of which is cognitive and one of which is implementational. Instead, neurocognitive systems contain many levels of organization, which may be more or less directly involved in explaining cognitive phenomena (Chapter 8). The activities of neurons and networks thereof are routinely invoked by neuroscientists to explain cognitive phenomena. Thus, the levels of neurons and networks are as cognitive as any higher level. Second, the idea that higher-level digital computation emerges from lower-level nondigital computation requires both an explicit account of how

[3] E.g., Eliasmith (2000) argues that neural signals are digital because spikes are all-or-none and spike timing is affected by noise. It's worth noting that Eliasmith (2003, 2007, 2013) and I agree that the theory of cognition needs to be grounded on neural computation and that, in his theoretical and modeling work, even Eliasmith makes no use of the mathematics of digital computation.

this occurs and some empirical support, both of which are lacking to date. I'll come back to this at the end of this chapter.[4]

Caveat 2: The present argument is not formulated in terms of the notion of representation. It rests on the notion of digits as discrete entities that can be concatenated into strings. The digits discussed in this paper may be vehicles for many types of representation. The present argument does not depend on which if any notion of representation is adopted. Specifically, it is indifferent to the distinctions between digital vs. analog, localized vs. distributed, and "symbolic" vs. "sub-symbolic" representation (Rumelhart and McClelland 1986; Smolensky 1989).[5] In fact, the present argument is consistent both with representationalism and its denial. That said, Chapter 12 made it clear that neurocomputational systems do build and manipulate representations, which model body and environment and guide behavior.

For a class of particular events to constitute digits within a computing mechanism, it must be possible to type them in an appropriate way. In the next subsection I discuss the difficulties in doing so for spikes. For a plurality of putative digits to constitute a string, it must be possible to determine, at the very least, which digits belong to which strings. Later I will discuss the difficulties in doing so for pluralities of spikes.

13.4.1 Typing Spikes as Digits

Digits belong to finitely many types, which must be unambiguously distinguishable by the system that manipulates them. So, for spikes and their absence to be digits, it must be possible to type them into finitely many types that are unambiguously distinguishable by neural mechanisms.

Spikes are all-or-none events; either they occur or they don't. So, there might be two types of digits: the *presence* or *absence* of a spike at some time. This was McCulloch and Pitts's proposal. For this proposal to be adequate, more needs to be said about the time during which the presence or absence of spikes is to be counted.

[4] The proposal just discussed should not be confused with the view defended by Aydede (1997). Aydede argues that classicism is not committed to concatenation between language-of-thought symbols because the syntactically complex symbolic structures may be encoded implicitly (by patterns of activation of neural network units) rather than explicitly (by concatenating primitive symbols). The implicit encoding schemes discussed by Aydede still involve concatenated digits in the form of patterns of active units, so Aydede's twist on classicism is yet another variant of digital CTC.

[5] Maley (2011) gives a helpful account of digital vs. analog representation. Digital and analog vehicles in our sense need not be representations, although they can be. When they are, the notions I use are more encompassing than Maley's. Maley's notions of digital and analog representations are species of digital and analog representational vehicles in the present sense.

The relevant time cannot be instantaneous. Otherwise, there would be uncountably many digits in any finite amount of time, which is incompatible with the notion of digit. Besides that, spikes do take some time—about 1 ms—to occur. McCulloch and Pitts divided time into intervals whose length was equal to synaptic delay. Their proposal suffers from a number of shortcomings. First, they assumed that synaptic delay is constant; recent studies show it to be variable (e.g., Lin and Faber 2002). Second, they assumed that any time interval separating two spikes is an exact multiple of the master time interval, whereas there is no empirical evidence of any such master time interval. Third, in order to count the presence or absence of spikes from different neurons as belonging to the same computation, they assumed that neurons are perfectly synchronized, in the sense that their relevant time intervals begin and end in unison. The third assumption is also relevant to the concatenation of digits into strings, so I will discuss it in subsequent subsections. In what follows, I will focus on the first two assumptions.

The lack of evidence for the first two assumptions could be remedied by choosing a fixed time interval of unambiguous physiological significance. But, even though spikes may occur fairly reliably within 1 millisecond of neural stimuli in vitro (Mainen and Sejnowski 1995), actual neural firing in vivo (and in many in vitro preparations as well) often occurs with high variability, due to a large number of factors in the fluctuating cellular environment, which includes thousands of neuronal connections and dozens of different cell types. Thus, spikes are often *highly stochastic* functions of a large number of highly variable conditions. While some neural firing can be quite temporally precise, it cannot be claimed that spikes occur only within fixed time intervals of physiological significance. To summarize, since spiking is affected in varying degrees by events (such as input spikes) that occur at different times in the neuron's past history (e.g., Shepherd 1999, 126), there is no known principled way of typing both relevant causes and the time intervals during which they occur into finitely many types.

By itself, that spikes are probabilistic need not undermine the analogy between spikes and digits, since digital computations may be probabilistic as well as deterministic. In a probabilistic digital computation, there are a finite number of possible outcomes, each of which has a certain probability of occurring. In other words, the sample space of a probabilistic digital computation is finite. This seems superficially analogous to neural spiking, which occurs with a certain probability at a time (given a certain input).

But there is a crucial disanalogy. Digital computational events, whether deterministic or probabilistic, are temporally discrete in a way that spikes are not. In the case of a probabilistic digital computation, events occur within fixed functionally significant time intervals; this makes it possible to specify the probabilities of the finitely many outcomes effectively, outcome by outcome, so that each constitutes a digit. By contrast, there appear to be no functionally significant time intervals during which spikes are to be counted. As a result, spikes must be counted over real time. Because of this, the probability that a spike occurs at any instant of real time is zero. In other words, the sample space for spiking events during a time interval is uncountably infinite. Spike probabilities need to be given by probability densities, which specify the probability that a spike occurs within an infinitesimal time interval (Dayan and Abbott 2001, 24). In conclusion, since there can only be finitely many digit types but there are an uncountable number of possible spiking events, spiking events cannot be digits.

There is another, independent and perhaps deeper reason for denying that the presence or absence of a spike can constitute a digit. Until now, the discussion has been based on the implicit assumption that the physiological significance of the presence or absence of *individual* spikes is functionally relevant to the processing of neural signals. But this assumption is unwarranted in many cases. Although the presence or absence of spikes does affect neural systems differentially, individual spikes rarely seem to have unambiguous functional significance for the processing of neural signals (though see VanRullen et al. 2005).

First, neurons and neural systems can exhibit spontaneous activity, not triggered by any sensory stimulus; indeed, spontaneous neural activity—including postsynaptic activity (cf. Alle et al. 2009)—has been suggested to account for at least 75% of the brain's energy consumption (Raichle and Mintun 2006). (This spontaneous activity need not be stochastic: spontaneous firing of neocortical neurons in cortical microcircuits has been shown to have regular, pacemaker-like activity (Le Bon-Jego and Yuste 2007), and pacemaker cells in the suprachiasmatic nuclei can fire regularly even in the absence of any neuronal input.) Some have suggested that neural signals consist in relatively small deviations from a background level of activity. This is what makes the hypothesis of a "default network," suggested by BOLD fMRI imaging studies, so compelling (Raichle et al. 2001; Zheng and Raichle 2010), although still controversial (Morcom and Fletcher 2007). Finally, even a noisy, stochastic spontaneous firing state can have some function; witness the role of noise in enhancing weak subthreshold signals in the nervous system via stochastic resonance (Douglass et al. 1993; Levin and Miller 1996; Moss et al. 2004).

Functional, noisy, or neither, spontaneous activity per se is not an objection to digital CTC. For digital computations can depend on both inputs and internal states. In principle, spontaneous activity could be analogous to the internal state of a digital computing mechanism. But the internal states of (fully) digital computing mechanisms are such that their contribution to a computation can be described as digits (discrete states) and specified independently of the contribution given by the input. By contrast, spontaneous neural activity has all the characteristics already noted for spikes in general, which led to the conclusion that spikes are not digits.

Second, regardless of whether a spike is caused by spontaneous activity, a sensory input, or both, what appears to have functional significance is often not any individual spike, but rather the firing *rate* of a neuron. Consider the simple case of the crayfish photoreceptor, which fires at a rate of about 5 Hz in the dark, and speeds up to 30 Hz when exposed to visible light (Bruno and Kennedy 1962). In such clear cases of rate coding, individual spikes have no functional significance on their own. For that matter, many individual spikes might in principle be added to or removed from a spike train without appreciably altering its functional significance (Dayan and Abbott 2001, chap. 1). In fact, neuroscientists typically assess the functional significance of neural activity by computing average firing rates, namely, by averaging the firing rates exhibited by a neuron over many similar trials.

Thus, there are principled reasons to doubt that spikes and their absence can be treated mathematically as digits, to be concatenated into strings. But individual spikes are not the only candidates to the role of digits. Before abandoning the issue of digits, we should briefly look at two other proposals.

One possibility is that digits are to be found in precise patterns of several spikes. There have been serious attempts to demonstrate that some repeating triplets or quadruplets of precisely timed ($\approx$1 ms) spikes from single neurons have functional significance. These attempts have failed: rigorous statistical analysis indicates that both the number and types of precisely timed spike patterns are those that should be expected by chance (Oram et al. 1999). In fact, in many cases signals coming from individual neurons are too noisy to constitute the minimal units of neural processing; according to some estimates, the minimal signaling units in the human brain are of the order of 50–100 neurons (Shadlen and Newsome 1998).

Another possibility is that spike rates themselves should be taken to be digits of sorts. This was von Neumann's proposal (von Neumann 1958). Unlike the previous proposal, this one is based on variables that are functionally significant for the processing of neural signals. But this proposal raises more problems than it solves. Again, there is the problem of finding a significant time interval. Spike rates vary continuously, so there seems to be no nonarbitrary way to parse them into time intervals and count the resulting units as digits. It's also unclear how spike trains from individual neurons, understood as digits, could be concatenated with other spike trains from other neurons to form strings. Perhaps most importantly, spike rates cannot be digits in the standard sense, because they cannot be typed into finitely many types of unambiguous physiological significance. As von Neumann noted, there is a limit to the precision of spike rates, so if we attempt to treat them as digits, there will be a significant margin of error as to which type any putative digit belongs to; the same problem occurs with individual spike timings. As a consequence, computability theory as normally formulated does not apply to spike rates—a different mathematical theory is needed. Von Neumann pointed out that he didn't know how to do logic or arithmetic over such entities as spike rates, and hence he didn't know how the brain could perform logic or arithmetic using such a system. As far as I know, no one after von

Neumann has addressed this problem. Without a solution to this problem, von Neumann's proposal remains an empty shell.

13.4.2 Fitting Spikes into Strings

In the previous section, we saw there is no way to treat spikes, or pluralities thereof, as the digits of a computation. Even if they could, it wouldn't follow that neural mechanisms perform digital computations in any useful sense. For although some trivial digital computations are defined over single digits or pairs of them (i.e., those performed by ordinary logic gates), nontrivial digital computations are defined over strings of digits. Strings of digits consist of multiple digits concatenated together, whereby there is no ambiguity as to which digits belong to a string and which position they occupy within a string. If pluralities of spikes were to count as strings of digits, there should be a way to unambiguously determine which spikes belong to which string and which position they occupy within the string. But there is no evidence that this can be done. The two natural candidates for strings are pluralities of spikes from synchronous neurons and spike trains from individual neurons. Let's examine them in turn.

13.4.2.1 Strings versus Synchronous Spikes

In the case of pluralities of synchronous spikes, the concatenation relation might be given by spatial and temporal contiguity within a structure defined by some combination of anatomical and functional properties. For instance, synchronous spikes from a cortical column may be seen as belonging to the same string. Implicitly, this is how inputs and outputs of many artificial neural networks are defined.

A first difficulty with this proposal is the looseness of the boundaries between neural structures. It doesn't appear that neural structures, such as cortical columns, are defined precisely enough that one can determine, neuron-by-neuron, whether each neuron belongs to one structure or another. The boundaries are fuzzy, so that neurons can be recruited to perform different functions as they are needed, and the functional organization of neural structures is reshaped as a result (Recanzone, Merzenich, and Schreiner 1992; Elbert and Rockstroh 2004). This first problem may not be fatal. Perhaps the boundaries between neural structures are fuzzy and yet, at any time, there is a way to assign each spike to one and only one string of spikes synchronous to it.

But there is a fatal problem: synchrony between spikes is a matter of degree. Just as there is no meaningful way to divide the activity of a neuron into discrete functionally significant time intervals (Section 13.4.1), there is no meaningful way to divide the activity of neural ensembles into discrete functionally significant time intervals to determine in an absolute manner whether their spikes are synchronous. Even if such absolute time determinations were possible, the inherent noisiness in neural firing times would result in a varying degree of synchrony over time. Synchrony between spikes has been studied intensely by the nonlinear dynamics community over the past few decades, in the context of the synchronization of coupled oscillators (Pikovsky et al. 2003). Synchrony appears to play an important role in normal brain functions such as attention (Steinmetz et al. 2000; Fries et al. 2001, 2008; Roy et al. 2007) and has been applied to the analysis of neural pathologies such as epilepsy (Wong et al. 1986; Uhlhaas and Singer 2006) and Parkinson's disease (Tass et al. 2003). But the interesting question about neural synchrony is not *whether*, in an absolute sense, there is synchrony in a neural process, but rather *how much* synchrony there is. The typical definition of synchrony used within the neurophysiological community is that of *stochastic phase synchronization*, which measures the degree to which the phase difference between neurons remains constant; thus, neurons can be synchronized even with a (relatively) constant phase lag (Pikovsky et al. 2003).

As a result, even if neural spikes were digits (which they aren't), pluralities of synchronous spikes would not be strings of digits.

13.4.2.2 Strings versus Spike Trains

In a spike train from a single neuron, the concatenation relation may be straightforwardly identified with the temporal order of the spikes. Nevertheless, in order to treat a spike train as a string, its beginning (first spike) and end (end spike) must be identified unambiguously. Again, two independent problems undermine this proposal.

The first problem is that the question of which spikes belong to a string appears to be ill defined. One reason is the inherent stochasticity present in much neuronal firing, discussed above. While spike timing can be exquisitely precise, much neural activity is subject to various fluctuations, not least the variability in neurotransmitter release at the synaptic terminal. This can result in a noisy background of "spontaneous" (i.e., not driven by external sensory input) activity. This can be seen, for example, in the typically broad power spectrum of spiking frequencies from the unstimulated crayfish mechanoreceptor-photoreceptor system. Under external periodic stimuli, the stimulus frequency appears "encoded" in the power spectrum, superimposed on the background of variable spontaneous activity (Bahar 2003; Bahar and Moss 2003). There does not seem to be any functionally significant way to parse *which* spikes are the ones encoding the stimulus, however, or to divide the neural responses to sensory inputs into subgroups with unambiguous beginnings and ends.

The second problem is that, given the amount of noisiness in neural activity, individual spike trains are not useful units of functional significance. What has functional significance is not a specific train of spikes, each of which occurs during a particular time interval, but a pattern shared by many different spike trains that are generated by the same neuron—or rather, typically, neuronal ensemble—in response to the same sensory stimulus. The only known way to find useful units of functional significance is to average spike trains over many trials and use average spike trains as units of functional significance.

For these reasons, even if neural spikes could be treated as digits (which they can't), spike trains could not be meaningfully treated as strings of digits.

13.4.3 What About the Neural Code?

A final relevant issue is the relationship between neural codes and (digital) computational codes. In both cases, the same term is used, which suggests an analogy between the two. Is there such an analogy?

The expression "neural code" usually refers to the functional relationship between properties of neural activity and variables external to the nervous system (e.g., see Dayan and Abbott 2001). Roughly speaking, the functional dependence of some property of a neural response on some property of an environmental stimulus is called neural encoding, whereas the inference from some property of a neural response to some property of the stimulus is called neural decoding. Neural encoding is a measure of how accurately a neural response reflects a stimulus; neural decoding is a measure of how accurately a neural mechanism can respond to a stimulus.

By contrast, in the theory of (digital) computation, "coding" refers to one-to-one (sometimes, many-to-one) mappings between classes of strings of digits (see, e.g., Rogers 1967, 27–9). Given each member of a (countably infinite) class of strings over some alphabet, any algorithm that produces a corresponding member of a different class of strings (possibly over a different alphabet) is called an encoding. This kind of encoding is usually reversible: for any encoding algorithm there is a decoding algorithm, which recovers the inputs of the encoding algorithm from its outputs.

Neural coding is only loosely analogous to computational coding. Both are relationships between two classes of entities, and both give procedures for retrieving one class of entities from the other. But the disanalogies are profound. Whereas neural encoding is only approximate and approximately reversible, (digital) computational coding is (by definition) exact and is usually exactly reversible. Most importantly, whereas neural encoding relates physical quantities in the environment to neural responses, neither one of which is likely to be accurately described as a class of digital strings, computational coding is a relationship between classes of digital strings. For these reasons, neural coding should not be confused with (digital) computational coding. The existence of a neural code does not entail that neural processes are digital computations.

13.4.4 Against Digital CTC

Given current evidence, the most functionally significant variables for the purpose of understanding the processing of neural signals are properties of neural spike trains such as firing rates and spike timing. I have argued that neither spike trains from single neurons nor pluralities of synchronous spikes from multiple neurons are viable candidates for strings of digits. Without strings, we can't identify operations defined over strings. And without operations defined over strings, we can't have digital computations.

The discrete nature of spikes was the main empirical motivation for digital CTC. Once the original assumptions about spikes that supported digital CTC are rejected, the burden is on supporters of digital CTC to find evidence that some neural variables are suited to being treated as strings of digits, and that this has something to do with explaining cognitive phenomena.

13.5 Neural Computation Is Sui Generis

Given the present state of the art, a defense of digital CTC (or analog CTC, for that matter) as a general account of neural computation is not only implausible—it is unmotivated. Unlike in 1943, when digital CTC was initially proposed, today we have an advanced science of neural mechanisms and processes, based on sophisticated mathematical theories and models that are, in turn, grounded in sophisticated experimental techniques. I am referring to what is often called theoretical and computational neuroscience (Dayan and Abbott 2001; Izhikevich 2007; Ermentrout and Terman 2010; Eliasmith 2013). Neurocomputational models can include, for example, spiking outputs from model cells, but also terms that rely on graded changes in parameters and on spike rates.

Theoretical neuroscientists build mathematical models of neural mechanisms and use them to explain neural and cognitive phenomena. Many philosophers and psychologists have unduly neglected the development of this discipline. Specifically, they have rarely noticed that in realistic mathematical models of neural processes, which can be tested against experiments, the explanatory role of (digital) computability theory and (digital) computer design is nil.

When a process is explained by digital computation, at a minimum the explanation includes a stream of digital inputs, internal states, and outputs, a set of rules for manipulating the digits, and a system of digital circuits that perform the manipulations. This is how computational explanation works in digital computers (Chapter 6, Piccinini 2015). But nothing of this sort has been found to explain neural processes. If the present argument is right, none is forthcoming.

In a nutshell, current evidence indicates that typical neural signals, such as spike trains, are graded like continuous signals but are constituted by discrete functional elements (spikes); thus, typical neural signals are neither continuous signals nor strings of digits. It follows that neural computation is sui generis.

Many neuroscientists will probably react with a yawn, because few if any neuroscientists take typical neural processes to be digital computations. Yet some psychologists and philosophers still theorize about cognition using constructs that rely either implicitly or explicitly on digital computation, and some of them still defend digital CTC (e.g., Fodor 2008; Gallistel and King 2009; Schneider 2011). As I've pointed out, however, understanding neural computation requires specially designed mathematical tools rather than the mathematics of analog or digital computation. Moreover, there remains considerable confusion in the literature—including the neuroscience literature—about what CTC does and does not entail. In the following section, I will examine some widespread fallacies. I will conclude with a plea to cognitive scientists who rely on nonneural notions of computation to replace or reinterpret their theories in terms of neural computation.

13.6 Some Fallacies to Avoid

CTC properly so called is often conflated with other theses that are weaker than or logically independent of it. At least two issues are relevant here.

The first issue has to do with the relationship between CTC and representationalism. Many authors believe that computation is the same as information processing, so that all computational systems process information (or, more strongly, process representations) (e.g., Churchland, Koch, and Sejnowski 1990; Fodor 2008; O'Brien and Opie 2006; Shagrir 2010a). By contrast, I argued that computation and information processing are conceptually distinct. In fact, the relevant sort of information processing—integrating and processing semantic information transmitted by different types of physical signal—does entail computing, but the reverse does not hold (Chapter 6). Equating computation with information processing obscures the structural (nonsemantic) differences between different types of computation, such as the differences between digital, analog, and neural computation that I have discussed in this chapter. To understand the nature of neural computation, it is essential to define computation independently of information processing (and, a fortiori, representation) so as to investigate its structural aspects. That said, neural computations do construct and manipulate representations (Chapter 12).

The second issue has to do with the relationship between CTC and mechanistic explanation. CTC and mechanism have been run together in the literature. For example, David Marr dubbed the mathematical analysis of the functions of neural mechanisms "computational theory," by which he meant theory of neural computations (Marr 1982). From this, it follows that neural processes are computations. But, thus understood, CTC ceases to be an empirical hypothesis about the specific functional nature of neural processes, to become a trivial consequence of the definition of mechanistic explanation. Suppose, instead, that we begin our investigation with a rich notion of mechanistic explanation (Chapters 2 and 3). Under such a notion, computation is one kind of mechanistic process among others (Chapter 6). Then, we may ask—with McCulloch and Pitts and other computationalists—whether neural processes are computations and what kind of computation they are. As I argued, we can find empirical evidence that neural processes are computations and that neural computation is sui generis.

CTC is often used to derive various consequences, which may or may not follow from some version of digital CTC but certainly do not follow from generic CTC: (i) that in principle we can capture what is functionally relevant to neural processes in terms of some formalism taken from (digital) computability theory, such as Turing machines, (ii) that it is possible to design computer programs that are functionally equivalent to neural processes in the sense in which computer programs can be computationally equivalent to each other, (iii) that the study of neural (or mental) computation is independent of the study of neural implementation, (iv) that the Church–Turing thesis applies to neural activity in the sense in which it applies to digital computers. None of these alleged consequences of CTC follow from the view that neural activity is computation in the generic sense.

If neural activity is not digital computation, as I have argued, there is no reason to suppose that computability theory formalisms can capture what is functionally relevant to neural processes. They might be used to build computational *models* of neural processes, but the explanation of neural processes will be given by the constructs of theoretical neuroscience, not the notion of digital computation. The computational models of computational neuroscience are computer programs that perform numerical integration of coupled nonlinear ordinary or partial differential equations that, in turn, describe aspects of what is known about neural computations.

The relationship between computer programs that model neural processes and the processes they model is significantly different from the relationship between computer programs that simulate each other (Chapter 10, Section 10.3). When a computer program simulates another program (some aspect of) the computations of the simulating program encode (in the sense of computer science) the computations of the simulated program. By contrast, when a computer program implements a mathematical model of a neural process, the computations of the simulating program are a means for generating approximate representations of the states that the mathematical model attributes to the neural process. This kind of model attempts to capture something essential to a neural process based on what is known about the process by employing suitable abstractions and idealizations. There is no sense in which such modeling work can proceed independently of the empirical study of neural processes. Computational modeling is a different activity, with different rules, from the simulation of one computer program by another program.[6]

Finally, the Church–Turing thesis is largely orthogonal to CTC. The Church–Turing thesis states that any function that is computable in an intuitive sense is computable by some Turing machine. Many authors have argued that the Church–Turing thesis proves CTC to be true (e.g., Churchland 2007). This argument commits the fallacy of conflating mechanism with a form of CTC restricted to the computations that fall under the Church–Turing thesis (Chapter 10).

When mechanism and CTC are kept distinct, as they should be, two separate questions about neural computation arise. One question—the most discussed in the literature—is whether neural computations are computable by Turing machines. If McCulloch and Pitts's theory of the brain (or a similar theory) were correct, then the answer would be positive. Another possibility is that neural computations are digital but more powerful than those of Turing machines—neural mechanisms might be hypercomputers (Copeland 2000; Siegelmann 2003).

The more basic question, however, is whether neural activity is digital computation in the first place. If, as I have argued here, neural activity is *not* digital computation, then the question of whether neural computation is computable by Turing machines doesn't even arise. Of course, there remains the question of whether computing humans can do more than Turing machines. Human beings can certainly perform digital computations, as anyone who has learned basic arithmetic can attest. And presumably there are neural processes that explain human digital computing. But there is no evidence that human beings are computationally more powerful than Turing machines. No human being has ever been able to compute functions, such as the halting function, that are not computable by Turing machines. In this limited sense, the Church–Turing thesis appears to be true.

13.7 From Cognitive Science to Cognitive *Neuro*science

If neural computation is sui generis, what happens to computational theories in psychology, many of which rely on digital computation? Mental models (e.g., Johnson-Laird 1983, 2010), production systems (e.g., Newell 1990), and "theories" (e.g., Murphy 2002) are just a few examples among many others of digital computational constructs employed by psychologists to explain cognitive phenomena. For over seventy years, various forms of *digital* CTC—either classical or connectionist—have provided the background assumption against which many

[6] For more on the methodology of computational modeling, see Humphreys 2004; Craver 2006; Piccinini 2007b; Winsberg 2010.

computational constructs in psychology have been interpreted and justified. Computational psychological theories are legitimate, the assumption goes, because the digital computations they posit can be realized by neural computations. But if neural computations are not digital, or if digital neural computations are an exception rather than a rule, what should we do with computational psychological theories that rely on digital computation?

One tempting response is that psychological explanations remain unaffected because they are *autonomous* from neuroscientific explanations. According to a traditional view, psychological theories are not directly constrained by neural structures; psychology offers functional analyses, not mechanistic explanations like neuroscience (Fodor 1968a; Cummins 1983). On closer examination, however, functional analysis turns out to be an elliptical or partial mechanistic explanation (Chapter 7). Thus, psychological explanation is a type of mechanistic explanation after all. Psychological explanation is not autonomous from neuroscientific explanation; instead, psychological explanation is directly constrained by neural structures.

Another tempting response is to eliminate psychological explanations. That would be premature. Computational theories in psychology are sometimes our current best way of capturing and explaining psychological phenomena.

What needs to be done, instead, is to take psychological constructs that rely on digital computation and gradually reinterpret or replace them with theoretical constructs that can be realized by known neural processes, such as the spike trains of neuronal ensembles. The shift that psychology is currently undergoing, from classical cognitive psychology to cognitive neuroscience, goes in this direction. Much groundbreaking work in psychology over the past two or three decades may be seen as contributing to this project (e.g., Anderson 2007; Barsalou 1999; Gazzaniga 2009; Kalat 2008; Kosslyn et al. 2006; O'Reilly et al. 2014; Posner 2004). We are on our way to explaining more and more neural and cognitive phenomena in terms of neurocomputational mechanisms and processes. There is no reason to restrict our rich discipline of theoretical neuroscience and psychology to the narrow and incorrect framework of digital CTC.

I am calling for abandoning the classical approach of just searching for computational explanations of human behavior without worrying much, if at all, about neural computation. It doesn't follow that we should consider only algorithms that can be implemented using spiking neurons and immediately drop any other research program. But it does follow that anyone seriously interested in explaining cognition should strive to show how the computations she posits may be carried out by neural processes, to the extent that this can be made plausible based on current neuroscience.

I would not rule out that some cognitive phenomena are explained by digital computation, or at least by neural computations that closely approximate many properties of digital computation. Specialized neural computations, which approximate at least some properties of digital computations, might help explain some of our linguistic and arithmetical capacities. If this is on the right track, the next frontier for serious digital computationalists is to develop an explicit account of how such specialized neural computations occur and find neuroscientific evidence to back it up.

A final objection may be raised. If computational theories of cognition must be grounded in neural computation, as I have argued, shouldn't neural computation be explained in turn by molecular processes, and those by atomic processes, and those by subatomic processes, all the way down to the fundamental physical level? Aren't I recommending a form of brute reductionism?

Ideally, every level of organization should be explained by the level below it. In an ideally complete mechanistic explanation of a phenomenon, the capacities of entities at each level are explained by the organized subcapacities of those entities' components. But no brute reductionism follows because each level is invariant over many changes at the lower levels, and each level's properties are only one aspect of their lower level realizers (Chapters 1–2). In fact, to understand a higher-level phenomenon, the most important task is to identify the higher-level invariants that are the primary explanandum of the phenomenon.

In constructing multilevel mechanistic explanations, it is at least as important to understand which aspect of a realizer a higher level is, and what the relevant higher-level invariants are, as it is to explain the higher level in terms of lower levels. In other words, an ideally complete mechanistic explanation of a phenomenon integrates all and only the information that is needed to explain the phenomenon at each level of mechanistic organization. This is why no level (e.g., the "cognitive" level, if there were a unique cognitive level) can be considered in isolation from the others.

This is also why each level of description of a mechanism yields specific predictions that cannot be made at other levels; because each level articulates essential information that is at best implicit at lower levels. The information most relevant to explaining and predicting cognitive phenomena in detail is likely found at the levels of neural systems, neural networks, and neurons—where neural computation takes place. So, any explanation of cognition worthy of that name needs to take neural computation into account.

This completes my neurocomputational account of cognition. But there is one loose end that needs to be tied up. Throughout the book, I have gone along with the mainstream assumption that cognition can be studied in terms of computation and information processing, without worrying too much about phenomenal consciousness. Even if this assumption is correct, the question of the relation between computation and consciousness remains to be addressed. Can consciousness be explained in terms of computation and information processing? If not, how can it be explained? That's what the last chapter is about.

14
Computation and the Function of Consciousness

14.1 The Function of Consciousness

The framework developed in this book sheds new light on the relation between phenomenal consciousness and computation. Phenomenal consciousness is a huge topic; I cannot do justice to it. This chapter's aim is simply to clarify the relation between consciousness and computation and make room for some hitherto overlooked possibilities about the nature of consciousness.

In Chapter 4, I articulated mechanistic functionalism. Functionalism about X says that X has a functional nature. Mechanistic functionalism about X says that the nature of X is the functional organization of the mechanism for X. Functional organization is the plurality of components, functions, and organizational relations that make up a mechanism and explain its capacities.

Functionalism about the mind is often assumed to entail that the mind has a computational nature—that the mind is the software of the brain (Putnam 1967a). Chapter 6 should have made it abundantly clear that this is a bad mistake. In the interesting sense of "computation"—the sense that matters for explaining mental capacities—computing mechanisms are a relatively small subset of all functional mechanisms. They are those mechanisms whose teleological function is manipulating medium-independent vehicles in accordance with a rule. Therefore, having a functional nature by no means entails having a computational nature.

In Chapter 9, I articulated the Computational Theory of Cognition (CTC), according to which cognition has a computational explanation or, more strongly, cognition has a computational nature. I also offered two powerful reasons for CTC: that those aspects of neural processes that are most functionally relevant—spike frequency and timing—are medium independent, and that information processing in the relevant sense requires computation.[1] Yet, even if cognition is computational, it doesn't follow that everything about the mind is computational.

[1] Perhaps aspects of the body and environment are also relevant to some extent. As always, I focus solely on what type of neural properties are relevant, remaining entirely neutral on the extent to which the mind is extended. In other words, I ignore the possibility of extended mind solely in order to simplify the exposition.

Neurocognitive Mechanisms: Explaining Biological Cognition. Gualtiero Piccinini, Oxford University Press (2020).

DOI: 10.1093/oso/9780198866282.001.0001

For present purposes, the mind can be divided into cognition and phenomenal consciousness. By "phenomenal consciousness" I mean the qualitative aspect of subjective experience. From now on, by "consciousness" I mean phenomenal consciousness.

The relationship between cognition and consciousness is complicated and controversial. What is clear and uncontroversial is that much of cognition does not require consciousness—that is, much of cognition can occur and often does occur in the absence of any consciousness. Thus, much of cognition can be explained without worrying about consciousness.

What about those aspects of cognition that are conscious? Is consciousness required for them to take place, for them to take place in biological organisms, or at least for them to function correctly in biological organisms? And what about consciousness itself? How does that work? This chapter clarifies the relation between computation and consciousness, in the service of clarifying what does and does not follow from both functionalism and CTC.

One possibility is that consciousness is reducible to cognition, which according to CTC boils down to (embodied and embedded) computation and information processing. For instance, according to the prominent global broadcast theory, consciousness arises when cognitive states are globally broadcast throughout the cognitive system (Baars 1988, 1997; Dehaene and Naccache 2001; Dehaene 2014). More generally, according to this type of view, being in a conscious state is the same as being in an appropriate kind of computational and information carrying state. If this is correct, then the whole mind is reducible to computation and information processing. I call this the Computational Theory of Mind (CTM).

CTM says that the whole mind—including consciousness—can be explained in terms of computation and information processing or, more strongly, that the nature of the whole mind—including consciousness—is computational and informational (Lycan 1987; Dennett 1991; Rey 2005).[2] The standardly acknowledged alternative to CTM is that consciousness is type-identical to lower-level physical properties—this is the type-identity theory of mind (e.g., Place 1956; Feigl 1958; Smart 1959). The type-identity theory of mind, thus understood, is consistent with CTC, though not with CTM.[3]

The dialectic between CTM and the type-identity theory misses one plausible account of phenomenal consciousness. That's the view that consciousness is due to higher-level properties that are functional yet noncomputational—properties that are neither computational nor type-identical to lower-level physical properties. I will refer to this view as *noncomputational functionalism*.

[2] This terminology may be confusing because "Computational Theory of Mind" is sometimes used for what I call Computational Theory of Cognition.

[3] A third option is property dualism, the view that consciousness is due to nonphysical properties (Chalmers 1996b). Property dualism poses a distinct set of problems that I will not discuss here. I discuss some of those problems in more detail in Piccinini 2017c.

According to noncomputational functionalism, the functions involved in producing consciousness are medium-*dependent* functions, whereas computational functions are medium-*independent*. This is an option worth exploring because both CTM and the type-identity theory face serious objections.[4]

More precisely, there are two types of theory according to which consciousness is due to noncomputational, higher-level properties. Type 1 theories assert that consciousness is due to purely functional yet noncomputational higher-level properties. According to Type 1 theories, consciousness has a purely functional yet not-wholly-computational and not-wholly-informational nature. Type 2 theories assert that consciousness is due to higher-level properties that are not wholly functional in nature because they have qualitative aspects that are not entirely reducible to functional properties. According to Type 2 theories, consciousness is a higher-level property not wholly functional in nature. Which type of theory is correct depends not only on the nature of consciousness but also on the metaphysics of properties—specifically, the relationship between qualities and causal powers. (Given the account of functions I gave in Chapter 3, functions are a kind of causal power.)

The connection between the theory of consciousness and the metaphysics of properties is an area ripe for exploration. As I mentioned in Chapter 1, however, I will not address the metaphysics of properties any further. To keep matters simple, in what follows I will elide the distinction between Type 1 and Type 2 theories and refer to both as noncomputational functionalism even though, strictly speaking, Type 2 theories are not pure functionalist theories.

I want to reiterate that noncomputational functionalism is neither computationalist nor reductive in the sense of the type-identity theory. It is not computationalist because the functions involved in producing phenomenal consciousness are *not* (wholly) computational functions. They are medium-*dependent* functions, whereas computational functions are medium-*independent*. At the same time, the functions involved in producing phenomenal consciousness *are* functions, so they are higher-level properties, which are aspects of their realizers (Chapters 1 and 2). Since they are aspects of their realizers, they are *not* identical to their realizers. Therefore, the type-identity theory does not hold. (Even the token-identity theory does not hold, for the same reason.)

Noncomputational functionalism is neutral on whether the functions that are involved in consciousness are teleological. If consciousness is tied to the performance of teleological functions (in the sense of Chapter 3), then consciousness provides a regular contribution to the goals of organisms. Since consciousness occurs spontaneously and is not under intentional control, the goals in question must be

[4] Type-identity theories are implausible largely because consciousness seems to arise from the organization of a very complex system, which makes it a higher-level functional property. For recent arguments against computational accounts of consciousness, see Piper 2012 and Bartlett 2012.

biological—that is, goals such as survival, reproduction, development, and helping others. Thus, if consciousness is tied to teleological functions being performed, then consciousness plays a valuable biological role—presumably to facilitate certain cognitive functions.

There are two kinds of role that consciousness could play with respect to cognition. The first kind is to enable aspects of cognition that would otherwise not occur within biological organisms. For example, nervous systems might be so constituted that without being conscious, they cannot integrate different sources of information into a unified percept. This is consistent with the hypothesis that those same cognitive functions can be performed unconsciously, although doing so may require other, nonbiological forms of cognition.

The second kind of possible role is to enable aspects of cognition that could not occur at all without consciousness—either within biological organisms or any other cognitive systems. For example, some forms of creative thought may be such that they can only take place within a conscious system. If this were the case, then an artificially intelligent agent could replicate those forms of cognition only if it possessed consciousness. So much for the possible teleological functions of consciousness.

It's also possible that consciousness is produced by functions that are not teleological. In that case, consciousness would be either a byproduct of other teleological functions—a so-called spandrel—or a frozen evolutionary accident (Robinson, Maley, and Piccinini 2015). Biological traits fall into two classes: traits with a teleological function, which are typically adaptations, and traits without a teleological function, which are typically either byproducts of other traits or frozen biological accidents due to genetic drift. If consciousness falls into the second class, then consciousness does not enhance cognition in any way. It might be either epiphenomenal or, more likely, have physical effects that play no useful biological role.

The bottom line is that functionalism does not entail that consciousness has a wholly computational or informational explanation, or that computation and information processing are sufficient for consciousness. In fact, contrary to what many assume, functionalism is consistent with the possibility that consciousness has a functional yet noncomputational and noninformational nature. In the rest of this chapter, I will embed this point within the classical literature on functionalism and CTM and show how distinguishing carefully between functions in general and computational functions in particular makes room for a noncomputational functionalist view of consciousness.

14.2 Functionalism and the Computational Theory of Mind

Philosophers often conjoin functionalism and CTM. Yet their relationship remained obscure for a long time. With Jerry Fodor, I am struck by "the widespread failure to distinguish the computational program in psychology from the functionalist program in metaphysics" (Fodor 2000, 104). For example, Paul Churchland (2005) argues that functionalism is false because the brain is not a classical (i.e., more or less Turing-machine-like) computing system but a connectionist one. His argument presupposes that functionalism entails *classical* CTC and, by extension, a form of CTM that entails classical CTC. But functionalism—properly understood—does *not* even entail CTM, let alone classical CTC.

Assessing functionalism and CTM requires clarity on what they amount to and what evidence counts for or against them. In the following sections, I will formulate them and their relationship clearly enough that we can determine which type of evidence is relevant to each. I aim to dispel some sources of confusion that surround functionalism and CTM and clarify how functionalism and CTM may or may not legitimately come together. One upshot of this chapter, anticipated in the previous section, is that functionalism may be combined with a noncomputational theory of consciousness.

To a first approximation, functionalism is the view that the mind is the "functional organization" of the brain, or any other organ or system that performs mental functions (cf. Putnam 1960, 149; 1967b, 200; 1967a, 32). Another formulation of functionalism is that mental states are "functional states" (Putnam 1967a, 30).[5] Putnam's main example of a description of functional organization is the machine table of a Turing machine. For him, a functional organization is a set of functional states with their functional relations, where a functional state is defined by its causal relations to inputs, outputs, and other functional states (under normal conditions). Thus, under Putnam's notion of functional organization, the two formulations of functionalism are equivalent.

Under the broader notion of functional organization that I defended in Chapter 4, there is more to functional organization than individual functional states and their relations. There are also aggregates of states, components bearing the states, functional properties of the components, and relations between the components and their properties. Since the first formulation of functionalism is more general than the second, I prefer the first, but most of what I say below applies to the second formulation as well.

Stronger or weaker versions of functionalism may be formulated depending on how much of the mind is taken to be functional—how many mental states, or which aspects thereof, are taken to be functional. Are all mental states functional,

[5] In the cited references, Putnam writes "functional organization of organisms" or "functional organization of the human being" rather than "functional organization of the brain," as I wrote. Since much of the functional organization of human beings—or organisms, for that matter—is irrelevant to the mind, I replaced "organism" and "human being" with "brain." Even so, the brain is unlikely to coincide with the exact boundaries of the mind. For present purposes, it will be convenient to understand "brain," whenever appropriate, as referring to whatever aspects of the functional organization of organisms perform mental functions.

or only some? Are all aspects of mental states functional, or only, say, their nonphenomenal aspects?

CTM is the view that the entire mind, including phenomenal consciousness, has a computational and informational nature, or at least a computational and informational explanation. For present purposes, what matters about CTM is that, since the brain is the organ of the mind, the functional organization of the brain is wholly computational and informational or, more narrowly, that mental states are computational states.

Functionalism plus CTM equals *computational functionalism*. In a well-known slogan, computational functionalism says that the mind is the software of the brain (or any other mental organ or system; I will omit this qualification from now on). Taken at face value, this slogan draws an analogy between the mind and the software of ordinary, program-controlled computers. But the same slogan is often understood to suggest, more modestly, that the mind is the computational organization of the brain—or that mental states are computational states—without the implication that such a computational organization is that of a program-controlled computer. As we shall see, the ambiguity between the strong and the weak reading is one source of confusion in this area.

Computational functionalism has been popular among functionalists who are sympathetic to CTC. It has also encountered ferocious resistance (Chapter 11). Before we attempt to assess whether the mind is the software of the brain, however, we should understand what this means and what counts as evidence for or against it.

An important caveat. The functionalism I am primarily concerned with descends from some early writings of Hilary Putnam and Jerry Fodor (Putnam 1960, 1967a, 1967b; Fodor 1965, 1968a, 1968b). It is a metaphysics of mind—the main alternatives being dualism, behaviorism, and the type-identity theory. It accounts for minds as they are described by scientific theories and explanations. It is also known as psychofunctionalism (Block 1980).[6]

14.3 The Analogy between Minds and Computers

At the origin of computational functionalism are analogies between some features of minds and some features of computers. Different analogies pull towards different versions of the view.

Putnam, the chief founder of computational functionalism, drew an analogy between the individuation conditions of mental states and those of Turing machine

[6] In Chapter 4 I reviewed where this type of functionalism fits within the broader functionalist landscape.

states (Putnam 1960, 1967a, 1967b).[7] Putnam noticed that the states of Turing machines are individuated by the way they affect and are affected by other Turing machine states, inputs, and outputs. By the same token, he thought, mental states are individuated by the way they affect and are affected by other mental states, stimuli, and behavior. At first, Putnam did not conclude that mental states are Turing machine states, because—he said—the mind might not be a causally closed system (Putnam 1960). A bit later, though, Putnam reckoned that mental states can be characterized functionally, like those of Turing machines, though he added that the mind might be something "quite different and more complicated" than a Turing machine (Putnam 1967b). Finally, Putnam went all the way to computational functionalism: mental states are (probabilistic) Turing machine states (Putnam 1967a).

As Putnam's trajectory illustrates, the analogy between the individuation of mental states and that of Turing machine states does not entail computational functionalism. The latter conclusion was reached by Putnam after drawing his analogy, on independent grounds. What grounds?

It's hard to know for sure. The relevant papers by Putnam contain references to the plausibility and success of computational models of mental phenomena, including McCulloch and Pitts's (1943) CTC cum theory of mind. McCulloch and Pitts's theory was widely interpreted to claim that, in essence, the brain is a Turing machine (without tape), and that mental states are what today we would call neurocomputational states. As we've seen, von Neumann interpreted McCulloch and Pitts's work as *proof* that "anything that can be exhaustively and unambiguously described, anything that can be exhaustively and unambiguously put into words, is ipso facto realizable by a suitable finite neural network" of the McCulloch and Pitts type (von Neumann 1951, 23).

But the nervous system described by their theory is a simplified and idealized version of the real thing (Chapter 5). More importantly, McCulloch and Pitts's theory does not entail that everything can be simulated by Turing machines or that everything is computational. Alas, neither von Neumann nor his early readers were especially careful on these matters. After von Neumann, fallacious arguments from the Church–Turing thesis—sometimes in conjunction with McCulloch and Pitts's actual or purported results—to the conclusion that the mind is computational began to proliferate (Chapter 10).

Since McCulloch and Pitts's networks can be simulated by digital computers, von Neumann's (unwarranted) assertion entails that anything that can be exhaustively and unambiguously described can be simulated by a digital computer. If you add to this a dose of faith in scientists' ability to describe phenomena—"exhaustively and unambiguously"—you obtain (limited) pancomputationalism: at a suitable level of

[7] For a more detailed reconstruction and discussion of Putnam's functionalism and computational functionalism, see Piccinini 2004b and Shagrir 2005.

description, everything is computational. Thus, pancomputationalism made its way into the literature, and Putnam was one of the earliest philosophers to pick it up. As he put it, "everything is a Probabilistic Automaton [i.e., a kind of Turing machine] under some Description" (Putnam 1967a, 31). Together with Putnam's analogy between mental states and Turing machine states and the alleged plausibility of computational psychology, (limited) pancomputationalism is the most likely ground for Putnam's endorsement of computational functionalism.

For present purposes, the most important thing to notice is that the resulting version of computational functionalism is a very weak thesis. This remains true if computational functionalism is disengaged from Putnam's appeal to Turing machine states in favor of the thesis that mental states are, more generally, computational states (Block and Fodor 1972). If mental states are computational simply because at some level, everything is computational, then computational functionalism tells us nothing specific about the mind. It is a trivial consequence of the purported general applicability of computational descriptions to the natural world. This version of computational functionalism does not tell us how the mind works or what is special about it. Such a weak thesis stands in sharp contrast with others, which derive from different analogies between features of minds and features of computers.[8]

Both minds and digital computing systems manipulate complex combinatorial structures. Minds produce natural language sentences and other complex

[8] David Chalmers has pointed out to me that the weak thesis may be strengthened by arguing that while computation is insufficient for the instantiation of most properties, computation is sufficient for the instantiation of mental properties. Unlike most properties, mental properties might be such that they are instantiated "in virtue of the implementation of computations" (Chalmers, personal correspondence). This strengthening makes a difference only insofar as we have good evidence that computation—in the sense in which every physical system purportedly computes—is sufficient for mentation.

Chalmers defends a thesis of computational sufficiency along these lines as follows. He defines a notion of abstract causal organization, which involves "the patterns of interaction among the parts of the system, abstracted away from the make-up of individual parts and from the way the causal connections are implemented," and yet includes "a level fine enough to determine the causation of behavior" (Chalmers 2011). He then argues that unlike most nonmental properties, all there is to mental properties is abstract causal organization, and abstract causal organization can be fully and explanatorily captured computationally. If Chalmers is right, then (the right kind of) computation is sufficient for mentation while being insufficient for most other properties.

Chalmers's argument does not establish computational sufficiency for mental properties in a way that makes a difference for present purposes. Chalmers faces a dilemma. If abstract causal organization is truly fine grained enough to determine the causation of a system's behavior, then—contrary to Chalmers's intent—abstract causal organization will capture (the causal aspects of) *any* property whatsoever (including digestion, combustion, etc.). If, instead, abstract causal organization excludes enough information about a system to rule out at least certain properties (such as digestion and combustion), then—again, contrary to Chalmers's intent—there is no reason to accept that abstract causal organization will capture every aspect of mental properties. Either way, the specific connection between mentation and computation is not strengthened. Ritchie (2011) provides a more detailed reply to Chalmers along these lines; in his response, Chalmers (2012) concedes that in his argument he did not distinguish between functionalism and computational functionalism, although he does not seem to notice how missing that distinction undermines his argument. Thus, Chalmers's argument does not affect our main discussion.

sequences of actions. Digital computers manipulate strings of digits such that there are recursive rules describing the structure of the inputs and outputs as well as recursive rules describing the relationship between inputs and outputs. Furthermore, for any (universal) formalism for specifying digital computations (e.g., Turing machines) and any recursive function, there is a program describing a way to compute that function within that formalism. Finally, whether a program computes a function correctly (e.g., whether it computes square roots correctly) is an open question—it's "open to rational criticism" (Putnam 1960, 149), as it were.

Thanks in part to the complexity of their structure and the sensitivity of digital computing systems to such structure, digital computational vehicles may be systematically interpreted. So, computing systems' activities are usually characterized by semantic descriptions. For example, we say that digital computers do arithmetic calculations, which is an activity individuated in terms of operations on numbers, which are possible referents of strings of digits. This interpretability of the vehicles manipulated by computers has often been seen as part of the analogy between mental states and computational states, because mental states are also typically seen as endowed with semantic content. This, in turn, has contributed to the attractiveness of CTM.[9] But matters of mental content are controversial, and without agreement on mental content, the putative semantic analogy between mental states and computational states is a shaky basis on which to explicate computational functionalism. In the next section, I will also argue that the semantic properties of computational states make little difference for our purposes. Setting semantic properties aside, let's go back to the complexity of the structures being manipulated.

The analogy between the structures manipulated by minds and those manipulated by computing mechanisms is stronger than the previous one: prima facie, most things cannot manipulate combinatorial structures of arbitrary recursive complexity according to arbitrary recursive rules. This stronger analogy is a likely source of some versions of CTM and, derivatively, of computational functionalism. But this analogy is still insufficient to underwrite the slogan "the mind is the software of the brain." The reason is that the analogy holds between minds and digital computing mechanisms in general, and many digital computing mechanisms (e.g., ordinary Turing machines) don't possess any software in the relevant sense. For the relevant notion of software, we need yet another analogy, which holds between certain mental capacities and certain capacities of program-controlled digital computers.

[9] According to Smith, "The *only* compelling reason to suppose that we (or minds or intelligence) might be computers stems from the fact that we, too, deal with representations, symbols, meanings, and the like" (1996, 11). Smith is exaggerating in calling this the *only* reason. But the semantic analogy between minds and computers does have a long and influential history. For a more detailed discussion, see Piccinini 2004a.

Program-controlled digital computers ("computers" from now on, unless otherwise noted), unlike other computing systems, have an endless versatility in manipulating strings of digits. If they are universal, they embody in a single entity the universality of whole programming systems. Computers are versatile because they can store, manipulate, and execute programs.[10] To a first approximation, a program is a list of instructions for executing a task defined over strings of digits. An instruction is also a string of digits, which affects a computer in a special way. Most computers can execute many different (appropriately written) programs—typically, within certain limits of time and memory, they can execute *any* program. Because of this, computers can acquire any number of new capacities simply by acquiring and switching between programs. They can also refine their capacities by altering their programs. Just as minds can learn to execute a seemingly endless number of tasks, computers can execute any task, defined over strings of digits, for which they are given an appropriate program.

This special property of computers—their capacity to store and execute programs—gives rise to the special form of explanation that we employ for their behavior. How is my computer letting me write this book? By executing a word-processing program. How does it allow me to search the World Wide Web? By executing an Internet browsing program. And so on for the myriad capacities of my computer. These are explanations by program execution:

> An *explanation by program execution* of a capacity *C* possessed by a system *S* is a program *P* for *C* such that *S* possesses *C* because *S* executes *P*.

Explanation by program execution applies only to systems, such as computers, that have the capacity to execute programs. Other relevant systems include certain automatic looms and music machines. Even though computers are not the only kind of system subject to explanation by program execution, computers have other interesting properties that other program-controlled mechanisms lack. The main difference is that the processes generated by *computer* programs depend not only on the program but also on input data and, more specifically, on the precise configuration of the input data (viz., the input strings of digits) for their application: to each string of input data there corresponds a different computation. By contrast, automatic looms have inputs but not input data, and produce the same output pattern regardless of the properties of the input. (Whether the pattern is visible is another matter, which depends in part on the color of the inputs, but the loom does not respond to input color.) Furthermore, typical computers have an internal memory in which they can store and manipulate their own data and

[10] For a more detailed and systematic discussion of computer kinds and their properties, see Piccinini 2015, chap. 11.

programs, and they can compute any recursive function for as long as they have time and memory space.

These remarkable capacities of computers—to manipulate strings of digits and to store and execute programs—suggest a bold hypothesis. Perhaps brains are computers, and perhaps minds are nothing but the programs running on neural computers. If so, then we can explain the multiple capacities minds exhibit by positing specific programs for those capacities. The versatility of minds would then be explained by assuming that brains have the same special power that computers have: the power to compute by storing and executing programs. Strictly speaking, this is the source of the computational functionalist slogan: the mind is the software of the brain.[11]

Compare this version of computational functionalism to the first one. Here we have a putative explanation of human behavior, based on an analogy with what explains computers' behavior. This version tells us how the mind works and what's special about it: the brain has the capacity of storing and executing different programs, and the brain's switching between programs explains its versatility. It is a strong thesis: of all the things we observe, only brains and computers exhibit such seemingly endless ability to switch between tasks and acquire new skills. Presumably, there are few if any other systems whose behavior is explained in terms of (this type of) program execution.

If we take this formulation of computational functionalism seriously, we ought to find an adequate explication of program execution. We ought to make explicit what differentiates systems that compute by executing programs from other kinds of system. For if minds are to be interestingly analogous to some aspect of computers, there must be something that minds and computers share and other systems lack—something that accounts for the versatility of minds and computers as well as the explanation of this versatility by program execution. Unfortunately, the received view of software implementation, which is behind the standard view of program execution, does not satisfy this condition of adequacy.

14.4 Troubles with Program Execution

Ever since Putnam (1967a) formulated computational functionalism, the received view of software implementation has been the mapping account (Chapter 6, Section 6.4). If there are two descriptions of a physical system, a physical

[11] An argument to this effect is in Fodor 1968b, which is one of the founding documents of computational functionalism and which Fodor 2000 singles out as conflating functionalism and CTM. An argument along similar lines is in Newell 1990, 113ff. Other influential authors offered similar considerations. For the role program execution played in Alan Turing's thinking about intelligence, see Turing 1950 and Piccinini 2003b. For the role program execution played in von Neumann's thinking about brains, see von Neumann 1958 and Piccinini 2003c.

description and a computational description, and if the computational description maps onto the physical description, then the physical system implements the computational description and the computational description is the system's software.[12]

The problem with the mapping account is that it turns every physical system into a computer executing its computational description. As I mentioned in the previous section, everything can be given computational descriptions. For instance, some cosmologists study the evolution of galaxies using cellular automata. According to the received view of software implementation, this turns galaxies into hardware running the relevant cellular automata programs. If satisfying computational descriptions is sufficient for implementing them in the sense in which ordinary computers execute their programs, then every physical system is a computer executing its computational descriptions. This is not only counterintuitive—it also trivializes the notion of computer as well as the analogy at the origin of computational functionalism. If the mind is the software of the brain in the sense in which certain cellular automata are the software of galaxies, then the analogy between minds and computers becomes an analogy between minds and everything else. As a consequence, the strong version of computational functionalism collapses into something very much like the weak one.

To make matters worse, the same physical system satisfies *many* computational descriptions. An indefinite number of cellular automata—using different state transition rules, different time steps, or cells that represent regions of different sizes—map onto the same physical dynamics. Furthermore, an indefinite number of formalisms different from cellular automata, such as Turing machines or C++ programs, can be used to compute the same functions computed by cellular

[12] Here is a case in point:

> [A] programming language can be thought of as establishing a mapping of the physical states of a machine onto sentences of English such that the English sentence assigned to a given state expresses the instruction the machine is said to be executing when it is in that state. (Fodor 1968b, 638)

Beginning in the 1970s, some authors attempted to go beyond the mapping account by imposing further constraints on implementation. Most prominently, Bill Lycan (1987) imposed a teleological constraint. Although this was a step in the right direction, Lycan and others used "software/hardware" and "role/realizer" interchangeably. They offered no account specific to *software* implementation as opposed to role realization, so the conflation between functionalism and CTM remained unaffected. When talking specifically about computation, philosophers continued to appeal to versions of the mapping view:

> [A] physical system is a computational system just in case there is an appropriate (revealing) *mapping between the system's physical states and the elements of the function computed.* (Churchland and Sejnowski 1992, 62; emphasis added)

> [C]omputational theories construe cognitive processes as formal operations defined over symbol structures . . . Symbols are just functionally characterized objects whose individuation conditions are specified by *a realization function f_g which maps equivalence classes of physical features of a system to what we might call "symbolic" features.* Formal operations are just those physical operations that are differentially sensitive to the aspects of symbolic expressions that under the realization function f_g are specified as symbolic features. The mapping f_g allows a causal sequence of physical state transitions to be interpreted as a *computation.* (Egan 1992, 446; first emphasis added)

automata. Given the received view of software implementation, it follows that galaxies are running all these programs at once.[13]

By the same token, brains implement all their indefinitely many computational descriptions. If the mind is the software of the brain, as computational functionalism maintains, then given the standard view of software implementation, we obtain either indeterminacy as to what the mind is, or that the mind is a collection of indefinitely many pieces of software. This is not a promising metaphysics of mind, nor is it a way of explaining mental capacities in terms of program execution.[14]

The problem under discussion should not be confused with the unlimited pancomputationalism described by Putnam (1988). He argues that any physical system implements a large number of computations, or perhaps every computation, because a large number of (or perhaps all) state transitions between computational states can be freely mapped onto the state transitions between the physical states of a system. For example, I can take the state transitions my web browser is going through and map them onto the state transitions my desk is going through. As a result, my desk implements my web browser. I can establish the same mapping relation between a large number of (or perhaps all) computations and an arbitrary physical system. From this, Putnam concludes that the notion of computation is observer-relative in a way that makes it useless to the philosophy of mind. His argument is based on an especially liberal version of the mapping account.

But the mapping account of software implementation can be strengthened to avoid unlimited pancomputationalism. As many authors have noted (e.g., Chrisley 1995; Copeland 1996; Chalmers 1996a; Bontly 1998; Scheutz 2001), the computational descriptions employed by Putnam are anomalous. In the case of kosher computational descriptions—the kind normally used in scientific modeling—the work of generating successive descriptions of a system's behavior is done by a computer running an appropriate program (e.g., a weather forecasting

[13] Matthias Scheutz has suggested an amendment to the standard explication of software implementation, according to which for a computational description to be considered relevant to software implementation, *all* its states and *all* its computational steps must map onto the system that is being described (Scheutz 2004). This proposal rules out many computational descriptions, such as computational models that employ C++ programs, as irrelevant to what software is being implemented by a system, and hence it improves on the standard view. But this proposal still leaves in place indefinitely many computational descriptions of any given system, so it doesn't fully solve the present problem.

[14] The thesis that everything has computational descriptions is more problematic than it may appear, in a way that adds a further difficulty for the mapping account of software implementation. For ordinary computational descriptions are only approximate descriptions rather than exact ones, and hence a further undesirable consequence of the standard view is that programs can only approximate the behavior of the systems that are supposed to be implementing them. I discuss the relevance of approximation to (limited) pancomputationalism and the way pancomputationalism trivializes CTM in more detail in Piccinini 2015, chap. 4.

program), not by the mapping relation. In the sort of descriptions employed in Putnam's argument, instead, the descriptive work is done by the mapping relation.

In our example, my web browser does not generate successive descriptions of the state of my desk. If I want a genuine computational description of my desk, I have to identify states and state transitions of the desk, represent them by a computational description (thereby fixing the mapping relation between the computational description and the desk), and then use a computer to generate subsequent representations of the state of the desk, while the mapping relation stays fixed. So, Putnam's unlimited pancomputationalism is irrelevant to genuine computational descriptions. Still, the problem under discussion remains: everything can be given an indefinite number of bona fide computational descriptions.

To solve this problem, we must conclude that being described computationally is insufficient for implementing software, which is to say, we must go beyond the mapping account of software implementation. The same point is supported by independent considerations. The word "software" was coined to characterize specific systems called "computers." Computers perform different activities from those performed by other systems such as drills or valves—let alone galaxies. We consider the invention of computers in the 1940s a major intellectual breakthrough—the discovery of something new. We have specific disciplines—computer science and computer engineering—that study the peculiar activities and characteristics of computers and only computers. For all these reasons, a good account of software implementation must draw a principled distinction between computers and other systems.

Philosophers largely ignored this problem until a few years ago, and the charitable reader may legitimately wonder why. A first part of the answer is that philosophers interested in computational theories used to devote most of their attention to explaining mental phenomena, leaving computation per se largely unanalyzed.

A second part of the answer is that computationalist philosophers typically endorse the semantic view of computation, according to which computational states are individuated, at least in part, by their content (e.g., Shagrir 2001, 2006). The semantic view appears to offer protection to the received view of software implementation because, independently of the semantic view, it is plausible that only some things, such as mental states, are individuated by their content. If computational states are individuated by their content and content is present only in few things, then explanation by program execution will apply at most to things that have content, and the trivialization of the notion of software is thereby avoided. Unfortunately, the protection offered by the semantic view is illusory.

For starters, even the semantic view of computation does not capture the notion of program execution that applies to ordinary computers. Computers (and some automatic looms, for that matter) execute programs whether or not the digits they manipulate have content, and there are mechanisms that perform computations defined over interpreted strings of digits just like those manipulated by computers but do so without executing programs (e.g., Boolean circuits). Second, there are computationalists who maintain that content—or at least cognitive content—plays no explanatory or individuative role in a computational theory of cognition (Stich 1983; Egan 1992, 2003, 2014). Conjoining computational functionalism with the semantic view of computation begs the question of whether content should play a role in a computational theory of mind. Finally, and most seriously, the semantic view of computation is incorrect, because computability theorists and computer designers—i.e., those to whom we should defer in individuating computational states—individuate computational states without appealing to their semantic properties (Chapter 6, Section 6.4; Piccinini 2015, chap. 3). For these reasons, the semantic view of computation needs to be rejected and cannot restore the relevant notion of software implementation to health.

To a first approximation, the distinction between computers and other systems can be drawn in terms of program execution, where program execution is understood informally as a special kind of activity pertaining to special mechanisms (cf. Fodor 1968b, 1975; Pylyshyn 1984). Computers are among the few systems whose behavior we normally explain by invoking the programs they execute. When we do so, we explain each activity of a computer by appealing to the unique program being executed. A program may be described in many different ways: instructions, subroutines, whole program in machine language, assembly language, or higher-level programming language. But, modulo the compositional and functional relations between programs and their components at different levels of description, a computer runs one and only one program at any given time. An expert could retrieve the unique program run by a computer and write it down, instruction by instruction.

True, modern computers can run more than one program "at once," but this has nothing to do with applying different computational descriptions to them at the same time. It has to do with computers' capacity to devote some time to running one program, quickly switch to another program, quickly switch back, and so forth, creating the impression that they are running several programs at the same time. (Some computers *can* execute many programs in parallel. This is because they have many different processors, i.e., program-executing components. Each processor executes one and only one program at any given time.)[15] All of this needs noncircular explication. A good account of software implementation must say why computers execute programs while most other systems don't, and hence what minds need to be like in order to be the putative software of brains. To prepare for that, it's time to clarify the relationship between functionalism and CTM.

[15] It's also true that infinitely many programs can compute the same function. But in any given computer processor, at any given time, there is a fact of the matter as to which program is generating the behavior in question: it's the one that the processor is executing.

14.5 Functional Organization

According to functionalism, the mind is the functional organization of the brain. According to CTM, the functional organization of the brain is computational. These theses are prima facie logically independent—it should be possible to accept one while rejecting the other. But, according to one construal, functional organizations are specified by computational descriptions connecting a system's inputs, internal states, and outputs (Putnam 1967a; Block and Fodor 1972). Under this construal, functional organizations are ipso facto computational, and hence functionalism entails CTM. This consequence makes it impossible to reject CTM without also rejecting functionalism, which may explain why attempts at refuting functionalism often address explicitly only its computational variety (e.g., Block 1978; Churchland 2005). The same consequence led to Fodor's admission that he and others conflated functionalism and CTM (2000, 104).

To avoid conflating functionalism and CTM, we need a notion of functional organization that doesn't beg the question of CTM. The broadest notion of functional organization is the purely causal one, according to which functional organization includes all causal relations between a system's internal states, inputs, and outputs. Given this notion, functionalism amounts to the thesis that the mind is the causal organization of the brain, or that mental states are individuated by their causal properties. Indeed, this is how functionalism is often formulated. The good news is, this version of functionalism is not obviously committed to CTM, because prima facie, causal properties are not ipso facto computational. The bad news is, the resulting version of functionalism is too weak to underwrite a theory of mind.

The causal notion of functional organization applies to all systems with inputs, outputs, and internal states. A liberal notion of input and output generates an especially broad causal notion of functional organization, which applies to all physical systems. For instance, every physical system may be said to take its state at time t_0 as input, go through a series of internal states between t_0 and t_n, and yield its state at t_n as output. A more restrictive notion of input and output generates more interesting notions of functional organization. For instance, opaque bodies may be said to take light of all wavelengths as input and yield light of only some wavelengths plus thermal radiation as output. Still, the purely causal notion of functional organization is too vague and broad to do useful work in the philosophy of mind (and computation, for that matter). How should the notions of input and output be applied? Which of the many causal properties of a system are relevant to explaining its capacities? To answer these questions, we need to restrict our attention to the causal properties of organisms and artifacts that are relevant to explaining their specific capacities.

To fulfill this purpose, we turn to the notion of functional analysis. Functional analysis was introduced in modern philosophy of mind by Fodor (1965, 1968a).

He used examples like the camshaft, whose function is to lift an engine's valve so as to let fuel into the piston. The camshaft has many causal properties, but only some of them, such as its capacity to lift valves, are functionally relevant—relevant to explaining an engine's capacity to generate motive power. Fodor argued that psychological theories are functional analyses, like our analysis of the engine's capacity in terms of the functions of its components.

When Fodor defined psychological functional analysis in general, however, he departed from his examples and assimilated psychological functional analyses to computational descriptions.[16] Several other authors developed a similar notion of functional analysis, retaining Fodor's assimilation of functional analyses to computational descriptions (Cummins 1975, 1983, 2000; Dennett 1978a; Haugeland 1978; Block 1995). If functional organizations are specified by functional analyses and functional analyses are computational descriptions, then functional organizations are ipso facto computational. The mongrel of functional analysis and computational description is another source of the conflation between functionalism and CTM.

To avoid this conflation, we need a notion of functional organization that has the relevant explanatory power—like Fodor et al.'s—but does not commit us to the view that every functionally organized system is computational. The mechanistic notion of functional organization that I defined in Chapter 4 offers us what we need. As I argued there, a formulation of functionalism in terms of mechanisms is also independently motivated. For an important lesson of recent philosophy of science is that (the relevant kind of) explanation in the special sciences, such as psychology and neuroscience, takes a mechanistic form (Chapters 6–9):

> A *mechanism M* with capacities **C** is a set of spatiotemporal components $A_1, \ldots, A_n$, their functions **F**, and **F**'s relevant causal and spatiotemporal relations **R**, such that *M* possesses **C** because (i) *M* contains $A_1, \ldots, A_n$, (ii) $A_1, \ldots, A_n$ have functions **F** organized in way **R**, and (iii) **F**, when organized in way **R**, constitute **C**.

A mechanism in the present sense exhibits its capacities thanks to its components, their functions, and their organization. Biologists ascribe functions to types of biological traits (e.g., the digestive function of stomachs) and engineers ascribe them to types of artifacts and their components (e.g., the cooling function of refrigerators). The functions ascribed to traits and artifacts are distinct from their accidental effects (e.g., making noise or breaking under pressure), and hence are only a subset of their causal powers. As a consequence, tokens of organs and artifacts that do not perform their functions may be said to malfunction or be

[16] Cf. "the paradigmatic psychological theory is a list of instructions for producing behavior" (Fodor 1968b, 630). For a more extended discussion, see Chapter 7, Section 7.2; Piccinini 2015, chap. 5.

defective. As I said in Chapter 4, a mechanism's functional organization includes the states and activities of components, the spatial relations between components, the temporal relations between the components' activities, and the specific ways the components' activities affect one another.

Different notions of mechanism may be generated by employing different notions of function. As I discussed in Chapter 3, several notions of function may be used. The narrowest notion is that of causal role, which is a regular effect that contributes to a capacity of a containing system. The strongest notion is that of a selected effect, which is a causal role that was selected for. The intermediate notion is that of a causal role that makes a regular contribution to a goal of organisms. Each of these notions of function gives rise to a notion of mechanism.

To serve our purposes, the notions of function and mechanism cannot depend on the notion of computation so that all mechanisms turn into computing mechanisms, on pain of begging the question of CTM again. Fortunately, computation plays no such role in my explication(s) of "function" and "mechanism." As a result, appealing to mechanisms does not beg the question of CTM.

This shows that we need not fasten together functions and computation the way Putnam, Fodor, and their followers did. When we appeal to the function of camshafts to explain the capacities of engines, our function ascription is part of a mechanistic explanation of the engine's capacities in terms of its components, their functions, and their organization. We do not appeal to programs executed by engines, nor do we attribute any computation to engines. In fact, most people would consider engines good examples of systems that do *not* work by executing programs or, more generally, by performing computations. The same point applies to the vast majority of mechanisms, with the notable exception of computers and other computing mechanisms—including, perhaps, brains.

As we saw in Chapter 4, mechanisms give us the notion of functional organization that is relevant to understanding theories in psychology, neuroscience, and computer science. On that basis, I offered a mechanistic formulation of functionalism, which does justice to the original motivations of functionalism without begging the question of CTM. Functionalism about a system *S* should be construed as the thesis that *S* is the functional organization of the mechanism that exhibits *S*'s capacities. On this basis, the claim that the mind is the functional organization of the brain amounts to the following: the brain is the relevant mechanism, and the mind is its functional organization. This is mechanistic functionalism.

Mechanistic functionalism has a further great advantage, which is especially relevant to the concerns of this chapter: it is based on a notion of mechanism that offers us the materials for explicating the notion of program execution and, more generally, computation.

14.6 Mechanisms, Computation, and Program Execution

A mechanism may or may not perform computations, and a mechanism that performs computations—a computing mechanism—may or may not do so by executing programs. To illustrate the latter distinction, consider Turing machines. Turing machines are made out of a tape of unbounded length, an active device that can take a finite number of states, letters from a finite alphabet, and functional relations (specified by a machine table) between tape, active device, states, and letters. Of course, Turing machines are usually thought of as abstract, in the same sense in which mathematically defined triangles and circles are abstract. Like triangles and circles, Turing machines can be physically implemented. Physically implemented Turing machines and other digital computing mechanisms operate on concrete counterparts of strings of letters, which I call strings of digits. Whether abstract or concrete, Turing machines are mechanisms, subject to mechanistic explanation no more and no less than other mechanisms.[17]

Some Turing machines compute only one function. Other Turing machines, called *universal*, can compute any computable functions. This difference in computation power has a mechanistic explanation. Universal Turing machines, unlike nonuniversal ones, treat some digits on their tape as programs; they manipulate their data by appropriately responding to the programs. Because of this, universal Turing machines—unlike nonuniversal ones—may be said to *execute* the programs written on their tape. The behavior of all Turing machines is explained by the computations they perform on their data, but only the behavior of universal Turing machines is explained by the execution of programs. Besides Turing machines, many other mechanisms compute: finite state automata,

[17] The distinction between the abstract and the concrete easily leads to confusion on these matters. For example, Thomas Polger argues that abstract computations, being abstract, are not causally individuated (2007, 239–44). (He also expresses some second thoughts; cf. fn. 18.) Polger slides between talk of abstract functions, computations, machines, states, algorithms, and programs, as if the same considerations applied to all. I take issue with that. Abstract functions are not realized in the same sense in which machines are. Functions are *computed* by machines—they are relations holding between the inputs and outputs of machines. Machines compute functions by *following* algorithms. As we shall see, some special machines not only follow algorithms, but also do so by *executing* programs. Algorithms may be seen as sequences of statements or as relations holding between machine states and actions; programs may be seen as sequences of strings, sequences of statements, or relations holding between machine states and actions. Abstract functions are not causally individuated, but machines as well as their states and computations are—at least insofar as they are physically realizable. Algorithms and programs may or may not be, depending on how they are conceived of.

On a different note, it is common to see references to different levels of description of computing mechanisms, some of which are said to be more abstract than others (e.g., Newell 1980; Marr 1982). In this sense, a description is more or less abstract depending on whether it includes less or more details about a system. Here I am not questioning the distinction between more abstract and more concrete levels of description and I am not focusing on the "implementation" or "physical" level at the expense of more "abstract" computational levels. Rather, I am offering a better way of understanding computational levels as levels of being (Chapter 1). Insofar as levels of description are relevant to the explanation of a system's capacities, they are all describing aspects of a mechanism—they are all part of a complete mechanistic explanation of the system, regardless of how abstract they are (Chapter 6).

pushdown automata, RAM machines, etc. Of these, some compute by executing programs and some don't. Of those that do, some are universal and some aren't.

Like Turing machines, the capacities of most biological systems and artifacts are explained mechanistically (Bechtel and Richardson 2010; Craver and Darden 2013). Unlike Turing machines, the capacities of most biological systems are not explained by appealing to putative computations they perform, let alone programs that they execute (except, of course, in the case of brains and other putative computing mechanisms).

So, explaining a capacity by program execution is not the same as explaining it computationally, which is not the same as explaining it mechanistically. Rather, explaining a (computational) capacity by program execution is a special *kind* of computational explanation, which is a special *kind* of mechanistic explanation. Computing mechanisms are subject to explanation by program execution because of their peculiar mechanistic properties. More specifically, computers are subject to computational explanation because of their peculiar mechanistic properties, some of which (though not all) they share with other computing mechanisms.

The rest of this section explicates the above distinctions by applying the mechanistic account of physical computation (Chapter 6). I will begin with the subclass of mechanisms that perform computations and whose (relevant) capacities are explained by the computations they perform. Then I will identify the subclass of computing mechanisms that execute programs and whose (relevant) activities are explained by the programs they execute. Once we have an account of these distinctions, we will have the resources to explicate computational functionalism.

Most functional mechanisms are partially individuated by their capacities. For instance, stomachs are things whose function is to digest food, and refrigerators are things whose function is to lower the temperature of certain regions of space. Capacities, in turn, may be analyzed in terms of inputs received from the environment and outputs delivered to the environment. Stomachs take undigested food as input and yield digested food as output; refrigerators take their inside at a certain temperature as input and deliver the same region at a lower temperature as output.

Their inputs and outputs may be taxonomized in many ways, which are relevant to the capacities to be explained. In our examples, foods and temperatures are taxonomized, respectively, in terms of whether and how they can be processed by stomachs and refrigerators in the relevant ways. Being a specific kind of functional mechanism, computing mechanisms are individuated by inputs, outputs, and internal states of a specific kind and by a specific way of processing those inputs, outputs, and internal states. The inputs, internal states, and outputs of computing mechanisms are medium-independent vehicles.

A special class of computing mechanisms are digital. Their inputs, outputs, and internal states are strings of digits. A digit is a state of a particular that belongs to one and only one of a finite number of relevant types. The digits' types are unambiguously distinguishable (and hence individuated) by the effects they have on the mechanism that manipulates them. That is, every digit of the same type affects a mechanism in the same way relative to generating the mechanism's output, and each type of digit affects the mechanism in a different way relative to generating the mechanism's output.

In other words, ceteris paribus, if Digit_1 and Digit_2 are two tokens of the same type, then substituting Digit_1 for Digit_2 results in the exact same computation (type) with the same output string (type), whereas if Digit_1 and Digit_2 are of different types, then substituting Digit_1 for Digit_2 results in a different computation, which may generate a different output string.[18] This property of digits differentiates them from many other classes of particulars, such as temperatures and bites of food, which belong to indefinitely many relevant types. (There is no well-defined functional classification of temperatures or foods such that every temperature or bite of food belongs to one among a finite number of relevant types.)

A string is a list of permutable digits individuated by the digits' types, their number, and their order within the string. Every finite string has a first and a last digit member, and each digit that belongs in a string (except for the last member) has a unique successor. A digit within a string can be substituted by another digit without affecting the other digits' types, number, or position within the string. In particular, when an input string is processed by a mechanism, ceteris paribus, the digits' types, their number, and their order within the string make a difference to what output string is generated.

The fact that digits are organized into strings further differentiates strings of digits from the inputs and outputs of other kinds of mechanism. Neither temperatures nor bites of food are organized into strings in the relevant sense. The comparison is unfair, because neither bites of food nor temperatures are digits to begin with. But let us suppose, for the sake of the argument, that we could find a way to unambiguously taxonomize bites of food into finitely many (functionally relevant) types. For instance, we could taxonomize bites of food into protein bites, fat bites, etc. If such a taxonomy were viable, it would turn bites of food into digits. Still, sequences of bites of food would not constitute *strings* of digits, because digestion—unlike computation—is not precisely sensitive to the order in which an organism bites its food.

Among systems that manipulate strings of digits, some do so in a special way: under normal conditions, they produce output strings of digits from input strings of digits in accordance with a general rule, which applies to all relevant strings and

[18] It is possible for two different computations to generate the same output from the same input. This simply shows that computations are individuated more finely than input–output mappings.

depends on the inputs (and perhaps the internal states) for its application.[19] The rule in question specifies the function *computed* by the system. Some systems manipulate strings without performing computations over them. For instance, a genuine random number generator yields strings of digits as outputs, but not on the basis of a general rule defined over strings. (If it did, its output would not be genuinely random.) Systems that manipulate strings of digits in accordance with the relevant kind of rule deserve to be called digital computing mechanisms.

The activities of digital computing mechanisms are explained by the computations they perform. For example, if you press the buttons marked "21," ":," "7," and "=," of a (well-functioning) calculator, after a short delay it will display "3." The explanation of this behavior includes the facts that 3 is 21 divided by 7, "21" represents 21, ":" represents division, "7" represents 7, "=" represents equality, and "3" represents 3. Most crucially, the explanation involves the fact that under those conditions, the calculator performs a specific calculation: to divide its first input datum by the second. The capacity to calculate is explained, in turn, by an appropriate mechanistic explanation.

Calculators have input devices, processing units, and output devices. The function of the input devices is to deliver input data and commands from the environment to the processing units, the function of the processing units is to perform the relevant operations on the data, and the function of the output devices is to deliver the results to the environment. By iterating this explanatory strategy, we can explain the capacities of a calculator's components in terms of its components' functions and their organization.

Until now, I've sketched an account of digital computing mechanisms in general: digital computing mechanisms are mechanisms whose function is manipulating strings of digits according to appropriate rules; their behaviors are explained by the computations they perform. There remains to explicate the more interesting notion of a (program-controlled) computer—a mechanism that computes by executing programs.

Computers have special processing units, usually called processors. Processors are capable of performing a finite number of primitive operations on input strings (of fixed length) called "data." Which operation a processor performs on its data is determined by further strings of digits, called "instructions." Different instructions cause different operations to be performed by a processor. The performance of the relevant operation in response to an instruction is what constitutes the execution of that instruction. A list of instructions constitutes a program. The execution of a program's instructions in the relevant order constitutes the execution of the

[19] Which strings are relevant? All the strings from the relevant alphabet. For each computing mechanism, there is a relevant finite alphabet. Notice that the rule need not define an output for all input strings (and perhaps internal states) from the relevant alphabet. If some outputs are left undefined, then under those conditions the mechanism should produce no output strings of the relevant type.

program. So, by executing a program's instructions in the relevant order, a computer processor executes the program. This is a brief mechanistic explanation of (program-controlled) computers and their capacity to execute programs in terms of their components, functions, and organization. The capacity of a processor to execute instructions can be further explained by a mechanistic explanation of the processor in terms of its components, their functions, and their organization.[20]

Only computing mechanisms of a specific kind, namely computers, have processors capable of executing programs (and memories for storing programs, data, and results). This is why only the capacities of computers, as opposed to the capacities of other computing mechanisms—let alone mechanisms that do not perform computations—are explained by program execution. Computational explanation by program execution says that there are strings of digits whose function is to determine a sequence of operations to be performed by a processor on its data.

In other words, program execution requires that some states of some components of the computer *function* as a program; in an explanation by program execution, "program" is used as a function term. The way a program determines what the computer does is cashed out in terms of the computer's mechanistic properties. A program-controlled computer is a very special kind of computing mechanism, which has the capacity to execute programs. This is why the appeal to program execution is explanatory for computers—because it posits programs and processors *inside* computers.

As a consequence, when the behavior of ordinary computers is explained by program execution, the program is not just a description. The program is also a (stable state of a) *physical component* of the computer, whose function is to generate the relevant capacity of the computer. Programs are physically present within computers, where they have a function to perform. Until recently, this simple and straightforward point had been almost entirely missed in the philosophical literature.[21]

[20] Cf. any standard textbook on computer organization and design, such as Patterson and Hennessy 1998.

[21] For an exception, see Moor 1978, 215. Robert Cummins, one of the few people to discuss this issue explicitly, maintained that "programs aren't causes but abstract objects or play-by-play accounts" (Cummins 1983, 34; see also Cummins 1977). Cummins's weaker notion of program execution is quite popular among philosophers and is yet another side of the fuzziness surrounding functionalism and CTM. This is because the weaker notion is not the one used in computer science and does not underwrite the strong analogy between mental capacities and computers' capacities that is behind the slogan "the mind is the software of the brain," and yet the weaker notion is often used in explicating CTM or even functionalism. The main reason for Cummins's view of program execution seems to be the way he mixes functional analysis and computational description. Roughly, Cummins thinks that explaining a capacity by program execution is the same as giving a functional analysis of it, and therefore the program is not a part of the computer but a description of it (see Section 14.4). This leads Cummins and his followers to the paradoxical conclusion that connectionist networks compute by executing algorithms or programs (Cummins and Schwarz 1991; Roth 2005). But it should be obvious

14.7 Computational Functionalism

We now have the means to explicate computational functionalism:

Computational functionalism: the mind is the software of the brain.

In the broadest sense, this may be interpreted to say that the mind is (an aspect of) the computational organization of the brain, where computational organization is the functional organization of a computing mechanism and the brain is a computing mechanism. In other words, systems that realize minds are mechanisms that manipulate medium-independent vehicles according to rules; the mind is the collection of computational states and properties such that the mechanism manipulates those states in accordance with those rules.

This broad interpretation incorporates a general analogy between minds and many artificial computing mechanisms, according to which both minds and artificial computing mechanisms manipulate medium-independent vehicles in accordance with appropriate rules. This version of computational functionalism is compatible with any nontrivial version of CTM, including connectionist and neurocomputational versions of CTM. If we restrict computation to *digital* computation, computational functionalism incorporates the more specific analogy between minds and digital computing systems based on the fact that both can manipulate complex combinatorial structures in accordance with rules. As I pointed out in Section 14.3, this analogy is neither as strong nor as explanatory as the analogy between minds and (program-controlled) computers. The more general analogy is not based on the notion of software used in computer science, which is the notion that explains the capacities of (program-controlled) computers and inspires the slogan "the mind is the software of the brain." Accordingly, in explicating computational functionalism we should give precedence to the literal notion of software. (I'll come back to more general formulations later.)

In its strongest and most literal form, computational functionalism says that (i) the brain contributes to the production of behavior by storing and executing programs on computational data, in the sense sketched in the previous section, and (ii) the mind is constituted by the programs stored and executed by the brain, plus, perhaps, the states and processes generated by executing those programs. As in the broader version of computational functionalism, the mind is an aspect of the computational organization of a computing mechanism; in addition, the computing mechanism is a program-controlled computer and the mind is its

that connectionist networks in general—just like Turing machines in general—do not store and execute programs in the sense I explicated in the main text, which is why their behavior is not as flexible as that of digital computers. For a detailed account of computation in neural networks, see Piccinini 2015, chap. 13. More recently, Cummins has agreed that "stored programs are certainly causes" (personal correspondence).

programs (plus, perhaps, the states and processes generated by executing the programs). This doctrine has some interesting consequences for the study of minds and brains.

Computational functionalism incorporates a strong version of classical CTC, which licenses explanations of mental capacities by program execution. This is a kind of mechanistic explanation, which explains mental capacities by positing a specific kind of mechanism with specific functional properties. Briefly, the posited mechanism includes memory components, which store programs, and at least one processor, which manipulates and executes them. Together, the interaction between memories and processors determines how the system processes its data and produces its outputs. The capacities of the system are explained as the result of the processing of data performed by the processor(s) in response to the program(s).

Computational functionalism entails that minds are medium-independent and hence multiply realizable, in the sense in which different tokens of the same type of computer program can run on different kinds of hardware. So, if computational functionalism is correct, then—pace foes of multiple realizability—mental programs can also be specified independently of how they are implemented in the brain, in the same way in which one can describe which programs are (or should be) run by digital computers without worrying much about how they are physically implemented. Under the computational functionalist hypothesis, this is the task of psychological theorizing. Psychologists may speculate on which programs are executed by brains when exhibiting certain mental capacities. The programs thus posited are how-possibly explanations for those capacities.

One surprise is that, when interpreted literally, computational functionalism entails that the mind is a (stable state of) a component of the brain, in the same sense in which computer program tokens are (stable states of) components of computers. As a consequence, even a brain that is not processing any data—analogously to an idle computer, or even a computer that is turned off—can still have a mind, provided that its programs are still physically present. This consequence seems to offend some people's intuitions about what it means to have a mind, but it's independently plausible. It corresponds to the sense in which people who are asleep or otherwise unconscious still have minds. Their minds are relatively "causally quiescent," as David Armstrong (1981) puts it.

Computational functionalism describes the mind as a program, which means that the function of the mind is to determine which sequences of operations the brain has to perform. This presupposes a very specific mechanistic explanation of the brain as a program-controlled computer, i.e., a mechanism with certain components that have certain functions and a certain organization. Whether a system is a particular kind of mechanism is an empirical question. In this important respect, computational functionalism turns out to incorporate a strong empirical hypothesis.

Philosophers of mind have usually recognized that CTM is an empirical hypothesis in two respects. On one hand, there is the empirical question of whether a computer can be programmed to exhibit all of the capacities that are peculiar to minds. This is one traditional domain of artificial intelligence. On the other hand, there is the empirical question of whether all mental capacities can be explained by program execution. This is one traditional domain of cognitive psychology. As to neuroscience, many CTM theorists have traditionally considered it irrelevant to testing their hypothesis, on the grounds that the same software can be implemented by different kinds of hardware. This attitude is unsatisfactory in two respects.

First, as we have seen, at least two historically influential construals of functionalism entail CTM. But if CTM is a logical consequence of the metaphysical doctrine of functionalism, then the empirical status of CTM is tied to that of functionalism: if functionalism is a priori true (as some philosophers believe), then CTM should need no empirical testing; conversely, any empirical disconfirmation of CTM should disconfirm functionalism too. An important advantage of my proposed reformulation of functionalism is that it does not entail CTM. This leaves CTM free to be an empirical hypothesis about the specific functional organization of the brain, which—when conjoined with functionalism—gives rise to computational functionalism.

But, second, if CTM is an empirical hypothesis to the effect that mental capacities are the result of program execution, it isn't enough to test it by programming computers and attempting to explain mental capacities by program execution. Indeed, to assume that this is enough for testing it begs the question of whether the brain is the sort of mechanism that could run mental programs at all—whether it is a (program-controlled) computer. Assuming that the mind is the software of the brain presupposes that the brain has components of the relevant kinds, with the relevant functional and organizational properties.

Whether the brain is a kind of program-controlled computer is itself an empirical question. If the brain were not functionally organized in the right way, the strong form of computational functionalism about the mind would turn out to be false. This shows that computational functionalism incorporates an empirical hypothesis that can be effectively tested only by neuroscience. Whether brains are one kind of mechanism or another can only be determined by studying brains.

This sense in which the strong form of computational functionalism embodies an empirical hypothesis is more fundamental than the other two. If the brain is a (program-controlled) computer, then both classical artificial intelligence and classical cognitive psychology have a fair chance of succeeding. But if the brain is *not* a (program-controlled) computer, then classical artificial intelligence and cognitive psychology will fail to explain how the mind actually works. Classical artificial intelligence and cognitive psychology may still succeed in reproducing mental capacities, depending on the extent to which it is possible to reproduce the capacities of systems that are not program-controlled computers by executing programs. It may be possible to reproduce all or many mental capacities by executing programs even though the brain is not a program-controlled computer, the mind is something other than the programs running on the brain, or both. The extent to which this is possible is a difficult question, which I briefly discussed in Chapter 11, Section 11.3.

That being said, there is only one way to find out the specific alphabet and computer code in which putative mental programs are written, the types of data they manipulate, the memory registers that store them, the processing components that process the data, and the operations by which processing components manipulate the data: to study neural circuits, the vehicles they manipulate, and the operations they perform. Surprise, surprise: this research has actually been done for decades by neurophysiologists in cooperation with theoretical and computational neuroscientists. In Chapter 13, I argued that the empirical evidence they've collected undermines all digital versions of CTC, and a fortiori the specific digital version of CTM that is behind the strong form of computational functionalism. In general, neural computations are not digital, let alone executions of programs.[22]

I have formulated and discussed computational functionalism primarily using the notion of program execution because the analogy between minds and program-controlled computers is the motivation behind the strong version of computational functionalism. There is no question that many of those who felt the pull of the analogy between some features of minds and some features of computers—such as Alan Turing, John von Neumann, Jerry Fodor, Allen Newell, and Herbert Simon—did so in part because of the explanatory power of program execution.

But, as I noticed in Section 14.2, computational functionalism is ambiguous between a strong and a weak reading. It is equally obvious that many other authors, who are (or were at one point) sympathetic to the analogy between some features of minds and some features of computers, such as Hilary Putnam, Robert Cummins, Paul and Patricia Churchland, Michael Devitt and Kim Sterelny (Devitt and Sterelny 1999), and even Warren McCulloch and Walter Pitts (at least in 1943), would resist the conclusion that brains store and execute programs. Is there a way to cash out their view without falling into the trivial conclusion that the mind can be described computationally in the sense in which anything else can? Indeed, there is. Their view is captured by one of the more general interpretations of computational functionalism mentioned at the beginning of this section.

The account of computational explanation I offered in Chapter 6 applies to all computing mechanisms, regardless of whether they are controlled by programs. In

[22] If so, how can we explain the kind of cognitive flexibility that is the very source of the analogy between minds and program-controlled computers? I sketched the beginning of a possible alternative explanation in Chapter 9, Section 9.4.

fact, I explicated computation by program execution in terms of the more general notion of digital computation, which I explicated, in turn, in terms of the more general notion of computation simpliciter. Brief recap: computation is the manipulation of medium-independent vehicles according to a rule, digital computation is computation over strings of digits, and computation by program execution is digital computation performed in response to instructions that encode the relevant rule. The digital computers we use every day compute by executing programs, but nonuniversal Turing machines, finite state automata, and many connectionist networks perform computations without executing programs. Analog computers and some other nonconventional computing systems perform computations on nondigital vehicles.

To cover theories that don't appeal to program execution, all we need to do is interpret computational functionalism in terms of computation simpliciter, without appealing to program execution. According to this generalized computational functionalism, the mind is (some aspect of) the computational organization of a (computing) mechanism, regardless of whether that mechanism is a program-controlled computer, a nonclassical neural network, or any other kind of computing mechanism. Given the generalized formulation, psychological explanations need not invoke the execution of programs—they can invoke either program execution or some other kind of neural computation that is presumed to generate the behavior to be explained. This kind of explanation is still a mechanistic explanation that appeals to the manipulation of medium-independent vehicles in accordance with an appropriate rule by appropriate components with appropriate functions. Therefore, this generalized formulation of computational functionalism still presupposes that the brain has the relevant mechanistic properties, which can be studied empirically by neuroscience. Given this generalization, computational functionalism is compatible with any version of CTC, including connectionism and computational neuroscience.

14.8 Functionalism, the Computational Theory of Mind, and Computational Functionalism

I have discussed three theses:

Functionalism: The mind is the functional organization of the brain.

CTM: Mental capacities have a computational explanation, which—in combination with the empirical fact that the brain is the organ of the mind—entails that the functional organization of the brain is computational.

Computational functionalism (generalized): The mind is the computational organization of the brain.

Computational functionalism is the conjunction of functionalism and CTM. I have formulated these doctrines within a mechanistic framework and exhibited some of their mutual relations.

Functionalism does not entail CTM, and by now it should be easy to see why. There are plenty of functional organizations that do not involve program execution or any other computational process. That the mind is functionally constituted is consistent with noncomputational mechanistic explanations applying to the mind or at least some recalcitrant aspects of the mind such as phenomenal consciousness. Thus, it is a fallacy to attack functionalism by impugning some computationalist hypothesis or another.

The thesis that the functional organization of the brain is computational, which is entailed by CTM, does not entail functionalism either. That thesis is compatible with the mind being the computational organization of the brain, but also with (some aspects of) the mind being a noncomputational but still functional property of the brain, or even some nonfunctional property of the brain, such as its physical composition, the speed of its action, its color, or, more plausibly, the intentional content or phenomenal quality of its states. In short, one may believe that the brain is a computing system while opposing or being neutral about functionalism, at least with respect to some aspects of the mind.

The view that the brain is a computing mechanism is an empirical, mechanistic hypothesis about the brain. Even if the brain is a computing mechanism, the mind may or may not be the brain's computational organization—perhaps there are aspects of the mind that have to do with other properties, e.g., the phenomenal qualities of mental states. But if brains turn out *not* to be computing mechanisms, then CTM (and hence computational functionalism) is false. So, whether or not we agree with computational functionalism, we can still focus on whether the brain is a computing mechanism and investigate CTM. This, of course, can only be done by studying the functional organization of the brain empirically.

The standard formulations of computational functionalism in philosophy of mind have made it difficult to discuss CTM as productively as it can be. They have convinced many philosophers that CTM is an a priori thesis, to be discussed by philosophical arguments and thought experiments and assessed by the extent to which it solves philosophical problems such as the mind–body problem. This has led philosophers to ignore that, in so far as it has empirical content, CTM embodies an empirical scientific hypothesis about the functional organization of the brain, which comes in several varieties that ought to be assessed by neuroscience. In Chapter 9, I argued on empirical grounds that cognition is largely explainable computationally, so the brain is a computing mechanism after all. In Chapter 13, I argued that neural computations are neither digital nor analog but sui generis. That covers the cognitive aspects of the mind. Consciousness is another matter.

14.9 Noncomputational Functionalism about Consciousness

In the early literature on computational approaches to the mind (ca. 1943–69), consciousness was rarely mentioned and even more rarely discussed in depth. Remarkably, the analogies between minds and computational systems that I've discussed in the previous sections—the analogies that motivate computational functionalism—have virtually nothing to do with consciousness. Or so it seems. We should appreciate that early defenders of computational approaches were primarily responding to behaviorism.[23] It was radical enough for them to explain behavior in terms of computations and representations. Consciousness was beyond the pale.

Things started changing in the 1970s. Daniel Dennett (1969) offered one of the first detailed accounts of consciousness in terms of computation and information processing. Block and Fodor (1972) made a side remark that computational functionalism does not account for consciousness, and Block (1978) followed up with his classic argument to the same effect and endorsed the type-identity theory about consciousness. Other authors responded that computational functionalism or some computational theory does accommodate consciousness after all (Lycan 1987; Dennett 1991, Rey 2005).

Beginning in the 1990s, consciousness became a hot topic, and a new twist on computational functionalism become popular. This is known as representationalism about consciousness. The idea is that phenomenal consciousness can be explained in terms of intentionality, and intentionality can be explained in terms of computation and information processing (Harman 1990; Tye 1995; Dretske 1995; Rosenthal 2005).

Intentionality is the ability to be about things. For example, the words in this chapter are meaningful and describe certain facts and, perhaps, so do our mental states. Mental intentionality is often explained in terms of representation, namely, by positing mental representations (intentional states) that refer to whatever we perceive, think about, or act on at any given time. We've seen that computational states need not be representations, but they often are (Chapter 12). That is, typical computational states represent things. By positing computational states with the appropriate representational properties, CTM might be able to account for intentionality.

Now consider that typical conscious states have representational properties. The orangeness you experience when you see an orange represents the orangeness of the orange. The roundness you experience represents the roundness of the orange. The size you experience represents the size of the orange. And so forth.

[23] In philosophy, they were also responding to the type-identity theory (Place 1956; Feigl 1958; Smart 1959). But the type-identity theory was not a rival *scientific* paradigm. The rival paradigm was behaviorism.

Many if not all aspects of conscious experience appear to represent something or other. Representationalism turns this point into a theory of consciousness. All there is to the quality of a conscious experience, says representationalism, is what it represents. A conscious experience is just a state with a certain representational content. Sometimes representationalists disagree about what, exactly, our conscious experiences represent. But representationalists agree that, as long as we have an adequate account of the representational content of our conscious states, we have an adequate account of phenomenal consciousness.

Most versions of CTM do assume that mental states are representations. Thus, if there is an adequate account of the content of mental representations, CTM plus representationalism (an account of phenomenal consciousness in terms of their representational content) yields an account of phenomenal consciousness. Since representationalism explains consciousness in terms of representation, and representation is consistent with CTM, representationalism is clearly compatible with CTM and computational functionalism.

Another wrinkle in the dialectic on the metaphysics of consciousness is property dualism—the view that consciousness is due to nonphysical properties. Some property dualists maintain that computation is sufficient for consciousness in the sense that, given certain computational states, the basic laws of nature guarantee that certain phenomenal states will be present (e.g., Chalmers 1996b, 2011). Setting property dualism aside, the two physicalistically respectable positions about consciousness remain computational functionalism and the type-identity theory.

Unless they resort to the type-identity theory, physicalists and even leading property dualists assume that phenomenal consciousness must be explained in terms of computation and information processing (e.g., Lycan 1987; Dennett 1991; Chalmers 2011; the same assumption is implicit in typical representationalist accounts of phenomenal consciousness). The main reason is the general plausibility of functionalism combined with the widespread conflation between functionalism and CTM, which I have debunked above. This gives rise to the following line of reasoning: conscious mental states are functional states (functionalism), functional states are computational (conflation); therefore, conscious mental states are computational.

I have argued at length, however, that computation is a special kind of mechanistic process among others. Therefore, computational states are a special class of functional states. Conversely, not all functional states are computational. This makes room for a theory of consciousness that does not require conscious states to be (entirely) computational in nature.

The distinction between functionalism and computational functionalism, and the corresponding distinction between multiple realizability and medium independence, opens up the possibility of a novel view about consciousness. This novel account combines what is appealing about (mechanistic) functionalism—that it fits within the multilevel mechanistic framework of cognitive neuroscience and allows for the multiple realizability of mental states—with what is appealing about the critique of computational functionalism about consciousness—that computation alone is too extrinsic to account for consciousness.

The central idea is that phenomenal consciousness is a kind of macroscopic physical state brought about by certain types of physical mechanisms, of which human brains are an instance. When the right type of physical system is in the relevant macroscopic state, consciousness ensues.[24] Brains are capable of bringing that state about, and other physical systems may be able to do it too if they have the right causal powers. (Whether other physical systems have the relevant causal powers is unknown because we do not know exactly what physical macrostates constitute consciousness.)

According to this noncomputational functionalism, consciousness has a functional nature in the sense of being an aspect of the functional organization of the brain, but it is not (entirely) computational in nature. This is still consistent with consciousness being physical, because the functional organization of the brain is still physical. This noncomputational functionalism is consistent with the multiple realizability of consciousness, although multiple realizability is not guaranteed because the relevant kind of functional organization might require a unique kind of structure.

The important difference between this view and computational functionalism is that, according to noncomputational functionalism, no amount of computation (cum information processing or representation) is sufficient for consciousness. Thus, consciousness is not medium independent, which means that consciousness is not a state whose sole function is processing variables based on differences between different portions of the variables. On the contrary, the conscious aspect of a mental state may be due to a global brain state whose function includes processing specific physical variables (conscious ones) in ways that are sensitive to some of their specific macroscopic physical properties (namely, their phenomenal character).

This account does not solve all the problems about consciousness but it does constitute substantive progress on this most intractable of philosophical problems, because it identifies a noncomputational and perhaps more plausible version of mechanistic functionalism than those previously available. Therefore, this account is worth investigating and pursuing. The progress is afforded by combining a multilevel mechanistic functionalism (Chapter 4) with the important yet generally unheeded distinction between computational and functional states. By relying on the functional organization of the brain, this noncomputational functionalism can address some classic objections to computational functionalism coming from the

[24] Cf. Piccinini 2007d for an early step in this direction.

physicalist and biological camps (e.g., Block 2007) and incorporate the insights of such views.

There is a whole space of possibilities worth exploring. For ease of reference, I have dubbed these views *noncomputational functionalism*. This label is slightly misleading because the theories I'm referring to can combine functions with qualities, where qualities may or may not reduce to functions. According to these theories, the nature of consciousness is a combination of qualities and not-entirely-computational functions.

Qualities are properties intrinsic to an object, like being round or triangular. They come in levels of being, where each higher-level quality is an aspect of its lower-level realizers. Functional properties are causal powers that also come in levels of being, where each higher-level causal power is an aspect of its lower-level realizers. Typically, higher-level qualities and functions are multiply realizable—they can be aspects of many different lower-level realizers.

The functions invoked here are not (entirely) medium-independent. Therefore, they are not (entirely) computational. This does not mean that they are identical to lower-level properties. Again, they are higher-level properties, which means they are aspects of their realizers and probably multiply realizable.

The relationship between qualities and causal powers is a complex issue that I will not address here. Options include that qualities reduce to powers (dispositionalism), powers to qualities (categoricalism), qualities and powers are the same thing (identity theory, not to be confused with type-identity reductionism), properties are a combination of qualities and powers (hybrid view), or there is some more complicated relationship between the two. Different versions of noncomputational functionalism are generated by different views about the relationship between qualities and powers. Only the theory that combines noncomputational functionalism and dispositionalism is a functionalist view in the strictest sense.

14.10 Summing Up: Explaining Biological Cognition

In this book, I have defended a neurocomputational theory of cognition grounded in an egalitarian ontology of levels, multilevel mechanisms, teleological functions, and mechanistic functionalism. Here are its main conclusions:

1. There are many levels of mechanistic organization. Parts compose wholes; properties of parts realize properties of wholes.
2. All levels are equally real. Higher-level objects are invariants under some changes in their parts; properties of wholes are aspects of their realizers (Chapter 1).
3. Higher-level functional properties are often multiply realizable, and some multiply realizable properties are medium independent (Chapter 2).

4. Some mechanisms, including cognitive and computational mechanisms, have teleological functions, which are regular contributions to goals of organisms (Chapter 3).
5. Some properties are functional—defined in terms of the functional organization of a mechanism. Functions and structures are not separable: they mutually constrain one another (Chapter 4).
6. The original Computational Theory of Cognition states that neurocognitive processes are digital computations (Chapter 5).
7. Digital computation is one kind of computation among others, including but not limited to analog computation. Computation and information processing are medium-independent mechanistic processes (Chapter 6).
8. Constitutive explanation is mechanistic (Chapter 7).
9. Cognition is constitutively explained in terms of multilevel neurocognitive mechanisms. This is the type of explanation cognitive neuroscience seeks to provide (Chapter 8).
10. Neurocognitive processes are computations that process information (Chapter 9).
11. The Computational Theory of Cognition does not follow from the Church–Turing thesis (Chapter 10).
12. There are no compelling objections to the Computational Theory of Cognition (Chapter 11).
13. Neurocognitive processes operate on representations (Chapter 12).
14. Neurocognitive processes are neither digital nor analog—they are a sui generis form of computation (Chapter 13).
15. Consciousness may or may not have a wholly functional nature. Even if consciousness does have a wholly functional nature, its functional nature may not be wholly computational (Chapter 14).

This account reconciles the Computational Theory of Cognition with recent developments in computational, theoretical, and experimental neuroscience. It avoids the pitfalls of both traditional reductionism and traditional antireductionism. It opens up a new direction in our understanding of consciousness. I hope some of you, kind readers, find it fruitful.

Bibliography

Abraham, T. H. (2004). "Nicolas Rashevsky's Mathematical Biophysics." *Journal of the History of Biology* 37(2): 333–85.

Abraham, T. H. (2016). *Rebel Genius: Warren S. McCulloch's Transdisciplinary Life in Science.* Cambridge, MA: MIT Press.

Adams, F. (1979). "A Goal-State Theory of Function Attributions." *Canadian Journal of Philosophy* 9(3): 493–518.

Adams, F. and K. Aizawa (2010). "Causal Theories of Mental Content." In E. N. Zalta (ed.), *The Stanford Encyclopedia of Philosophy* (Spring 2010 Edition), URL = <https://plato.stanford.edu/archives/spr2010/entries/content-causal/>. Stanford, CA: The Metaphysics Research Lab, Centre for the Study of Language and Information, Stanford University.

Adrian, E. D. (1928). *The Basis of Sensation: The Action of the Sense Organs.* New York: Norton.

Adrian, E. D. and Y. Zotterman (1926). "The Impulses Produced by Sensory Nerve-endings. Part 2. The Response of a Single End-Organ." *Journal of Physiology* 61(2): 151–71.

Aizawa, K.: (1996). "Some Neural Network Theorizing before McCulloch: Nicolas Rashevsky's Mathematical Biophysics." In R. Moreno Díaz and J. Mira (eds.), *Brain Processes, Theories, and Models: An International Conference in Honor of W. S. McCulloch 25 Years after His Death* (64–70). Cambridge, MA: MIT Press.

Aizawa, K. (2003). *The Systematicity Arguments.* Boston: Kluwer.

Aizawa, K. (2013). "Multiple Realization by Compensatory Differences." *European Journal for Philosophy of Science* (3)1: 1–18.

Aizawa, K. (2017). "Multiple Realization and Multiple 'Ways' of Realization: A Progress Report." *Studies in the History and Philosophy of Science* 68: 3–9.

Aizawa, K. and C. Gillett (2009). "The (Multiple) Realization of Psychological and Other Properties in the Sciences." *Mind and Language* 24(2): 181–208.

Aizawa, K. and C. Gillett (2011). "The Autonomy of Psychology in the Age of Neuroscience." In P. M. Illari, F. Russo, and J. Williamson (eds.), *Causality in the Sciences* (202–23). Oxford: Oxford University Press.

Albright, T. D. (1984). "Direction and Orientation Selectivity of Neurons in Visual Area MT of the Macaque." *Journal of Neurophysiology* 52(6): 1106–30.

Alexander, G. E. and M. D. Crutcher (1990). "Preparation for Movement: Neural Representations of Intended Direction in Three Motor Areas of the Monkey." *Journal of Neurophysiology* 64(1): 133–50.

Alle, H., A. Roth, and J. R. P. Geiger (2009). "Energy-Efficient Action Potentials in Hippocampal Mossy Fibers." *Science* 325: 1405–8.

Amundson, R. and G. V. Lauder (1994), "Function without Purpose: The Uses of Causal Role Function in Evolutionary Biology." *Biology and Philosophy* 9(4): 443–69.

Andersen, H. (2018). "Complements, Not Competitors: Causal and Mathematical Explanations." *British Journal for the Philosophy of Science* 69(2): 485–508.

Anderson, J. R. (1978). "Arguments Concerning Representations for Mental Imagery." *Psychological Review* 85: 249–77.

Anderson, J. R. (1983). *The Architecture of Cognition*. Cambridge, MA: Harvard University Press.

Anderson, J. R. (1993). *Rules of the Mind.* Hillsdale, NJ: Erlbaum.

Anderson, J. R. (2007). *How Can the Human Mind Occur in the Physical Universe?* Oxford: Oxford University Press.

Anscombe, E. (1957). *Intention*. Ithaca, NY: Cornell University Press.

Arbib, M. A. (1989). "Comments on 'A Logical Calculus of the Ideas Immanent in Nervous Activity.'" In R. McCulloch (ed.), *Collected Works of Warren S. McCulloch* (341–2). Salinas, CA: Intersystems.

Arbib, M. A. (2000). "Warren McCulloch's Search for the Logic of the Nervous System." *Perspectives in Biology and Medicine* 43(2): 193–216.

Armstrong, D. (1997). *A World of States of Affairs*. Cambridge: Cambridge University Press.

Armstrong, D. M. (1968). *A Materialistic Theory of the Mind.* New York, NY: Routledge.

Armstrong, D. M. (1973). *Belief, Truth, and Knowledge*. Cambridge: Cambridge University Press.

Armstrong, D. M. (1981). "What Is Consciousness?" In D. M. Armstrong, *The Nature of Mind.* Ithaca, NY: Cornell University Press.

Aspray, W. (1985). "The Scientific Conceptualization of Information: A Survey." *Annals of the History of Computing* 7(2): 117–40.

Attneave, F. (1961). "In Defense of Homunculi." In W. Rosenblith, *Sensory Communication* (777–82). Cambridge, MA: MIT Press.

Audi, P. (2012). "Properties, Powers, and the Subset Account of Realization." *Philosophy and Phenomenological Research* 84(3): 654–74.

Awater, H., J. R. Kerlin, K. K. Evans, and F. Tong (2005). "Cortical Representation of Space Around the Blind Spot." *Journal of Neurophysiology* 94(5): 3314–24.

Ayala, F. J. (1970). "Teleological Explanations in Evolutionary Biology." *Philosophy of Science* 37: 1–15.

Azim, E., J. Jiang, B. Alstermark, and T. M. Jessell (2014). "Skilled Reaching Relies on a V2a Propriospinal Internal Copy Circuit." *Nature* 508(7496): 357–63.

Azzi, J. C., R. Gattass, B. Lima, J. G. Soares, and M. Fiorani (2015). "Precise Visuotopic Organization of the Blind Spot Representation in Primate V1." *Journal of Neurophysiology* 113(10): 3588–99.

Aydede, M. (1997). "Language of Thought: The Connectionist Contribution." *Minds and Machines* 7(1): 57–101.

Baars, B. (1988). *A Cognitive Theory of Consciousness*. Cambridge: Cambridge University Press.

Baars, B. (1997). *In the Theater of Consciousness*. Oxford: Oxford University Press.

Baars, B. J., W. P. Banks, and J. B. Newman (eds.) (2003). *Essential Sources in the Scientific Study of Consciousness*. Cambridge, MA: MIT Press.

Baddeley, R., Hancock, P., and Földiák, P. (eds.) (2000). *Information Theory and the Brain.* Cambridge: Cambridge University Press.

Bahar, S. (2003). "Effect of Light on Stochastic Phase Synchronization in the Crayfish Caudal Photoreceptor." *Biological Cybernetics* 89(3): 200–13.

Bahar, S. and F. Moss (2003). "Stochastic Phase Synchronization in the Crayfish Mechanoreceptor/photoreceptor System." *Chaos* 13(1): 138–44.

Bair, W. and C. Koch (1996). "Temporal Precision of Spike Trains in Extrastriate Cortex of the Behaving Macaque Monkey." *Neural Computation* 8: 1185–1202.

Barabási, A.-L. (2002). *Linked: The New Science of Networks.* Cambridge, MA: Perseus.

Barberis, S. D. (2013). "Functional Analyses, Mechanistic Explanations, and Explanatory Tradeoffs." *Journal of Cognitive Science* 14(3): 229–51.

Barrett, D. (2014). "Functional Analysis and Mechanistic Explanation." *Synthese* 191(12): 2695–2714.

Barsalou, L. W. (1999). "Perceptual Symbol Systems." *Behavioral and Brain Sciences* 22(4): 577–660.

Bartels, A. (2006). "Defending the Structural Concept of Representation." *Theoria* 21(55): 7–19.

Bartlett, G. C. (2012). "Computational Theories of Conscious Experience: Between and Rock and a Hard Place." *Erkenntnis* 76: 195–209.

Bartlett, G. C. (2017). "Functionalism and the Problem of Occurrent States." *The Philosophical Quarterly* 68(270): 1–20

Bastian, J. (1996). "Plasticity in an Electrosensory System. I. General Features of a Dynamic Sensory Filter." *Journal of Neurophysiology* 76(4): 2483–96.

Batterman, R. (2000). "Multiple Realizability and Universality." *British Journal for the Philosophy of Science* 51: 115–45.

Batterman, R. (2002). *The Devil in the Details: Asymptotic Reasoning in Explanation, Reduction, and Emergence*. New York: Oxford University Press.

Bauer, R. H. and J. M. Fuster (1976). "Delayed-Matching and Delayed-Response Deficit from Cooling Dorsolateral Prefrontal Cortex in Monkeys." *Journal of Comparative and Physiological Psychology* 90(3): 293–302.

Baum, E. B. (2004). *What Is Thought?* Cambridge, MA: MIT Press.

Baumgartner, M. (2013). "Rendering Interventionism and Non-Reductive Physicalism Compatible." *Dialectica* 67: 1–27.

Baumgartner, M. (2017). "The Inherent Empirical Underdetermination of Mental Causation," *The Australasian Journal of Philosophy* 96(2): 335–50.

Baxter, D. and A. Cotnoir (eds.) (2014). *Composition as Identity*. Oxford: Oxford University Press.

Baylor, D. A. (1987). "Photoreceptor Signals and Vision. Proctor Lecture." *Investigative Ophthalmology and Visual Science* 28(1): 34–49.

Baysan, U. (2015). "Realization Relations in Metaphysics." *Minds and Machines* 25: 247–60.

Baysan, U. (2017). "Causal Powers and the Necessity of Realization." *International Journal of Philosophical Studies* 25 (4): 525–31.

Bechtel, W. (2001). "Cognitive Neuroscience: Relating Neural Mechanisms and Cognition." In P. Machamer, P. McLaughlin, and R. Grush (eds.), *Philosophical Reflections on the Methods of Neuroscience*. Pittsburgh, PA: University of Pittsburgh Press.

Bechtel, W. (2008). *Mental Mechanisms: Philosophical Perspectives on Cognitive Neuroscience*. London: Routledge.

Bechtel, W. (2009). "Constructing a Philosophy of Science of Cognitive Science." *Topics in Cognitive Science* 1: 548–69.

Bechtel, W. (2013). "Addressing the Vitalist's Challenge to Mechanistic Science: Dynamic Mechanistic Explanation." In S. Normandin and C. T. Wolfe (eds.), *Vitalism and the Scientific Image in Post-Enlightenment Life Science* 1800–2010 (345–70). Dordrecht: Springer.

Bechtel, W. (2016). "Investigating Neural Representations: The Tale of Place Cells." *Synthese* 193(5): 1287–1321.

Bechtel, W. and A. Abrahamsen (2002). *Connectionism and the Mind: Parallel Processing, Dynamics, and Evolution in Networks*. Malden, MA: Blackwell.

Bechtel, W. and A. Abrahamsen (2010). "Dynamic Mechanistic Explanation: Computational Modeling of Circadian Rhythms as an Exemplar for Cognitive Science." *Studies in History and Philosophy of Science Part A* 41: 321–33.

Bechtel, W. and A. Abrahamsen (2013). "Thinking Dynamically about Biological Mechanisms: Networks of Coupled Oscillators." *Foundations of Science* 18: 707–23.

Bechtel, W. and J. Mundale (1999). "Multiple Realizability Revisited: Linking Cognitive and Neural States." *Philosophy of Science* 66(2): 175–207.

Bechtel, W. and R. C. Richardson (2010). *Discovering Complexity: Decomposition and Localization as Strategies in Scientific Research, Second Edition.* Cambridge, MA: MIT Press/Bradford Books.

Bechtel, W. and O. Shagrir (2015). "The Non-Redundant Contributions of Marr's Three Levels of Analysis for Explaining Information Processing Mechanisms." *Topics in Cognitive Science* 7(2): 312–22.

Beer, R. D. and P. L. Williams (2015). "Information Processing and Dynamics in Minimally Cognitive Agents." *Cognitive Science* 39: 1–38.

Bell, C. C. (1981). "An Efference Copy Which Is Modified by Reafferent Input." *Science* 214 (4519): 450–3.

Bell, C. C. (1982). "Properties of a Modifiable Efference Copy in an Electric Fish." *Journal of Neurophysiology* 47(6): 1043–56.

Bell, C. C., S. Libouban, and T. Szabo (1983). "Pathways of the Electric Organ Discharge Command and Its Corollary Discharges in Mormyrid Fish." *Journal of Comparative Neurology* 216(3): 327–38.

Bennett, M. R. and J. L. Kearns (2000). "Statistics of Transmitter Release at Nerve Terminals." *Progressive Neurobiology* 60(6): 545–606.

Bernstein, S. (2016). "Overdetermination Undermined." *Erkenntnis* 81: 17–40.

Bialek, W. and F. Rieke (1992). "Reliability and Information Transmission in Spiking Neurons." *Trends in Neurosciences* 15(11): 428–34.

Bickle, J. (2003). *Philosophy and Neuroscience: A Ruthlessly Reductive Approach.* Dordrecht: Kluwer.

Bickle, J. (2006). "Reducing Mind to Molecular Pathways: Explicating the Reductionism Implicit in Current Cellular and Molecular Neuroscience." *Synthese* 151: 411–34.

Bird, A. (2007). *Nature's Metaphysics: Laws and Properties.* Oxford: Oxford University Press.

Blasdel, G. G. (1992). "Orientation Selectivity, Preference, and Continuity in Monkey Striate Cortex." *Journal of Neuroscience* 12(8): 3139–61.

Block, N. (1978). "Troubles with Functionalism." In C. W. Savage (ed.), *Perception and Cognition: Issues in the Foundations of Psychology, Vol. 6* (261–325). Minneapolis, MN: University of Minnesota Press.

Block, N. (1980). "Introduction: What is Functionalism?" In Block, N (ed.), *Readings in Philosophy of Psychology* (171–84). London: Methuen.

Block, N. (1986). "Advertisement for a Semantics for Psychology." *Midwest Studies in Philosophy* 10(1): 615–78.

Block, N. (1995). "The Mind as the Software of the Brain." In Osherson, D., L. Gleitman, S. Kosslyn, E. Smith, and S. Sternberg (eds.), *An Invitation to Cognitive Science.* Cambridge, MA: MIT Press.

Block, N. (1997). "Anti-reductionism Slaps Back." *Noûs* 31(11): 107–32.

Block, N. (2006). "Max Black's Objection to Mind-Body Identity." In D. Zimmerman (ed.), *Oxford Studies in Metaphysics II (Vol. 2).* Oxford: Oxford University Press.

Block, N. (2007). *Consciousness, Function, and Representation: Collected Papers, Volume 1.* Cambridge, MA: MIT Press.

Block, N. J. and J. A. Fodor (1972). "What Psychological States Are Not." *The Philosophical Review* 81(2): 159–81.

Blum, L., F. Cucker, M. Shub, and S. Smale (1998). *Complexity and Real Computation.* New York, NY: Springer.

Boden, M. A. (1988). *Computer Models of Mind: Computational Approaches in Theoretical Psychology.* Cambridge: Cambridge University Press.

Boden, M. A. (1991). "Horses of a Different Color?" In W. Ramsey, S. P. Stich, and D. E. Rumelhart (eds.), *Philosophy and Connectionist Theory* (3–19). Hillsdale: LEA.

Boden, M. A. (2006). *Mind as Machine: A History of Cognitive Science.* Oxford: Oxford University Press.

Bogen, J. (2005). "Regularities and Causality: Generalizations and Causal Explanations." *Studies in History and Philosophy of Biological and Biomedical Sciences* 36: 397–420.

Bolhuis, J. J. and M. Everaert (2013). *Birdsong, Speech, and Language: Exploring the Evolution of Mind and Brain.* Cambridge, Mass.: MIT Press.

Bolkan, S. S., J. M. Stujenske, S. Parnaudeau, T. J. Spellman, C. Rauffenbart, A. I. Abbas, A. Z. Harris, J. Gordon, and C. Kellendonk (2017). "Thalamic Projections Sustain Prefrontal Activity During Working Memory Maintenance." *Nature Neuroscience* 20 (7): 987–96.

Bontly, T. (1998). "Individualism and the Nature of Syntactic States." *British Journal for the Philosophy of Science* 49: 557–74.

Boone, W. and G. Piccinini (2016a). "Mechanistic Abstraction." *Philosophy of Science* 83(5): 686–97.

Boone, W. and G. Piccinini (2016b). "The Cognitive Neuroscience Revolution." *Synthese* 193(5): 1509–34.

Breidbach, O. (2001). "The Origin and Development of the Neurosciences." In P. Machamer, R. Grush, and P. McLaughlin (eds.), *Theory and Method in the Neurosciences* (7–29). Pittsburgh, PA: University of Pittsburgh Press.

Bringsjord, S. (1995). "Computation, among Other Things, Is Beneath Us." *Minds and Machines* 4: 469–88.

Boorse, C. (1977). "Health as a Theoretical Concept." *Philosophy of Science* 44: 542–73.

Boorse, C. (1997). "A Rebuttal on Health." In J. M. Humber and R. F. Almeder (eds.), *What Is Disease?* (1–134). Totowa: Humana Press.

Boorse, C. (2002). "A Rebuttal on Functions." In A. Ariew, R. Cummins, and M. Perlman (eds.), *Functions: New Essays in the Philosophy of Psychology and Biology* (63–112). Oxford: Oxford University Press.

Boorse, C. (2014). "A Second Rebuttal on Health." *Journal of Medicine and Philosophy* 39: 683–724.

Born, R. T. and D. C. Bradley (2005). "Structure and Function of Visual Area MT." *Annual Review of Neuroscience* 28: 157–89.

Boshernitzan, M. (1986). "Universal Formulae and Universal Differential Equations." *The Annals of Mathematics*, 2nd Series 124(2): 273–91.

Brainard, M. S. and A. J. Doupe (2002). "What Songbirds Teach Us About Learning." *Nature* 417(6886): 351–8.

Brecht, M., M. Schneider, B. Sakmann, and T. W. Margrie (2004). "Whisker Movements Evoked by Stimulation of Single Pyramidal Cells in Rat Motor Cortex." *Nature* 427 (6976): 704–10.

Briggman, K. L., M. Helmstaedter, and W. Denk (2011). "Wiring Specificity in the Direction-Selectivity Circuit of the Retina." *Nature* 471(7337): 183–8.

Bringsjord, S. and K. Arkoudas (2007). "On the Provability, Veracity, and AI-Relevance of the Church-Turing Thesis." In A. Olszewski, J. Wolenski, and R. Janusz (eds.), *Church's Thesis after 70 Years* (66–118). Heusenstamm: Ontos.

Brooks, R. A. (1991). "Intelligence without Representation." *Artificial Intelligence* 47(1–3): 139–59.

Brooks, J. X., J. Carriot and K. E. Cullen (2015). "Learning to Expect the Unexpected: Rapid Updating in Primate Cerebellum During Voluntary Self-motion." *Nature Neuroscience* 18(9): 1310–17.

Bruno, M. S. and D. Kennedy (1962). "Spectral Sensitivity of Photoreceptor Neurons in the Sixth Ganglion of the Crayfish." *Computational Biochemical Physiology* 6: 41–6.

Buchanan, T. S., D. G. Lloyd, K. Manal, and T. F. Besier (2004). "Neuromusculoskeletal Modeling: Estimation of Muscle Forces and Joint Moments and Movements from Measurements of Neural Command." *Journal of Applied Biomechanics* 20(4): 367–95.

Buckner, C. (2018). "Empiricism without Magic: Transformational Abstraction in Deep Convolutional Neural Networks." *Synthese* 195: 5339–5372.

Buechner, J. (2018). "Does Kripke's Argument Against Functionalism Undermine the Standard View of What Computers Are?" *Minds and Machines* 28: 491–513.

Buhr, E. D. and J. S. Takahashi (2013). "Molecular Components of the Mammalian Circadian Clock." In A. Kramer and M. Merrow (eds.), *Circadian Clocks* (Vol. 217, 3–27). Berlin, Heidelberg: Springer.

Bullock, T. H. (1982). "Electroreception." *Annual Review of Neuroscience* 5: 121–70.

Burge, T. (1986). "Individualism and Psychology." *Philosophical Review* 95: 3–45.

Burge, T. (2010). *Origins of Objectivity*. Oxford: Oxford University Press.

Burnston, D. C. (2016a). "Computational Neuroscience and Localized Neural Function." *Synthese* 193(12): 3741–62.

Burnston, D. C. (2016b). "A Contextualist Approach to Functional Localization in the Brain." *Biology & Philosophy* 31(4): 527–50.

Butterfill, S. A. and C. Sinigaglia (2014). "Intention and Motor Representation in Purposive Action." *Philosophy and Phenomenological Research* 88(1): 119–45.

Caianiello, E. R. (1961). "Outline of a Theory of Thought Processes and Thinking Machines." *Journal of Theoretical Biology* 1: 204–35.

Calvo, P. (2016). "The Philosophy of Plant Neurobiology: A Manifesto." *Synthese* 193: 1323–43.

Cameron, R. (2010). "How to Have a Radically Minimal Ontology." *Philosophical Studies* 151: 249–64.

Cannon, W. B. (1932). *The Wisdom of the Body*. New York: Norton.

Cao, R. (2018). "Computational Explanation and Neural Coding." In M. Sprevak and M. Colombo (eds.), *The Routledge Handbook of the Computational Mind* (283–96). London: Routledge.

Carandini, M. and D. J. Heeger (2012). "Normalization as a Canonical Neural Computation." *Nature Reviews Neuroscience*, 13: 51–62.

Carlson, B. A. (2002). "Neuroanatomy of the Mormyrid Electromotor Control System." *Journal of Comparative Neurology* 454(4): 440–55.

Carruth, A. 2016. Powerful Qualities, Zombies and Inconceivability, *The Philosophical Quarterly* 66(262): 25–46.

Casini, L. (2017). "Malfunctions and Teleology: On the (Dim) Chances of Statistical Accounts of Functions." *European Journal for Philosophy of Science* 7(2): 319–35.

Castellani, E. (1998). "Galilean Particles: An Example of Constitution of Objects." In Castellani, E. (ed.), *Interpreting Bodies: Classical and Quantum Objects in Modern Physics* (181–94). Princeton: Princeton University Press.
Castro, A. J. (1972). "The Effects of Cortical Ablations on Digital Usage in the Rat." *Brain Research* 37(2): 173–85.
Cerminara, N. L., R. Apps, and D. E. Marple-Horvat (2009). "An Internal Model of a Moving Visual Target in the Lateral Cerebellum." *The Journal of Physiology* 587(2): 429–42.
Chalmers, D. J. (1996a). "Does a Rock Implement Every Finite-State Automaton?" *Synthese* 108(3): 309–33.
Chalmers, D. J. (1996b). *The Conscious Mind: In Search of a Fundamental Theory*. Oxford: Oxford University Press.
Chalmers, D. J. (2011). "A Computational Foundation for the Study of Cognition." *Journal of Cognitive Science* 12(4): 323–57.
Chalmers, D. J. (2012). "The Varieties of Computation: A Reply." *Journal of Cognitive Science* 13: 211–48.
Chang, H. (2004). *Inventing Temperature: Measurement and Scientific Progress*. New York: Oxford University Press.
Chang, H. T., T. C. Ruch, and A. A. Ward, Jr. (1947). "Topographical Representation of Muscles in Motor Cortex of Monkeys." *Journal of Neurophysiology* 10(1): 39–56.
Chang, L. and D. Y. Tsao (2017). "The Code for Facial Identity in the Primate Brain." *Cell* 169(6): 1013–28.
Chemero, A. (2009). *Radical Embodied Cognitive Science*. Cambridge, MA: MIT Press.
Chemero, A. and F. Faries (2018). "Dynamic Information Processing." In M. Sprevak and M. Colombo (eds.), *The Routledge Handbook of the Computational Mind* (134–48). London: Routledge.
Chemero, A. and M. Silberstein (2008). "After the Philosophy of Mind: Replacing Scholasticism with Science." *Philosophy of Science* 75: 1–27.
Chirimuuta, M. (2014). "Minimal Models and Canonical Neural Computations: The Distinctness of Computational Explanation in Neuroscience." *Synthese* 191(2): 127–54.
Chirimuuta, M. (2018). "Explanation in Computational Neuroscience: Causal and Non-Causal." *British Journal for the Philosophy of Science* 69(3): 849–80.
Chirimuuta, M. (2019). "Synthesis of Contraries: Hughlings Jackson on Sensory-Motor Representation in the Brain." *Studies in the History and Philosophy of Science Part C: Studies in the History and Philosophy of the Biological and Biomedical Sciences* 75: 34–44.
Chisholm, R. (1969). "The Loose and Popular and the Strict and Philosophical Senses of Identity." In Care, N and H. Grimm (eds.), *Perception and Identity* (82–106). Cleveland: Case Western Reserve University Press.
Chomsky, N. (1995). "Language and Nature." *Mind* 104(413): 1–61.
Chrisley, R. L. (1994). "Why Everything Doesn't Realize Every Computation." *Minds and Machines* 4: 403–30.
Church, A. (1936). "An Unsolvable Problem in Elementary Number Theory." *The American Journal of Mathematics* 58: 345–63.
Churchland, M. M., J. P. Cunningham, M. T. Kaufman, S. I. Ryu, and K. V. Shenoy (2010). "Cortical Preparatory Activity: Representation of Movement or First Cog in a Dynamical Machine?" *Neuron* 68(3): 387–400.
Churchland, P. M. (1981). "Eliminative Materialism and the Propositional Attitudes." *Journal of Philosophy* 78: 67–90.
Churchland, P. M. (1989). *A Neurocomputational Perspective*. Cambridge, MA: MIT Press.

Churchland, P. M. (2005). "Functionalism at Forty: A Critical Retrospective." *The Journal of Philosophy* 102: 33–50.

Churchland, P. M. (2007). *Neurophilosophy at Work*. Cambridge: Cambridge University Press.

Churchland, P. M. (2012). *Plato's Camera: How the Physical Brain Captures a Landscape of Abstract Universals*. Cambridge, MA: MIT Press.

Churchland, P. S. (1986). *Neurophilosophy: Toward a Unified Science of the Mind-Brain*. Cambridge, MA: The MIT Press.

Churchland, P. M. and P. S. Churchland (1990). "Could a Machine Think?" *Scientific American* 262: 32–9.

Churchland, P. S., C. Koch, and T. J. Sejnowski (1990). "What Is Computational Neuroscience?" In E. L. Schwartz (ed.), *Computational Neuroscience* (46–55). Cambridge, MA: MIT Press.

Churchland, P. S. and T. J. Sejnowski (1992). *The Computational Brain*. Cambridge, MA: MIT Press.

Cisek, P. and J. F. Kalaska (2010). "Neural Mechanisms for Interacting with a World Full of Action Choices." *Annual Review of Neuroscience* 33: 269–98.

Clapp, L. (2001). "Disjunctive Properties: Multiple Realizations." *Journal of Philosophy* 98 (3): 111–36.

Clark, A. (1993). *Associative Engines: Connectionism, Concepts, and Representational Change*. Cambridge, MA: MIT Press.

Clark, A. (2016). *Surfing Uncertainty: Prediction, Action, and the Embodied Mind*. Oxford and New York: Oxford University Press.

Clark, A. and D. J. Chalmers (1998). "The Extended Mind." *Analysis* 58(1): 7–19.

Cleland, C. E. (1993). "Is the Church–Turing Thesis True?" *Minds and Machines* 3: 283–312.

Coelho Mollo, D. (2018). "Functional Individuation, Mechanistic Implementation: The Proper Way of Seeing the Mechanistic View of Concrete Computation." *Synthese* 195(8): 3477–97.

Collins, T. (2010). "Extraretinal Signal Metrics in Multiple-saccade Sequences." *Journal of Vision* 10(14): 7.

Colombo, M. (2014). "Neural Representationalism, the Hard Problem of Content and Vitiated Verdicts. A Reply to Hutto & Myin (2013)." *Phenomenology and the Cognitive Sciences* 13(2): 257–74.

Confais, J., G. Kim, S. Tomatsu, T. Takei, and K. Seki (2017). "Nerve-Specific Input Modulation to Spinal Neurons during a Motor Task in the Monkey." *Journal of Neuroscience* 37(10): 2612–26.

Conant, R. C. and W. R. Ashby (1970). "Every Good Regulator of a System Must Be a Model of That System." *International Journal of Systems Science* 1(2): 89–97.

Copeland, B. J. (1996). "What is Computation?" *Synthese* 108: 224–359.

Copeland, B. J. (1998). "Turing's O-machines, Searle, Penrose, and the Brain." *Analysis* 58(2): 128–38.

Copeland, B. J. (2000). "Narrow versus Wide Mechanism: Including a Re-Examination of Turing's Views on the Mind–Machine Issue." *The Journal of Philosophy* 97: 5–32.

Copeland, B. J. (2002). "The Church–Turing Thesis." In E. N. Zalta (ed.), *The Stanford Encyclopedia of Philosophy* (Fall 2002 Edition), URL = <http://plato.stanford.edu/archives/fall2002/entries/church-turing/>. Stanford, CA: The Metaphysics Research Lab, Centre for the Study of Language and Information, Stanford University.

Couch, M. (2009a). "Multiple Realization in Comparative Perspective," *Biology and Philosophy* 24 (4): 505–19.

Couch, M. (2009b). "Functional Explanation in Context," *Philosophy of Science*, April 2009, vol. 76.
Cover, T. M. and J. A. Thomas (2006). *Elements of Information Theory* (2nd ed.). Hoboken, N.J.: Wiley-Interscience.
Cowan, J. D. (1990a). "McCulloch-Pitts and Related Neural Nets from 1943 to 1989." *Bulletin of Mathematical Biology* 52(1/2): 73–97.
Cowan, J. D. (1990b). "Neural Networks: The Early Days." In D. S. Touretzky (ed.), *Advances in Neural Information Processing Systems* 2 (829–42). San Mateo, CA: Morgan Kaufmann.
Craik, K. (1943). *The Nature of Explanation*. Cambridge: Cambridge University Press.
Crapse, T. B. and M. A. Sommer (2008). "Corollary Discharge across the Animal Kingdom." *Nature Reviews Neuroscience* 9(8): 587–600.
Craver, C. (2003). "The Making of a Memory Mechanism." *Journal of the History of Biology* 36: 153–95.
Craver, C. F. (2001). "Role Functions, Mechanisms, and Hierarchy." *Philosophy of Science* 68: 53–74.
Craver, C. F. (2006). "When Mechanistic Models Explain." *Synthese* 153: 355–76.
Craver, C. F. (2007). *Explaining the Brain*. Oxford: Oxford University Press.
Craver, C. F. (2008). "Physical Law and Mechanistic Explanation in the Hodgkin and Huxley Model of the Action Potential." *Philosophy of Science* 75: 1022–33.
Craver, C. F. (2010). Prosthetic Models. *Philosophy of Science* 77(5): 840–51.
Craver, C. F. (2013). "Functions and Mechanisms: A Perspectivalist View." In P. Huneman (ed.), *Function: Selection and Mechanisms* (133–58). Dordrecht: Springer.
Craver, C. F. (2014). "The Ontic Account of Scientific Explanation." In M. I. Kaiser, O. R. Scholz, D. Plenge, and A. Hüttemann (eds.), *Explanation in the Special Sciences: The Case of Biology and History* (27–52). Dordrecht: Springer.
Craver, C. and L. Darden (2001). "Discovering Mechanisms in Neurobiology." In P. Machamer, R. Grush, and P. McLaughlin (eds.), *Theory and Method in the Neurosciences* (112–37). Pittsburgh, PA: University of Pittsburgh Press.
Craver, C. F. and L. Darden (2013). *In Search of Mechanisms: Discoveries Across the Life Sciences*. Chicago: University of Chicago Press.
Craver, C. F. and D. M. Kaplan (2020). "Are More Details Better? On the Norms of Completeness for Mechanistic Explanations." *British Journal for the Philosophy of Science* 71(1): 287–319.
Craver, C. F. and M. Povich (2017). "The Directionality of Distinctively Mathematical Explanations." *Studies in History and Philosophy of Science* 63: 31–8.
Cubitt, T. S., Perez-Garcia, D., and Wolf, M. M. (2015). "Undecidability of the Spectral Gap." *Nature* 528, 207–11.
Cummins, R. (1975). "Functional Analysis." *Journal of Philosophy* 72(20): 741–65.
Cummins, R. (1977). "Programs in the Explanation of Behavior." *Philosophy of Science* 44: 269–87.
Cummins, R. (1983). "Analysis and Subsumption in the Behaviorism of Hull." *Philosophy of Science* 50(1): 96–111.
Cummins, R. (1989). *Meaning and Mental Representation*. Cambridge, MA: MIT Press.
Cummins, R. (1996). *Representations, Targets, and Attitudes*. Cambridge, MA: MIT Press.
Cummins, R. (2000). '"How Does It Work?" vs. "What Are the Laws?" Two Conceptions of Psychological Explanation.' In F. C. Keil and R. A. Wilson (eds.), *Explanation and Cognition*. Cambridge, MA: MIT Press.

Cummins, R. and G. Schwarz (1991). "Connectionism, Computation, and Cognition." In T. Horgan and J. Tienson (eds.), *Connectionism and the Philosophy of Mind* (60–73). Dordrecht: Kluwer.

Dacey, M. (2016). "Rethinking Associations in Psychology." *Synthese* 193(12): 3763–86.

Dacey, M. (2019). "Simplicity and the Meaning of Mental Association." *Erkenntnis* 84(6): 1207–28.

Daniel, P. M. and D. Whitteridge (1961). "The Representation of the Visual Field on the Cerebral Cortex in Monkeys." *Journal of Physiology* 159: 203–21.

Darling, W. G., M. A. Pizzimenti, and R. J. Morecraft (2011). "Functional Recovery Following Motor Cortex Lesions in Non-human Primates: Experimental Implications for Human Stroke Patients." *J Integr Neurosci* 10(3): 353–84.

Daugman, J. G. (1990). "Brain Metaphor and Brain Theory." In E. L. Schwartz (ed.), *Computational Neuroscience* (9–18). Cambridge, MA: MIT Press.

Daunizeau, A. K. (2015). "The Cybernetic Bayesian Brain – From Interoceptive Inference to Sensorimotor Contingencies." In T. Metzinger and J. M. Windt (eds).), *Open MIND: 35 (T)*. Frankfurt am Main: MIND Group. doi: 10.15502/9783958570108

Daunizeau, J., David, O., and K. Stephan (2011). "Dynamic Causal Modelling: A Critical Review of the Biophysical and Statistical Foundations." *NeuroImage* 58 (2): 312–22.

Davidson, D. (1970). "Events and Particulars." *Noûs* 4(1): 25–32.

Davis, M., R. Sigal, and E. J. Weyuker (1994). *Computability, Complexity, and Languages.* Boston: Academic Press.

Dayan, P. and L. F. Abbott (2001). *Theoretical Neuroscience: Computational and Mathematical Modeling of Neural Systems.* Cambridge, MA: MIT Press.

DeAngelis, G. C. and W. T. Newsome (1999). "Organization of Disparity-selective Neurons in Macaque Area MT." *Journal of Neuroscience* 19(4): 1398–1415.

Dehaene, S. (2014). *Consciousness and the Brain*. New York: Viking Press.

Dehaene, S. and Naccache, L. (2001). "Towards a Cognitive Neuroscience of Consciousness: Basic Evidence and a Workspace Framework." *Cognition* 79: 1–37.

Denk, W. and H. Horstmann (2004). "Serial Block-Face Scanning Electron Microscopy to Reconstruct Three-dimensional Tissue Nanostructure." *PLoS Biology* 2(11): e329.

Dennett, D. C. (1969). *Content and Consciousness*. London: Routledge.

Dennett, D. C. (1971). "Intentional Systems." *The Journal of Philosophy* 68(4): 87–106.

Dennett, D. C. (1978a). *Brainstorms.* Cambridge, MA: MIT Press.

Dennett, D. C. (1978b). "Artificial Intelligence as Philosophy and as Psychology." In D. C. Dennett, *Brainstorms* (109–26). Cambridge, MA: MIT Press.

Dennett, D. C. (1987). *The Intentional Stance*. Cambridge, MA: MIT Press.

Dennett, D. C. (1988). "Quining Qualia." In A. J. Marcel and E. Bisiach (eds.), *Consciousness in Contemporary Science* (42–77). Oxford: Clarendon Press.

Dennett, D. C. (1991). *Consciousness Explained.* Boston: Little, Brown and Co.

Devitt, M. and K. Sterelny (1999). *Language and Reality: An Introduction to the Philosophy of Language.* Cambridge, MA: MIT Press.

Dewhurst, J. (2018a). "Computing Mechanisms without Proper Functions." *Minds and Machines* 28: 569–88.

Dewhurst, J. (2018b). "Individuation without Representation." *British Journal for the Philosophy of Science* 69: 103–16.

Dewhurst, J., and Villalobos, M. (2017). "The Enactive Automaton as a Computing Mechanism." *Thought: A Journal of Philosophy*, 6(3): 185–92.

Dewhurst, J., and Villalobos, M. (2018). "Enactive Autonomy in Computational Systems." *Synthese*, 195(5): 1891–1908.

Ding, H., R. G. Smith, A. Poleg-Polsky, J. S. Diamond, and K. L. Briggman (2016). "Species-specific Wiring for Direction Selectivity in the Mammalian Retina." *Nature* 535(7610): 105–10.

Dorr, C. and G. Rosen (2002). "Composition as a Fiction." In R. Gale (ed.), *The Blackwell Guide to Metaphysics* (151–74). Oxford: Blackwell.

Douglass, J. K., L. Wilkens, E. Pantazelou, and F. Moss (1993). "Noise Enhancement of Information Transfer in Crayfish Mechanoreceptors by Stochastic Resonance." *Nature 365* (6444): 337–40.

Doupe, A. J. and P. K. Kuhl (1999). "Birdsong and Human Speech: Common Themes and Mechanisms." *Annual Review of Neuroscience* 22: 567–631.

Dow, B. M., A. Z. Snyder, R. G. Vautin, and R. Bauer (1981). "Magnification Factor and Receptive Field Size in Foveal Striate Cortex of the Monkey." *Experimental Brain Research* 44(2): 213–28.

Dowling, J. E. (2012). *The Retina: An Approachable Part of the Brain* (Rev. ed.). Cambridge, Mass.: Belknap Press of Harvard University Press.

Downey, A. (2018). "Predictive Processing and the Representation Wars: A Victory for the Eliminativist (via Fictionalism)." *Synthese* 195: 5115–39.

Dretske, F. (1981). *Knowledge and the Flow of Information.* Cambridge, MA: MIT Press.

Dretske, F. I. (1988). *Explaining Behavior: Reasons in a World of Causes.* Cambridge, MA: MIT Press.

Dretske, F. (1994). "If You Can't Make One, You Don't Know How It Works." *Midwest Studies in Philosophy* 19(1): 468–82.

Dretske, F. (1995). *Naturalizing the Mind.* Cambridge, MA: MIT Press.

Dreyfus, H. L. (1979). *What Computers Can't Do.* New York: Harper & Row.

Dreyfus, H. L. (1998). "Response to My Critics." In T. W. Bynum and J. H. Moor (eds.), *The Digital Phoenix: How Computers Are Changing Philosophy* (193–212). Oxford, UK: Blackwell.

Duffin, R. J. (1981). "Rubel's Universal Differential Equation." *Proceedings of the National Academy of Sciences USA* 78(8 [Part 1: Physical Sciences]): 4661–2.

Duwell, A. (2017). "Exploring the Frontiers of Computation: Measurement Based Quantum Computers and the Mechanistic View of Computation." In A. Bokulich and J. Floyd (eds.), *Turing 100: Philosophical Explorations of the Legacy of Alan Turing*, Boston Studies in the Philosophy and History of Science (vol. 324, 219–32). New York: Springer.

Edelman, G. M. (1992). *Bright Air, Brilliant Fire: On the Matter of the Mind.* New York: Basic Books.

Egan, F. (1992). "Individualism, Computation, and Perceptual Content." *Mind* 101(403): 443–59.

Egan, F. (1999). "In Defence of Narrow Mindedness." *Mind and Language* 14(2): 177–94.

Egan, F. (2003). "Naturalistic Inquiry: Where Does Mental Representation Fit In?" In L. M. Antony and N. Hornstein (eds.), *Chomsky and His Critics* (89–104). Malden, MA: Blackwell.

Egan, F. (2014). "How to Think about Mental Content." *Philosophical Studies* 170: 115–35.

Egan, F. and R. Matthews (2006). "Doing Cognitive Neuroscience: A Third Way." *Synthese* 153: 377–91.

Elbert, T. and B. Rockstroh (2004). "Reorganization of Human Cerebral Cortex: The Range of Changes Following Use and Injury." *Neuroscientist* 10(2): 129–41.

Eliasmith, C. (2000). "Is the Brain Analog or Digital? The Solution and its Consequences for Cognitive Science." *Cognitive Science Quarterly* 1(2): 147–70.

Eliasmith, C. (2003). "Moving Beyond Metaphors: Understanding the Mind for What It Is." *Journal of Philosophy* 100(10): 493–520.

Eliasmith, C. (2007). "How to Build a Brain: From Function to Implementation." *Synthese 159*(3): 373–88.

Eliasmith, C. (2013). *How to Build a Brain: A Neural Architecture for Biological Cognition.* Oxford: Oxford University Press.

Eliasmith, C. and C. H. Anderson (2003). *Neural Engineering: Computation, Representation and Dynamics in Neurobiological Systems.* Cambridge, MA: MIT Press.

Eliasmith, C. and O. Trujillo (2014). "The Use and Abuse of Large-Scale Brain Models." *Current Opinion in Neurobiology* 25: 1–6.

Ellis, B. (2010). "Causal Powers and Categorical Properties." In A. Marmodoro (Ed.), *The Metaphysics of Powers: Their Grounding and Their Manifestation* (133–42), New York: Routledge.

Elsayed, G. F., A. H. Lara, M. T. Kaufman, M. M. Churchland, and J. P. Cunningham (2016). "Reorganization between Preparatory and Movement Population Responses in Motor Cortex." *Nat Commun* 7: 13239.

Ermentrout, G. B. and D. H. Terman (2010). *Mathematical Foundations of Neuroscience.* New York, Springer.

Erneling, C. E. and D. M. Johnson (2005). *The Mind as a Scientific Object: Between Brain and Culture.* Oxford: Oxford University Press.

Etesi, G. and I. Németi (2002). "Non-Turing Computations via Malament-Hogarth Spacetimes." *International Journal of Theoretical Physics* 41: 342–70.

Fales, E. (1990). *Causation and Universals.* New York: Routledge.

Fazekas, P., and G. Kertesz (2019). "Are Higher Mechanistic Levels Causally Autonomous?" *Philosophy of Science* 86(5): 847–57.

Feigl, H. (1958). "The 'Mental' and the 'Physical.'" In H. Feigl, M. Scriven, and G. Maxwell (eds.), *Concepts, Theories and the Mind-Body Problem (Minnesota Studies in the Philosophy of Science, Volume 2)* (370–497). Minneapolis: University of Minnesota Press.

Feldman, D. E. (2012). "The Spike-Timing Dependence of Plasticity." *Neuron* 75(4): 556–71.

Feldman, J. A. and D. H. Ballard (1982). "Connectionist Models and their Properties." *Cognitive Science* 6: 205–54.

Felleman, D. J. and D. C. Van Essen (1991). "Distributed Hierarchical Processing in the Primate Cerebral Cortex." *Cerebral Cortex* 1(1): 1–47.

Ferretti, G. (2016). "Through the Forest of Motor Representations." *Consciousness and Cognition* 43: 177–96.

Fetz, E. E. (1992). "Are Movement Parameters Recognizably Coded in the Activity of Single Neurons?" *Behavioral and Brain Sciences* 15: 679–90.

Fetzer, J. H. (2001). *Computers and Cognition: Why Minds Are Not Machines.* Dordrecht: Kluwer.

Field, H. H. (1978). "Mental Representation." *Erkenntnis* 13: 9–61.

Fodor, J. A. (1965). "Explanations in Psychology." In M. Black (ed.) *Philosophy in America* (161–79). London: Routledge and Kegan Paul.

Fodor, J. A. (1968a). *Psychological Explanation.* New York: Random House.

Fodor, J. A. (1968b). "The Appeal to Tacit Knowledge in Psychological Explanation." *Journal of Philosophy* 65: 627–40.

Fodor, J. A. (1974). "Special Sciences (or: The Disunity of Science as a Working Hypothesis)." *Synthese* 28(2): 97–115.

Fodor, J. A. (1975). *The Language of Thought.* Cambridge, MA: Harvard University Press.

Fodor J. A. (1980). "Methodological Solipsism Considered as a Research Strategy in Cognitive Psychology." *Behavioral and Brain Sciences* 3(1): 63–73.

Fodor, J. A. (1981). "The Mind–Body Problem." *Scientific American* 244.

Fodor, J. A. (1983). *The Modularity of Mind*. Cambridge, MA: MIT Press.

Fodor, J. A. (1987). *Psychosemantics: The Problem of Meaning in the Philosophy of Mind*. Cambridge, MA: MIT Press.

Fodor, J. A. (1990). *A Theory of Content and Other Essays*. Cambridge, MA: MIT Press.

Fodor, J. A. (1997). "Special Sciences: Still Autonomous After All These Years." *Philosophical Perspectives* 11 Mind, Causation, and the World: 149–63.

Fodor, J. A. (1998). *Concepts*. Oxford: Clarendon Press.

Fodor, J. A. (2000). *The Mind Doesn't Work That Way*. Cambridge, MA: MIT Press.

Fodor, J. A. (2008). *LOT 2: The Language of Thought Revisited*. Oxford: Oxford University Press.

Fodor, J. A. and Z. W. Pylyshyn (1988). "Connectionism and Cognitive Architecture." *Cognition* 28: 3–71.

Folina, J. (1998). "Church's Thesis: Prelude to a Proof." *Philosophia Mathematica* 6: 302–23.

Forssberg, H., H. Kinoshita, A. C. Eliasson, R. S. Johansson, G. Westling, and A. M. Gordon (1992). "Development of Human Precision Grip. II. Anticipatory Control of Isometric Forces Targeted for Object's Weight." *Exp Brain Res* 90(2): 393–8.

Fox, P. T., M. A. Minton, M. E. Raichle, F. M. Miezin, J. M. Allman, and D. C. Van Essen (1986). "Mapping Human Visual Cortex with Positron Emission Tomography." *Nature* 323: 806–9.

Franklin, A. (2002). *Selectivity and Discord: Two Problems of Experiment*. Pittsburgh: University of Pittsburgh Press.

Franklin, A. (2013). *Shifting Standards: Experiments in Particle Physics in the Twentieth Century*. Pittsburgh: University of Pittsburgh Press.

Franklin, A. and S. Perovic (2016). "Experiment in Physics." In E. N. Zalta (ed.), *The Stanford Encyclopedia of Philosophy* (Winter 2016 Edition), URL = <https://plato.stanford.edu/archives/win2016/entries/physics-experiment/>. Stanford, CA: The Metaphysics Research Lab, Centre for the Study of Language and Information, Stanford University.

Freeman, W. J. (2001). *How Brains Make Up Their Minds*. New York: Columbia University Press.

Fredkin, E. (1990). "Digital Mechanics: An Information Process Based on Reversible Universal Cellular Automata." *Physica D* 45: 254–70.

Frege, G. (1892). "Über Sinn und Bedeutung." In *Zeitschrift für Philosophie und philosophische Kritik*, 100: 25–50. Translated as "On Sense and Reference" by M. Black in P. Geach and M. Black (eds. and trans.), *Translations from the Philosophical Writings of Gottlob Frege*, Oxford: Blackwell, 3rd edition, 1980.

French, S. (2015). "Identity and Individuality in Quantum Theory." In E. N. Zalta (ed.), *The Stanford Encyclopedia of Philosophy* (Fall 2015 Edition), URL = <https://plato.stanford.edu/archives/fall2015/entries/qt-idind/>. Stanford, CA: The Metaphysics Research Lab, Centre for the Study of Language and Information, Stanford University.

Fresco, N. (2014). *Physical Computation and Cognitive Science*. New York, NY: Springer.

Fresco, N., M. J. Wolf, and J. B. Copeland (2016). "On the Indeterminacy of Computation. In Methodological Issues in Philosophy of Computer Science Symposium." Presented at the 2016 Annual Meeting of the International Association for Computing and Philosophy, University of Ferrara, Italy.

Freud, S. (1895/1966). "Project for a Scientific Psychology." In E. Jones (ed.) and J. Strachey (trans.), *The Standard Edition of the Complete Psychological Works of Sigmund Freud* Vol. 1 (295–397). London: Hogarth Press.

Fries, P., J. H. Reynolds, A. E. Rorie, and R. Desimone (2001). "Modulation of Oscillatory Neuronal Synchronization by Selective Visual Attention." *Science* 291(5508): 1560–3.

Fries, P., T. Womelsdorf, R. Oostenveld, and R. Desimone (2008). "The Effects of Visual Stimulation and Selective Visual Attention on Rhythmic Neuronal Synchronization in Macaque Area V4." *Journal of Neuroscience* 28(18): 4823–35.

Funahashi, S., C. J. Bruce, and P. S. Goldman-Rakic (1989). "Mnemonic Coding of Visual Space in the Monkey's Dorsolateral Prefrontal Cortex." *Journal of Neurophysiology* 61(2): 331–49.

Funahashi, S., C. J. Bruce, and P. S. Goldman-Rakic (1993). "Dorsolateral Prefrontal Lesions and Oculomotor Delayed-response Performance: Evidence for Mnemonic 'Scotomas.'" *Journal of Neuroscience* 13(4): 1479–97.

Funahashi, S., M. V. Chafee, and P. S. Goldman-Rakic (1993). "Prefrontal Neuronal Activity in Rhesus Monkeys Performing a Delayed Anti-Saccade Task." *Nature* 365 (6448): 753–6.

Fuster, J. M. and G. E. Alexander (1970). "Delayed Response Deficit by Cryogenic Depression of Frontal Cortex." *Brain Research* 20(1): 85–90.

Fuster, J. M. and G. E. Alexander (1971). "Neuron Activity Related to Short-term Memory." *Science* 173(3997): 652–4.

Galison, P. (1987). *How Experiments End.* Chicago: University of Chicago Press.

Galison, P. (1997). *Image and Logic.* Chicago: University of Chicago Press.

Gallistel, C. R. (1990). "Representations in Animal Cognition: An Introduction." *Cognition* 37(1–2): 1–22.

Gallistel, C. R. (2008). "Learning and Representation." In J. Byrne (ed.), *Learning and Memory: A Comprehensive Reference* (227–42). Amsterdam: Elsevier.

Gallistel, C. R. and A. P. King (2009). *Memory and the Computational Brain: Why Cognitive Science Will Transform Neuroscience.* New York: Wiley/Blackwell.

Gallistel, C. R. (2017a). "The Coding Question." *Trends in Cognitive Science*, 21(7): 498–508.

Gallistel, C. R. (2017b). "The Neurobiological Bases for the Computational Theory of Mind." In R. G. d. Almeida and L. Gleitman (eds.), *On Concepts, Modules, and Language* (275–96). New York: Oxford University Press.

Gallois, A. (2016). "Identity over Time." In E. N Zalta (ed.), *The Stanford Encyclopedia of Philosophy* (Winter 2016 Edition), URL = <https://plato.stanford.edu/archives/win2016/entries/identity-time/>. Stanford, CA: The Metaphysics Research Lab, Centre for the Study of Language and Information, Stanford University.

Gandhi, N. J. and H. A. Katnani (2011). "Motor Functions of the Superior Colliculus." *Annual Review of Neuroscience* 34: 205–31.

Gandy, R. (1980). "Church's Thesis and Principles for Mechanism." In J. Barwise, H. J. Keisler, and K. Kuhnen (eds.), *The Kleene Symposium* (123–48). Amsterdam: North-Holland.

Ganguly, K. and J. M. Carmena (2009). "Emergence of a Stable Cortical Map for Neuroprosthetic Control." *PLoS Biology* 7(7): e1000153.

Gardenfors, P. (2005). "The Detachment of Thought." In C. Erneling and D. Johnson (eds.), *Mind as a Scientific Subject: Between Brain and Culture* (323–41). Oxford: Oxford University Press.

Garson, J. (2003). "The Introduction of Information into Neurobiology." *Philosophy of Science* 70 (5): 926–36.

Garson, J. (2013). "The Functional Sense of Mechanism." *Philosophy of Science* 80: 317–33.
Garson, J. (2016). *A Critical Overview of Biological Functions.* Dordrecht: Springer.
Garson, J. (2019a). *What Biological Functions Are and Why They Matter.* Cambridge: Cambridge University Press.
Garson, J. (2019b). "There Are No Ahistorical Theories of Function." *Philosophy of Science* 86(5): 1146–56.
Garson, J. and G. Piccinini (2014). "Functions Must Be Performed at Appropriate Rates in Appropriate Situations." *British Journal for the Philosophy of Science* 65: 1–20.
Garzon, F. C. (2008). "Towards a General Theory of Antirepresentationalism." *British Journal for the Philosophy of Science* 59(3): 259–92.
Gazzaley, A. and A. C. Nobre (2012). "Top-down Modulation: Bridging Selective Attention and Working Memory." *Trends in Cognitive Science* 16(2): 129–35.
Gazzaniga, M. (ed.), (2009). *The Cognitive Neurosciences.* Cambridge, MA: MIT Press.
Gerard, R. W. (1951). "Some of the Problems Concerning Digital Notions in the Central Nervous System. Cybernetics: Circular Causal and Feedback Mechanisms in Biological and Social Systems." In H. V. Foerster, M. Mead, and H. L. Teuber (eds.), *Transactions of the Seventh Conference* (11–57). New York: Macy Foundation.
Georgopoulos, A. P. and J. Ashe (2000). "One Motor Cortex, Two Different Views." *Nature Neuroscience* 3(10): 963; author reply 964–965.
Georgopoulos, A. P., J. F. Kalaska, R. Caminiti, and J. T. Massey (1982). "On the Relations between the Direction of Two-dimensional Arm Movements and Cell Discharge in Primate Motor Cortex." *Journal of Neuroscience* 2(11): 1527–37.
Georgopoulos, A. P., A. B. Schwartz, and R. E. Kettner (1986). "Neuronal Population Coding of Movement Direction." *Science* 233(4771): 1416–19.
Gibson, J. J. (1979). *The Ecological Approach to Visual Perception.* Boston, Houghton Mifflin.
Gilbert, C. D. and W. Li (2013). "Top-Down Influences on Visual Processing." *Nature Reviews Neuroscience* 14(5): 350–63.
Gillett, C. (2002). "The Dimensions of Realization: A Critique of the Standard View." *Analysis* 62: 316–23.
Gillett, C. (2003). "The Metaphysics of Realization, Multiple Realizability, and the Special Sciences." *The Journal of Philosophy* 100(11): 591–603.
Gillett, C. (2010). "Moving Beyond the Subset Model of Realization: The Problem of Qualitative Distinctness in the Metaphysics of Science." *Synthese* 177: 165–92.
Gillett, C. (2013). Understanding the Sciences through the Fog of "Functionalism(s)." In P. Huneman (ed.), *Functions: Selection and Mechanisms* (159–81). Dordrecht: Springer.
Gillett, C. (2016). *Reduction and Emergence in Science and Philosophy.* Cambridge: Cambridge University Press.
Giunti, M. (1997). *Computation, Dynamics, and Cognition.* New York, Oxford University Press.
Gładziejewski, P. (2016). "Predictive Coding and Representationalism." *Synthese* 193: 559–82.
Gładziejewski, P. and M. Miłkowski (2017). "Structural Representations: Causally Relevant and Different from Detectors." *Biology & Philosophy* 32(3): 337–55.
Glennan, S. (1996). "Mechanisms and the Nature of Causation." *Erkenntnis* 44: 49–71.
Glennan, S. (2002). "Rethinking Mechanistic Explanation." *Philosophy of Science* 69(3): S342–53.
Glennan, S. (2017). *The New Mechanical Philosophy.* Oxford: Oxford University Press.
Glennan, S. and P. Illari (2017). *The Routledge Handbook of Mechanisms and Mechanical Philosophy.* London: Routledge.

Glimcher, P. W. and E. Fehr (eds.) (2014). *Neuroeconomics: Decision Making and the Brain, Second edition.* Amsterdam: Elsevier.

Globus, G. G. (1992). "Towards a Noncomputational Cognitive Neuroscience." *Journal of Cognitive Neuroscience* 4(4): 299–310.

Gödel, K. (1965). "Postscriptum." In M. Davis (ed.), *The Undecidable* (71–3). New York: Raven.

Guttenplan, S. (ed.), (1994). *A Companion to the Philosophy of Mind.* Cambridge, MA: Blackwell.

Godfrey-Smith, P. (1994). "A Modern History Theory of Functions." *Nous* 28: 344–62.

Godfrey-Smith, P. (1996). *Complexity and the Function of Mind in Nature.* Cambridge: Cambridge University Press.

Goldberg, M. E. and R. H. Wurtz (1972). "Activity of Superior Colliculus in Behaving Monkey. I. Visual Receptive Fields of Single Neurons." *Journal of Neurophysiology* 35(4): 542–59.

Gollisch, T. (2009). "Throwing a Glance at the Neural Code: Rapid Information Transmission in the Visual System." *HFSP Journal* 3(1): 36–46.

Goodale, M. A. and A. D. Milner (1992). "Separate Visual Pathways for Perception and Action." *Trends in Neuroscience* 15(1): 20–5.

Gould, S. J. and R. Lewontin (1979). "The Spandrels of San Marco and the Panglossian Paradigm." *Proceedings of the Royal Society of London* 205: 281–8.

Graziano, M. S. (2016). "Ethological Action Maps: A Paradigm Shift for the Motor Cortex." *Trends in Cognitive Science* 20(2): 121–32.

Graziano, M. S., C. S. Taylor, T. Moore, and D. F. Cooke (2002). "The Cortical Control of Movement Revisited." *Neuron* 36(3): 349–62.

Grice, H. P. (1957). "Meaning." *The Philosophical Review* 66(3): 377–88.

Griffin, D. M., D. S. Hoffman, and P. L. Strick (2015). "Corticomotoneuronal Cells Are 'Functionally Tuned.'" *Science* 350(6261): 667–70.

Griffiths, P. E. (1993). "Functional Analysis and Proper Function." *British Journal for the Philosophy of Science* 44: 409–22.

Grush, R. (2003). "In Defense of Some 'Cartesian' Assumptions Concerning the Brain and Its Operation." *Biology and Philosophy* 18: 53–93.

Grush, R. (2004). "The Emulation Theory of Representation: Motor Control, Imagery, and Perception." *Behavioral and Brain Sciences* 27(3): 377–96; discussion 396–442.

Hacking, I. (1983). *Representing and Intervening.* Cambridge: Cambridge University Press.

Hahnloser, R. H. and A. Kotowicz (2010). "Auditory Representations and Memory in Birdsong Learning." *Curr Opin Neurobiol* 20(3): 332–9.

Haimovici, S. 2013. "A Problem for the Mechanistic Account of Computation." *Journal of Cognitive Science* 14: 151–81.

Hallett, M. (2000). "Transcranial Magnetic Stimulation and the Human Brain." *Nature* 406 (6792): 147–50.

Hartmann, K., E. E. Thomson, I. Zea, R., Yun, P. Mullen, J. Canarick, A. Huh, and M. A. Nicolelis (2016). "Embedding a Panoramic Representation of Infrared Light in the Adult Rat Somatosensory Cortex through a Sensory Neuroprosthesis." *Journal of Neuroscience* 36(8): 2406–24.

Harman, G. (1973). *Thought.* Princeton, Princeton University Press.

Harman, G. (1990). "The Intrinsic Quality of Experience." *Philosophical Perspectives* 4: 31–52.

Harman, G. (1999). *Reasoning, Meaning and Mind.* Oxford: Clarendon Press.

Harnad, S. (1996). "Computation Is Just Interpretable Symbol Manipulation; Cognition Isn't." *Minds and Machines* 4: 379–90.

Harvey, W. (1628/1889). *On the Motion of the Heart and Blood in Animals.* London: George Bell and Sons.
Haugeland, J. (1978). "The Nature and Plausibility of Cognitivism." *Behavioral and Brain Sciences* 2: 215–60.
Haugeland, J. (1981). "Analog and Analog." *Philosophical Topics* 12: 213–25.
Haugeland, J. (1985). *Artificial Intelligence: The Very Idea.* Cambridge, MA: MIT Press.
Haugeland, J. (1998). *Having Thought.* Cambridge, MA: Harvard University Press.
Hausman (2012). "Health, Naturalism, and Functional Efficiency." *Philosophy of Science* 79: 519–41.
Hausman, D. (2011). "Is an Overdose of Paracetamol Bad for One's Health?" *British Journal for the Philosophy of Science* 62: 657–68.
Healey, R. (2013). "Physical Composition." *Studies in History and Philosophy of Modern Physics* 44: 48–62.
Hebb, D. O. (1949). *The Organization of Behavior: A Neuropsychological Theory.* New York: Wiley.
Heil, J. (2003). *From an Ontological Point of View.* Oxford: Oxford University Press.
Heil, J. (2012). *The Universe as We Find It.* Oxford: Oxford University Press.
Heiligenberg, W. and J. Bastian (1984). "The Electric Sense of Weakly Electric Fish." *Annual Review of Physiology* 46: 561–83.
Heims, S. J. (1991). *Constructing a Social Science for Postwar America: The Cybernetics Group, 1946–1953.* Cambridge, MA: MIT Press.
Hemmo, M. and O. Shenker. 2015. "The Emergence of Macroscopic Regularity." *Mind and Society* 14(2): 221–44.
Herculano-Houzel, S. (2010). "Coordinated Scaling of Cortical and Cerebellar Numbers of Neurons." *Frontiers in Neuroanatomy* 4: 12.
Hewitt, A. L., L. S. Popa, S. Pasalar, C. M. Hendrix, and T. J. Ebner (2011). "Representation of Limb Kinematics in Purkinje Cell Simple Spike Discharge Is Conserved Across Multiple Tasks." *Journal of Neurophysiology* 106(5): 2232–47.
Hilbert, D. and W. Ackermann (1928). *Grundzüge der theoretischen Logik.* Berlin: Springer.
Hille, B. (2001). *Ion Channels of Excitable Membranes.* Sunderland, MA: Sinauer.
Hochberg, L. R., D. Bacher, B. Jarosiewicz, N. Y. Masse, J. D. Simeral, J. Vogel, S. Haddadin, J. Liu, S. S. Cash, P. van der Smagt, and J. P. Donoghue (2012). "Reach and Grasp by People with Tetraplegia Using a Neurally Controlled Robotic Arm." *Nature* 485(7398): 372–5.
Hochstein, E. (2016). "Giving Up on Convergence and Autonomy: Why the Theories of Psychology and Neuroscience are Codependent as well as Irreconcilable." *Studies in History and Philosophy of Science Part A*, 56: 135–44.
Hodges, A. (1983). *Alan Turing: The Enigma.* New York: Simon & Schuster.
Hodgkin, A. L. and A. F. Huxley (1952). "A Quantitative Description of Membrane Current and Its Application to Conduction and Excitation in Nerve." *Journal of Physiology* 117: 500–44.
Hodgkin, A. L. and A. F. Huxley (1939). "Action Potentials Recorded from Inside a Nerve Fibre." *Nature* 144: 710–11.
Hoffmann-Kolss, V. (2014). "Interventionism and Higher-Level Causation." *International Studies in the Philosophy of Science* 28(1): 49–64.
Hogarth, M. L. (1994). "Non-Turing Computers and Non-Turing Computability." *PSA 1994* (1): 126–38.
Hohwy J. (2013). *The Predictive Mind.* New York: Oxford University Press.
Holdefer, R. N. and L. E. Miller (2002). "Primary Motor Cortical Neurons Encode Functional Muscle Synergies." *Experimental Brain Research* 146(2): 233–43.

Hopfield, J. J. (1982). "Neural Networks and Physical Systems with Emergent Collective Computational Abilities." *Proceedings of the National Academy of Sciences USA* 79: 2554–8.

Horgan, T. and J. Tienson (1996). *Connectionism and the Philosophy of Psychology.* Cambridge, MA: MIT Press.

Horsley, V. (1907). "Dr. Hughlings Jackson's Views of the Functions of the Cerebellum, as Illustrated by Recent Research." *The British Medical Journal* 1(2414): 803–8.

Horsley, V. (1909). The Linacre Lecture on the Function of the So-called Motor Area of the Brain. *The British Medical Journal* 2(2533): 121–32.

Horst, S. W. (1996). *Symbols, Computation, and Intentionality: A Critique of the Computational Theory of Mind.* Berkeley, CA: University of California Press.

Houk, J. C. and S. P. Wise (1995). "Distributed Modular Architectures Linking Basal Ganglia, Cerebellum, and Cerebral Cortex: Their Role in Planning and Controlling Action." *Cerebral Cortex* 5(2): 95–110.

Householder, A. S. (1941). "A Theory of Steady-State Activity in Nerve-Fiber Networks: I. Definitions and Preliminary Lemmas." *Bulletin of Mathematical Biophysics* 3: 63–9.

Householder, A. S. and H. D. Landahl (1945). *Mathematical Biophysics of the Central Nervous System.* Bloomington, Principia.

Hubel, D.H. and T. N. Wiesel (1962). "Receptive Fields, Binocular Interaction and Functional Architecture in the Cat's Visual Cortex." *Journal of Physiology* 160: 106–54.

Hubel, D. H. and T. N. Wiesel (1968). "Receptive Fields and Functional Architecture of Monkey Striate Cortex." *Journal of Physiology* 195(1): 215–43.

Hubel, D. H. and T. N. Wiesel (2005). *Brain and Visual Perception: The Story of a 25-year Collaboration.* New York, N.Y.: Oxford University Press.

Hughlings Jackson, J. (1867) "Remarks on the Disorderly Movements of Chorea and Convulsion, and on Localisation." *Medical Times and Gazette*, 669–70.

Hughlings Jackson, J. (1868) "Notes on the Physiology and Pathology of the Nervous System." *Medical Times and Gazette* 696.

Hughes, R. I. G. (1999). "The Ising Model, Computer Simulation, and Universal Physics." In M. S. Morgan and M. Morrison (eds.), *Models as Mediators* (97–145). Cambridge: Cambridge University Press.

Humphreys, P. (2004). *Extending Ourselves: Computational Science, Empiricism, and Scientific Method.* Oxford: Oxford University Press.

Humphreys, P. (2016). *Emergence: A Philosophical Account.* Oxford: Oxford University Press.

Hupe, J. M., A. C. James, B. R. Payne, S. G. Lomber, P. Girard, and J. Bullier (1998). "Cortical Feedback Improves Discrimination between Figure and Background by V1, V2 and V3 Neurons." *Nature* 394(6695): 784–7.

Hütteman, A. (2004). *What's Wrong with Microphysicalism?* London: Routledge.

Hütteman, A. and D. Papineau (2005). "Physicalism Decomposed." *Analysis* 65(1): 33–9.

Hutto, D. D. and E. Myin (2013). *Radicalizing Enactivism*. Cambridge, MA: MIT Press.

Hutto, D. D. and E. Myin (2014). "Neural Representations Not Needed—No More Pleas, Please." *Phenomenology and the Cognitive Sciences* 13(2): 241–56.

Hutto, D. D. and E. Myin (2017). *Evolving Enactivism: Basic Minds Meet Content.* Cambridge, MA: MIT Press.

Illari, P. M. and J. Williamson (2012). "What is a Mechanism? Thinking about Mechanisms Across the Sciences." *European Journal of Philosophy of Science* 2: 119–35.

Inan, M. and M. C. Crair (2007). "Development of Cortical Maps: Perspectives from the Barrel Cortex." *Neuroscientist* 13(1): 49–61.

Inoue, S. and T. Matsuzawa (2007). "Working Memory of Numerals in Chimpanzees." *Curr Biol* 17(23): R1004–5.

Isaac, A. (2013). "Objective Similarity and Mental Representation." *Australasian Journal of Philosophy* 91(4): 683–704.

Isaac, A. (2018). "Embodied Cognition as Analog Computation." *Reti, Saperi, Linguaggi: Italian Journal of Cognitive Sciences* 2/2018: 239–62.

Ito, M. (1970). "Neurophysiological Aspects of the Cerebellar Motor Control System." *International Journal of Neurology* 7(2): 162–76.

Izhikevich, E. (2007). *Dynamical Systems in Neuroscience: The Geometry of Excitability and Bursting*. Cambridge, MA: The MIT Press.

James, W. (1890/1983). *The Principles of Psychology*. Cambridge, MA: Harvard University Press.

Jensen, M. O., V. Jogini, D. W. Borhani, A. E. Leffler, R. O. Dror, and D. E. Shaw (2012). "Mechanism of Voltage Gating in Potassium Channels." *Science Signaling* 336(6078): 229.

Jilk, D., C. Lebiere, R. O'Reilly and J. Anderson (2008) "SAL: An Explicitly Pluralistic Cognitive Architecture." *Journal of Experimental and Theoretical Artificial Intelligence* 20(3): 197–218.

Johansson, R. S. and K. J. Cole (1992). "Sensory-motor Coordination during Grasping and Manipulative Actions." *Current Opinions in Neurobiology* 2(6): 815–23.

Johansson, R.S. and I. Birznieks (2004). "First Spikes in Ensembles of Human Tactile Afferents Code Complex Spatial Fingertip Events." *Nature Neuroscience* 7: 170–7.

Johnson-Laird, P. N. (2010). "Mental Models and Human Reasoning." *Proceedings of the National Academy of Sciences USA* 107(43): 18243–50.

Johnson, D. M. and C. E. Erneling (eds.) (1997). *The Future of the Cognitive Revolution*. New York: Oxford University Press.

Johnson-Laird, P. N. (1983). *Mental Models: Towards a Cognitive Science of Language, Inference and Consciousness*. New York: Cambridge University Press.

Kakei, S., D. S. Hoffman, and P. L. Strick (1999). "Muscle and Movement Representations in the Primary Motor Cortex." *Science* 285(5436): 2136–9.

Kalaska, J. F. (2009). "From Intention to Action: Motor Cortex and the Control of Reaching Movements." *Advances in Experimental Medical Biology* 629: 139–78.

Kalat, J. W. (2008). *Biological Psychology*. Belmont, CA: Wadsworth Publishing.

Kálmar, L. (1959). "An Argument Against the Plausibility of Church's Thesis." In A. Heyting (eds.), *Constructivity in Mathematics* (72–80). Amsterdam: North-Holland.

Kaminski, J., J. Call, and J. Fischer (2004). 'Word Learning in a Domestic Dog: Evidence for "Fast Mapping."' *Science* 304(5677). 1682–3.

Kandel, E. R. (2006). *In Search of Memory: The Emergence of a New Science of Mind*. New York: W. W. Norton & Company.

Kandel, E. R. (2013). *Principles of Neural Science* (5th ed.). New York: McGraw-Hill.

Kaplan, D. M (2011) "Explanation and Description in Computational Neuroscience." *Synthese* 183(3): 339–73.

Kaplan, D. M. (2015). "Moving Parts: The Natural Alliance Between Dynamical and Mechanistic Modeling Approaches." *Biology and Philosophy* 30: 757–86.

Kaplan, D. M., and C. F. Craver (2011). "The Explanatory Force of Dynamical and Mathematical Models in Neuroscience: A Mechanistic Perspective." *Philosophy of Science* 78: 601–27.

Kaplan, D. M., and C. Hewitson (unpublished). "Modelling Bayesian Computation in the Brain: Unification, Explanation, and Constraints."

Kästner, L. (2017). *Philosophy of Cognitive Neuroscience: Causal Explanations, Mechanisms and Experimental Manipulations*. Berlin: de Gruyter.

Kauffman, S. (1993). *The Origins of Order: Self-Organization and Selection in Evolution*. Oxford: Oxford University Press.

Kauffman, S. (2002). *Investigations*. Oxford: Oxford University Press.

Kearns, J. T. (1997). "Thinking Machines: Some Fundamental Confusions." *Minds and Machines* 7: 269–87.

Keeley, B. L. (2000). "Shocking Lessons from Electric Fish: The Theory and Practice of Multiple Realization." *Philosophy of Science* 67(3): 444–65.

Keijzer, F. A. (1998). "Doing Without Representations Which Specify What to Do." *Philosophical Psychology* 11(3): 269–302

Kennedy, A., G. Wayne, P. Kaifosh, K. Alvina, L. F. Abbott, and N. B. Sawtell (2014). "A Temporal Basis for Predicting the Sensory Consequences of Motor Commands in an Electric Fish." *Nature Neuroscience* 17(3): 416–22.

Kiefer, A. and J. Hohwy (2018). "Content and Misrepresentation in Hierarchical Generative Models." *Synthese* 195(6): 2387–2415.

Kim, J. (1992). "Multiple Realization and the Metaphysics of Reduction." *Philosophy and Phenomenological Research* 52(1): 1–26.

Kim, J. (1998). *Mind in a Physical World: An Essay on the Mind–Body Problem and Mental Causation*. Cambridge, MA: MIT Press.

Kim, J. (2005). *Physicalism, or Something Near Enough*. Princeton, NJ: Princeton University Press.

Kim, J. (2006). "Emergence: Core Ideas and Issues." *Synthese* 151: 547–59.

Kingma, E. (2010). "Paracetamol, Poison, and Polio: Why Boorse's Account of Function Fails to Distinguish Health and Disease." *British Journal for the Philosophy of Science* 61: 241–64.

Kingma, E. (2016). "Situation-Specific Disease and Dispositional Function." *British Journal for the Philosophy of Science* 67: 391–404.

Kirsh, D. (2006). "Implicit and Explicit Representation." In L. Nadel (ed.), *Encyclopedia of Cognitive Science*. New York: Wiley.

Kitamura, T., S. K. Ogawa, D. S. Roy, T. Okuyama, M. D. Morrissey, L. M. Smith, R. L. Redondo, and S. Tonegawa (2017). "Engrams and Circuits Crucial for Systems Consolidation of a Memory." *Science* 356(6333): 73–8.

Kleene, S. C. (1952). *Introduction to Metamathematics*. Princeton, NJ: Van Nostrand.

Kleene, S. C. (1956). "Representation of Events in Nerve Nets and Finite Automata." In C. E. Shannon and J. McCarthy (eds.), *Automata Studies* (3–42). Princeton, NJ: Princeton University Press.

Klein, C. (2008). "An Ideal Solution to Disputes about Multiply Realized Kinds." *Philosophical Studies* 140(2): 161–77.

Klein, C. (2013). "Multiple Realizability and the Semantic View of Theories." *Philosophical Studies* 163(3): 683–95.

Klein, C. (2018). "What Do Predictive Coders Want?" *Synthese* 195: 2541–57.

Knight, B. W. (1972). "Dynamics of Encoding in a Population of Neurons." *Journal of General Physiology* 59(6): 734.

Knoll, A. (2018). "Still Autonomous After All." *Minds & Machines* 28:7–27.

Koch, C. (2004). *The Quest for Consciousness: A Neurobiological Approach*. Denver, Colo.: Roberts and Co.

Koch, C. and I. Segev (2000). "The Role of Single Neurons in Information Processing." *Nature Neuroscience Supplement* 3: 1171–7.

Komatsu, H., M. Kinoshita, and I. Murakami (2000). "Neural Responses in the Retinotopic Representation of the Blind Spot in the Macaque V1 to Stimuli for Perceptual Filling-In." *Journal of Neuroscience* 20(24): 9310–19.

Konishi, M. (1965). "The Role of Auditory Feedback in the Control of Vocalization in the White-Crowned Sparrow." *Zeitschrift für Tierpsychologie* 22(7): 770–83.

Koulakov, A. A., Raghavachari, S., Kepecs, A., and Lisman, J. E. (2002). "Model for a Robust Neural Integrator." *Nature Neuroscience* 5: 775–82.

Korenbrot, J. I. (2012). "Speed, Sensitivity, and Stability of the Light Response in Rod and Cone Photoreceptors: Facts and Models." *Progress in Retinal and Eye Research* 31(5): 442–66.

Kosslyn, S. (1980). *Image and Mind.* Cambridge, MA: Harvard University Press.

Kosslyn, S. (1994). *Image and Brain: The Resolution of the Imagery Debate.* Cambridge, MA: MIT Press.

Kosslyn, S., W. L. Thompson, and G. Ganis (2006). *The Case for Mental Imagery.* New York: Oxford University Press.

Kosslyn, S. and Van Kleeck (1990). "Broken Brains and Normal Minds: Why Humpty-Dumpty Needs a Skeleton." *Computational Neuroscience.* Cambridge, MA: MIT Press.

Kraemer, D. M. (2013). "Statistical Theories of Functions and the Problem of Epidemic Disease." *Biology and Philosophy* 28: 423–38.

Kraemer, D. M. (unpublished). "The Plurality of Uses of the Term 'Representation' in Brain Science from 1867–1990."

Krahe, R. and L. Maler (2014). "Neural Maps in the Electrosensory System of Weakly Electric Fish." *Current Opinions in Neurobiology* 24(1): 13–21.

Krauzlis, R. J., L. P. Lovejoy, and A. Zenon (2013). "Superior Colliculus and Visual Spatial Attention." *Annual Reviews Neuroscience* 36: 165–82.

Krickel, B. (2018). *The Mechanical World: The Metaphysical Commitments of the New Mechanistic Approach.* Berlin: Springer.

Kripke, S. (1982). *Wittgenstein on Rules and Private Language.* Cambridge: Harvard University Press.

Kustov, A. A. and D. L. Robinson (1996). "Shared Neural Control of Attentional Shifts and Eye Movements." *Nature* 384(6604): 74–7.

Lacquaniti, F., N. A. Borghese, and M. Carrozzo (1992). "Internal Models of Limb Geometry in the Control of Hand Compliance." *Journal of Neuroscience* 12(5): 1750–62.

Laird, J. E. (2012). *The Soar Cognitive Architecture.* Cambridge, MA: MIT Press.

Laird, J. E., A. Newell, and P. S. Rosenbloom (1987). "Soar: An Architecture for General Intelligence." *Artificial Intelligence* 33: 1–64.

Lamme, V. A., H. Super, and H. Spekreijse (1998). "Feedforward, Horizontal, and Feedback Processing in the Visual Cortex." *Current Opinions in Neurobiology* 8(4): 529–35.

Lange, M. (2017). *Because Without Cause.* New York: Oxford University Press.

Lange, M. (2018). "A Reply to Craver and Povich on the Directionality of Distinctively Mathematical Explanations." *Studies in the History and Philosophy of Science* 67: 85–8.

Lapicque, L. (1907/2007). "Quantitative Investigation of Electrical Nerve Excitation Treated as Polarization." Biological Cybernetics 97: 341–9. Translation from the French by N. Brunel and M. van Rossum.

Lashley, K. S. (1958). "Cerebral Organization and Behavior." In F. A. Beach, D. O. Hebb, C. T. Morgan, and H. V. Nissen (eds.), *The Brain and Human Behavior, Proceedings of the Association for Research of Nervous and Mental Disorders 36*: 1–18. *The Neuropsychology of Lashley; Selected Papers of K. S. Lashley* (pp. 529–43). New York: McGraw-Hill (1960) (reprint).

Leavitt, M. L., D. Mendoza-Halliday, and J. C. Martinez-Trujillo (2017). "Sustained Activity Encoding Working Memories: Not Fully Distributed." *Trends in Neuroscience* 40(6): 328–46.

Le Bon-Jego, M. and R. Yuste (2007). "Persistently Active, Pacemaker-Like Neurons in Neocortex." *Frontiers in Neuroscience* 1: 123–9.

Leopold, D. A. and N. K. Logothetis (1996). "Activity Changes in Early Visual Cortex Reflect Monkeys' Percepts During Binocular Rivalry." *Nature* 379(6565): 549–53.

Lettvin, J. L. (1989). "Strychnine Neuronography." In R. McCulloch (ed.), *Collected Works of Warren S. McCulloch, Vol. 1.* (50–58). Salinas, CA: Intersystems.

Levin, J. E. and J. P. Miller (1996). "Broadband Neural Encoding in the Cricket Cercal Sensory System Enhanced by Stochastic Resonance." *Nature* 380(6570): 165–8.

Levy, A. (2014). "What Was Hodgkin and Huxley's Achievement?" *British Journal for the Philosophy of Science* 65(3): 469–92.

Levy, A. and W. Bechtel (2013). "Abstraction and the Organization of Mechanisms." *Philosophy of Science* 80(2): 241–61.

Lewis, J. E. and W. B. Kristan, Jr. (1998a). "A Neuronal Network for Computing Population Vectors in the Leech." *Nature* 391(6662): 76–9.

Lewis, J. E. and W. B. Kristan, Jr. (1998b). "Representation of Touch Location by a Population of Leech Sensory Neurons." *Journal of Neurophysiology* 80(5): 2584–92.

Lewis, D. K. (1966). "An Argument for the Identity Theory." *The Journal of Philosophy* 63 (1): 17–25.

Lewis, D. K. (1970). "How to Define Theoretical Terms." *The Journal of Philosophy* 67(13): 427–46.

Lewis, D. K. (1972). "Psychophysical and Theoretical Identifications." *Australasian Journal of Philosophy* 50(3): 249–58.

Lewis, D. K. (1983). "New Work for a Theory of Universals." *Australasian Journal of Philosophy* 61(4): 343–77.

Lewis, D. K. (1986). *On the Plurality of Worlds.* Malden, MA: Blackwell.

Li, C. S., C. Padoa-Schioppa, and E. Bizzi (2001). "Neuronal Correlates of Motor Performance and Motor Learning in the Primary Motor Cortex of Monkeys Adapting to an External Force Field." *Neuron* 30(2): 593–607.

Li, P. H., G. D. Field, M. Greschner, D. Ahn, D. E. Gunning, K. Mathieson, A. Sher, A. M. Litke, and E. J. Chichilnisky (2014). "Retinal Representation of the Elementary Visual Signal." *Neuron* 81(1): 130–9.

Lin, J.-W. and D. S. Faber (2002). "Modulation of Synaptic Delay during Synaptic Plasticity." *Trends in Neurosciences* 25(9): 449–55.

Lipshitz, L. and L. A. Rubel (1987). "A Differentially Algebraic Replacement Theorem, and Analog Computability." *Proceedings of the American Mathematical Society* 99(2): 367–72.

Lisberger, S. G. (2009). "Internal Models of Eye Movement in the Floccular Complex of the Monkey Cerebellum." *Neuroscience* 162(3): 763–76.

List, C. and P. Menzies (2009). "Nonreductive Physicalism and the Limits of the Exclusion Principle." *Journal of Philosophy* 106(9): 475–502.

Liu, L. D. and C. C. Pack (2017). "The Contribution of Area MT to Visual Motion Perception Depends on Training." *Neuron* 95(2): 436–46.

Lloyd, D. G. and T. F. Besier (2003). "An EMG-driven Musculoskeletal Model to Estimate Muscle Forces and Knee Joint Moments *in Vivo*." *Journal of Biomechanics* 36(6): 765–76.

Lloyd, S. (2006). *Programming the Universe: A Quantum Computer Scientist Takes on the Cosmos.* New York: Knopf.

Loar, B. (1981). *Mind and Meaning*. Cambridge: Cambridge University Press.
Loewer, B. (2007). "Mental Causation or Something Near Enough." In B. McLaughlin and J. Cohen (eds.), *Contemporary Debates in Philosophy of Mind* (243–64). Malden, MA: Wiley-Blackwell.
London, M., A. Roth, L. Beeren, M. Häusser, and P. E. Latham (2010). "Sensitivity to Perturbations *in Vivo* Implies High Noise and Suggests Rate Coding in Cortex." *Nature* 466: 123–7.
Lovejoy, A. O. (1936). *The Great Chain of Being: A Study of the History of an Idea*. Cambridge, MA: Harvard University Press.
Lucas, J. R. (1961). "Minds, Machines, and Gödel." *Philosophy* 36: 112–37.
Lucas, J. R. (1996). "Minds, Machines, and Gödel: A Retrospect." In P. J. R. Millikan and A. Clark (eds.), *Machines and Thought: The Legacy of Alan Turing*. Oxford: Clarendon.
Lycan, W. (1987). *Consciousness*. Cambridge, MA, MIT Press.
Lycan, W. (1990). "The Continuity of Levels of Nature." In W. Lycan (ed.), *Mind and Cognition* (77–96). Malden, MA: Blackwell.
Lycan, W. G. (1981). "Form, Function, and Feel." *The Journal of Philosophy* 78(1): 24–50.
Lyre, H. (2017). "Structures, Dynamics and Mechanisms in Neuroscience: An Integrative Account." *Synthese* 195(12): 5141–58.
Machamer, P., L. Darden, and C. F. Craver (2000). "Thinking about Mechanisms." *Philosophy of Science* 67(1): 1–25.
Mainen, Z. F. and T. J. Sejnowski (1995). "Reliability of Spike Timing in Neocortical Neurons." *Science* 268(5216): 1503–6.
Makino, H., C. Ren, H. Liu, A. N. Kim, N. Kondapaneni, X. Liu, D. Kuzum, and T. Komiyama (2017). "Transformation of Cortex-wide Emergent Properties during Motor Learning." *Neuron* 94(4): 880–90.
Maley, C. J. (2011). "Analog and Digital, Continuous and Discrete." *Philosophical Studies* 115: 117–31.
Maley, C. J. (2018). "Toward Analog Neural Computation." *Minds and Machines* 28(1): 77–91.
Maley, C. J. and G. Piccinini (2013). "Get the Latest Upgrade: Functionalism 6.3.1." *Philosophia Scientiae* 17(2): 135–49.
Maley, C. J. and Piccinini, G. (2016). "Closed Loops and Computation in Neuroscience: What It Means and Why It Matters," in Ahmed El Hady (ed.), *Closed Loop Neuroscience* (271–7). London: Elsevier.
Maley, C. J. and Piccinini, G. (2017). "A Unified Mechanistic Account of Teleological Functions for Psychology and Neuroscience." In D. Kaplan (ed.), *Explanation and Integration in Mind and Brain Science* (236–56). Oxford: Oxford University Press.
Mandik, P. (2003). "Varieties of Representation in Evolved and Embodied Neural Networks." *Biology and Philosophy* 18: 95–130.
Mangel, S. C. (1991). "Analysis of the Horizontal Cell Contribution to the Receptive Field Surround of Ganglion Cells in the Rabbit Retina." *Journal of Physiology* 442: 211–34.
Marcus, G. F. (2001). *The Algebraic Mind: Integrating Connectionism and Cognitive Science*. Cambridge, MA: MIT Press.
Markram, H. (2006). "The Blue Brain Project." *Nature Reviews Neuroscience* 7: 153–60.
Markram, H., E. Muller, S. Ramaswamy, Michael W. Reimann, M. Abdellah, Carlos A. Sanchez, A. Ailamaki, L. Alonso-Nanclares, N. Antille, S. Arsever, Guy Antoine A. Kahou, Thomas K. Berger, A. Bilgili, N. Buncic, A. Chalimourda, G. Chindemi, J.-D. Courcol, F. Delalondre, V. Delattre, S. Druckmann, R. Dumusc, J. Dynes, S. Eilemann, E. Gal, Michael E. Gevaert, J.-P. Ghobril, A. Gidon, Joe W. Graham,

A. Gupta, V. Haenel, E. Hay, T. Heinis, Juan B. Hernando, M. Hines, L. Kanari, D. Keller, J. Kenyon, G. Khazen, Y. Kim, James G. King, Z. Kisvarday, P. Kumbhar, S. Lasserre, J.-V. Le Bé, Bruno R.C. Magalhães, A. Merchán-Pérez, J. Meystre, Benjamin R. Morrice, J. Muller, A. Muñoz-Céspedes, S. Muralidhar, K. Muthurasa, D. Nachbaur, Taylor H. Newton, M. Nolte, A. Ovcharenko, J. Palacios, L. Pastor, R. Perin, R. Ranjan, I. Riachi, J.-R. Rodríguez, Juan L. Riquelme, C. Rössert, K. Sfyrakis, Y. Shi, Julian C. Shillcock, G. Silberberg, R. Silva, F. Tauheed, M. Telefont, M. Toledo-Rodriguez, T. Tränkler, W. Van Geit, Jafet V. Díaz, R. Walker, Y. Wang, Stefano M. Zaninetta, J. DeFelipe, Sean L. Hill, I. Segev, and F. Schürmann (2015). "Reconstruction and Simulation of Neocortical Microcircuitry." *Cell* 163 (2): 456–92.

Marler, P. and S. Peters (1981). "Sparrows Learn Adult Song and More from Memory." *Science* 213(4509): 780–2.

Marr, D. (1982). *Vision.* San Francisco, CA: W. H. Freeman and Company.

Martin, C. B. (2007). *The Mind in Nature.* Oxford: Oxford University Press.

Masland, R. H. (2012). "The Neuronal Organization of the Retina." *Neuron* 76(2): 266–80.

Matsumoto, M. and H. Komatsu (2005). "Neural Responses in the Macaque V1 to Bar Stimuli with Various Lengths Presented on the Blind Spot." *Journal of Neurophysiology* 93(5): 2374–87.

Matyas, F., V. Sreenivasan, F. Marbach, C. Wacongne, B. Barsy, C. Mateo, R. Aronoff, and C. H. Petersen (2010). "Motor Control by Sensory Cortex." *Science* 330(6008): 1240–3.

Maunsell, J. H. and D. C. van Essen (1983). "The Connections of the Middle Temporal Visual Area (MT) and Their Relationship to a Cortical Hierarchy in the Macaque Monkey." *Journal of Neuroscience* 3(12): 2563–86.

McClelland, J. L. and M. A. Lambon Ralph (eds.), (2013). "Cognitive Neuroscience: Emergence of Mind from Brain." *The Biomedical & Life Sciences Collection.* London: Henry Stewart Talks Ltd.

McCulloch, W. S. (1974). "Recollections of the Many Sources of Cybernetics." *ASC Forum* VI(2): 5–16.

McCulloch, W. S. and W. H. Pitts: (1943). "A Logical Calculus of the Ideas Immanent in Nervous Activity." *Bulletin of Mathematical Biophysics* 7, 115–33. Reprinted in McCulloch 1964, 16–39.

McDonnell, N. (2017). "Causal Exclusion and the Limits of Proportionality." *Philosophical Studies* 174(6): 1459–74.

McFarland, J. M., A. G. Bondy, R. C. Saunders, B. G. Cumming, and D. A. Butts (2015). "Saccadic Modulation of Stimulus Processing in Primary Visual Cortex." *Nature Communications* 6: 8110.

McGee, V. (1991). "We Turing Machines Aren't Expected-Utility Maximizers (Even Ideally)." *Philosophical Studies* 64: 115–23.

McLaughlin, B. (1992). "The Rise and Fall of British Emergentism." In A. Berckermann, J. Kim, and H. Flohr (eds.), *Emergence or Reduction?* (49–93). De Gruyter.

McShea, D. W. (2012). "Upper-Directed Systems: A New Approach to Teleology in Biology." *Biology and Philosophy* 27: 663–84.

Meister, M. and M. J. Berry (1999). "The Neural Code of the Retina." *Neuron* 22: 435–50.

Mellor, D. H. (1989). "How Much of the Mind is a Computer?" In P. Slezak and W. R. Albury (eds.), *Computers, Brains and Minds* Dordrecht (47–69). Boston: Kluwer.

Merricks, T. (2001). *Objects and Persons.* Oxford: Oxford University Press.

Milkowski, M. (2013). *Explaining the Computational Mind.* Cambridge, MA: MIT Press.

Milkowski, M. (2015). "Explanatory Completeness and Idealization in Large Brain Simulations: A Mechanistic Perspective." *Synthese* 193 (5): 1457–78.

Milkowski, M. (2017). "Situatedness and Embodiment of Computational Systems." *Entropy* 19 (4), 162.
Milkowski, M. (2018). "Objections to Computationalism: A Survey." *Roczniki Filozoficzne*, 66(3), 57–75.
Milkowski, M., R. Clowes, S. Rucinska, A. Przegalinska, T. Zawidzki, J. Krueger, A. Gies, M. McGann, L. Afeltowicz, W. Wachowski, F. Stjernberg, V. Loughlin, and M. Hohol (2018). "From Wide Cognition to Mechanisms: A Silent Revolution." *Frontiers in Psychology* 9 (2392): 1–17.
Miller, E. K., C. A. Erickson, and R. Desimone (1996). "Neural Mechanisms of Visual Working Memory in Prefrontal Cortex of the Macaque." *Journal of Neuroscience* 16(16): 5154–67.
Miller, G. A., E. H. Galanter, and K. Pribram (1960). *Plans and the Structure of Behavior.* New York: Holt.
Miller, H., R. Rayburn-Reeves and T. Zentall (2009). "What Do Dogs Know About Hidden Objects?" *Behavioral Processes* 81(3): 439–46.
Millikan, R. G. (1984). *Language, Thought, and Other Biological Categories: New Foundations for Realism.* Cambridge, MA: MIT Press.
Millikan, R. G. (1989). "Biosemantics." *The Journal of Philosophy* 86(6): 281–97.
Millikan, R. G. (1993). *White Queen Psychology and Other Essays for Alice.* Cambridge, MA: MIT Press.
Mills, J. W. (2008). "The Nature of the Extended Analog Computer." *Physica D: Nonlinear Phenomena* 237(9): 1235–56.
Mishkin, M. and F. J. Manning (1978). "Non-spatial Memory after Selective Prefrontal Lesions in Monkeys." *Brain Research* 143(2): 313–23.
Mitchell, S. (2009). *Unsimple Truths: Science, Complexity and Policy.* Chicago: University of Chicago Press.
Montévil, M. and M. Mossio (2015). "Biological Organisation as Closure of Constraints." *Journal of Theoretical Biology* 372: 179–91.
Mooney, R. (2009). "Neural Mechanisms for Learned Birdsong." *Learning and Memory* 16(11): 655–69.
Moor, J. H. (1978). "Three Myths of Computer Science." *British Journal for the Philosophy of Science* 29: 213–22.
Moran, D. W. and A. B. Schwartz (2000). "One Motor Cortex, Two Different Views." *Nature Neuroscience* 3(10): 963; author reply 963–5.
Morcom, A.M. and P. C. Fletcher (2007). "Does the Brain Have a Baseline? Why We Should Be Resisting a Rest." *NeuroImage* 37(4): 1073–82.
Morgan, A. (2014). "Representations Gone Mental." *Synthese* 191(2): 213–44.
Morgan, A. and G. Piccinini (2017). "Towards a Cognitive Neuroscience of Intentionality." *Minds and Machines* 28: 119–39.
Mori, S. and J. Zhang (2006). "Principles of Diffusion Tensor Imaging and Its Applications to Basic Neuroscience Research." *Neuron* 51(5): 527–39.
Morris, K. (2011). "Subset Realization, Parthood, and Causal Overdetermination." *Pacific Philosophical Quarterly* 92: 363–79.
Morris, R. G., P. Garrud, J. N. Rawlins, and J. O'Keefe (1982). "Place Navigation Impaired in Rats with Hippocampal Lesions." *Nature* 297(5868): 681–3.
Moss, F., L. M. Ward and W. G. Sannita (2004). "Stochastic Resonance and Sensory Information Processing: A Tutorial and Review of Application." *Clinical Neurophysiology* 115(2): 267–81.

Movshon, J. A. and W. T. Newsome (1996). "Visual Response Properties of Striate Cortical Neurons Projecting to Area MT in Macaque Monkeys." *Journal of Neuroscience* 16(23): 7733–41.

Muckli, L. and L. S. Petro (2013)." Network Interactions: Non-Geniculate Input to V1." *Current Opinions in Neurobiology* 23(2): 195–201.

Murphy, G. L. (2002). *The Big Book of Concepts*. Cambridge, MA: MIT Press.

Murray, J. D., J. Jaramillo, and X. J. Wang (2017). "Working Memory and Decision-Making in a Frontoparietal Circuit Model." *Journal of Neuroscience* 37(50): 12167–86.

Mylopoulos, M. and E. Pacherie (2017). "Intentions and Motor Representations: The Interface Challenge." *Review of Philosophy and Psychology* 8(2): 317–36.

Nabavi, S., R. Fox, C. D. Proulx, J. Y. Lin, R. Y Tsien, and R. Malinow (2014). "Engineering a Memory with LTD and LTP." *Nature* 511(7509): 348–52.

Nagel, E. (1953). "Teleological Explanation and Teleological Systems." In S. Ratner (ed.), *Vision and Action* (537–58). New Brunswick, NJ: Rutgers University Press.

Nagel, E. (1977). "Teleology Revisited: Goal Directed Processes in Biology and Functional Explanation in Biology." *Journal of Philosophy* 74: 261–301.

Neander, K. (1983). *Abnormal Psychobiology*. Ph.D. dissertation, La Trobe.

Neander, K. (2012). "Teleological Theories of Mental Content." In E. N. Zalta (ed.), *The Stanford Encyclopedia of Philosophy* (Spring 2012 Edition), URL = <https://plato.stanford.edu/archives/spr2012/entries/content-teleological/>. Stanford, CA: The Metaphysics Research Lab, Centre for the Study of Language and Information, Stanford University.

Neander, K. (2017). *A Mark of the Mental: In Defense of Informational Teleosemantics*. Cambridge, MA: MIT Press.

Nelson, R. J. (1987). "Church's Thesis and Cognitive Science." *Notre Dame Journal of Formal Logic* 28(4): 581–614.

Newell, A. (1980). "Physical Symbol Systems." *Cognitive Science* 4: 135–83.

Newell, A. (1990). *Unified Theories of Cognition*. Cambridge, MA: Harvard University Press.

Newell, A. and H. A. Simon (1972). *Human Problem Solving*. Englewood Cliffs, NJ., Prentice Hall.

Newell, A. and H. A. Simon (1976). "Computer Science as an Empirical Enquiry: Symbols and Search." *Communications of the Association for Computing Machinery* 19: 113–26.

Newsome, W. T., K. H. Britten, and J. A. Movshon (1989). "Neuronal Correlates of a Perceptual Decision." *Nature* 341(6237): 52–4.

Newsome, W. T. and E. B. Pare (1988). "A Selective Impairment of Motion Perception Following Lesions of the Middle Temporal Visual Area (MT)." *Journal of Neuroscience* 8 (6): 2201–11.

Ney, A. (2010). "Convergence on the Problem of Mental Causation: Shoemaker's Strategy for (Nonreductive?) Physicalists." *Philosophical Issues* 20(1): 438–45.

Nietszche, F. W. (1882/1974). *The Gay Science: With a Prelude in Rhymes and an Appendix of Songs*. Translated, with commentary, by Walter Kaufmann. New York: Vintage Books.

Nietszche, F. W. (1883/2006). *Thus Spoke Zarathustra*. Translated by Adrian del Caro and edited by Robert Pippin. Cambridge: Cambridge University Press.

Northcott, R., and G. Piccinini (2018). "Conceived This Way: Innateness Defended," *Philosophers' Imprint* 18 (18): 1–16.

O'Brien, G. and J. Opie (2004). "Notes Toward a Structuralist Theory of Mental Representation." In H. Clapin, P. Staines, and P. Slezac (eds.), *Representation in Mind* (1–20). Amsterdam: Elsevier.

O'Brien, G. and J. Opie (2006). "How Do Connectionist Networks Compute?" *Cognitive Processing* 7: 30–41.

O'Doherty, J., A. Hampton and H. Kim (2007). "Model-Based fMRI and Its Application to Reward Learning and Decision Making." *Annals of the New York Academy of Sciences* 1104: 35–53.

Ó Nualláin, S., P. McKevitt, and E. MacAogáin (eds.) (1997). *Two Sciences of Mind: Readings in Cognitive Science and Consciousness*. Philadelphia: John Benjamins.

Oram, M. W., M. C. Wiener, R. Lestienne, and B. J. Richmond (1999). "Stochastic Nature of Precisely Timed Spike Patterns in Visual System Neuronal Responses." *Journal of Neurophysiology* 81: 3021–33.

O'Reilly, R. C. and Y. Munakata (2000). *Computational Explorations in Cognitive Neuroscience: Understanding the Mind by Simulating the Brain*. Cambridge, MA: MIT Press.

O'Reilly, R. C., Y. Munakata, M. J. Frank, T. E. Hazy, and Contributors (2014). *Computational Cognitive Neuroscience*. Wiki Book, 2nd Edition. URL: http://ccnbook.colorado.edu.

Oppenheim, P., and H. Putnam (1958). "Unity of Science as a Working Hypothesis." In H. Feigl, M. Scriven, and G. Maxwell (eds.), *Concepts, Theories, and the Mind–Body Problem*, Minnesota Studies in the Philosophy of Science, II, University of Minnesota Press, Minneapolis, pp. 3–36.

Orilia, F. and Swoyer, C. (2020). "Properties" *The Stanford Encyclopedia of Philosophy* (Summer 2020 Edition), Edward N. Zalta (ed.), URL = <https://plato.stanford.edu/archives/sum2020/entries/properties/>. Stanford, CA: The Metaphysics Research Lab, Centre for the Study of Language and Information, Stanford University.

Patterson, D. A. and J. L. Hennessy (1998). *Computer Organization and Design: The Hardware/Software Interface*. San Francisco: Morgan Kauffman.

Papineau, D. (1984). "Representation and Explanation." *Philosophy of Science* 51: 550–72.

Papineau, D. (1993). *Philosophical Naturalism*. Oxford: Blackwell.

Papineau, D. (2001). "The Rise of Physicalism." In C. Gillett and B. Loewer (eds.), *Physicalism and Its Discontents* (3–36). Cambridge: Cambridge University Press.

Papineau, D. (2008). "Must a Physicalist be a Microphysicalist?" In J. Hohwy and J. Kallestrup (eds.), *Being Reduced*. Oxford: Oxford University Press.

Penfield, W. and E. Boldrey (1937). "Somatic Motor and Sensory Representation in the Cerebral Cortex of Man as Studied by Electrical Stimulation." *Brain* 60: 389–443.

Penrose, R. (1994). *Shadows of the Mind*. Oxford: Oxford University Press.

Penrose, R. (1989). *The Emperor's New Mind: Concerning Computers, Minds, and The Laws of Physics*. Oxford: Oxford University Press.

Pereboom, D. and H. Kornblith (1991). "The Metaphysics of Irreducibility." *Philosophical Studies* 63: 125–45.

Pereboom, D. (2002). "Robust Nonreductive Materialism." *Journal of Philosophy* 99: 499–531.

Perkel, D. H. (1988). "Logical Neurons: The Enigmatic Legacy of Warren McCulloch." *Trends in Neurosciences* 11(1): 9–12.

Perkel, D. H. (1990). "Computational Neuroscience: Scope and Structure." In E. L. Schwartz (ed.), *Computational Neuroscience* (38–45). Cambridge, MA: MIT Press.

Perrett, D. I., E. T. Rolls, and W. Caan (1982). "Visual Neurones Responsive to Faces in the Monkey Temporal Cortex." *Experimental in Brain Research* 47(3): 329–42.

Pfenning, A. R., E. Hara, O. Whitney, M. V. Rivas, R. Wang, P. L. Roulhac, J. T. Howard, M. Wirthlin, P. V. Lovell, G. Ganapathy, J. Mountcastle, M. A. Moseley, J. W. Thompson, E. J. Soderblom, A. Iriki, M. Kato, M. T. P. Gilbert, G. Zhang, T. Bakken, A. Bongaarts, A. Bernard, E. Lein, C. V. Mello, A. J. Hartemink, and E. D. Jarvis (2014). "Convergent

Transcriptional Specializations in the Brains of Humans and Song-learning Birds." *Science* 346(6215): 1256846.

Piccinini, G. (2003a). *Computations and Computers in the Sciences of Mind and Brain.* Doctoral dissertation, Department of History and Philosophy of Science, University of Pittsburgh, Pittsburgh, PA. URL = <http://etd.library.pitt.edu/ETD/available/etd-08132003-155121/>

Piccinini, G. (2003b). "Alan Turing and the Mathematical Objection." *Minds and Machines* 13(1): 23–48.

Piccinini, G. (2003c). "Review of John von Neumann's The Computer and the Brain." *Minds and Machines* 13(2): 327–32.

Piccinini, G. (2004a). "Functionalism, Computationalism, and Mental Contents." *Canadian Journal of Philosophy* 34(3): 375–410.

Piccinini, G. (2004b). "Functionalism, Computationalism, and Mental States." *Studies in the History and Philosophy of Science* 35(4): 811–33.

Piccinini, G. (2004c). "The First Computational Theory of Mind and Brain: A Close Look at McCulloch and Pitts's 'Logical Calculus of Ideas Immanent in Nervous Activity.'" *Synthese* 141(2): 175–215.

Piccinini, G. (2007a). "Computing Mechanisms." *Philosophy of Science* 74(4): 501–26.

Piccinini, G. (2007b). "Computational Modeling vs. Computational Explanation: Is Everything a Turing Machine, and Does it Matter to the Philosophy of Mind?" *Australasian Journal of Philosophy* 85(1): 93–115.

Piccinini, G. (2007c). "Computationalism, the Church–Turing Thesis, and the Church–Turing Fallacy." *Synthese* 154(1): 97–120.

Piccinini, G. (2007d). "The Ontology of Creature Consciousness: A Challenge for Philosophy." (Commentary on "Consciousness without a Cerebral Cortex: A Challenge for Neuroscience and Medicine," by Björn Merker), *Behavioral and Brain Sciences* 30(1): 103–4.

Piccinini, G. (2008a). "Computers." *Pacific Philosophical Quarterly* 89(1): 32–73.

Piccinini, G. (2008b). "Computation without Representation." *Philosophical Studies* 137(2): 205–241.

Piccinini, G. (2008c). "Some Neural Networks Compute, Others Don't." *Neural Networks* 21(2–3).

Piccinini, G. (2009). "Computationalism in the Philosophy of Mind." *Philosophy Compass* 4: 1–18.

Piccinini, G. (2010a). "The Resilience of Computationalism." *Philosophy of Science* 77(5): 852–61.

Piccinini, G. (2010b). "The Mind as Neural Software? Understanding Functionalism, Computationalism, and Computational Functionalism." *Philosophy and Phenomenological Research* 81(2): 269–311.

Piccinini, G. (2011). "Two Kinds of Concept: Implicit and Explicit." *Dialogue* 50:179–93.

Piccinini, G. (2015). *Physical Computation: A Mechanistic Account.* Oxford: Oxford University Press.

Piccinini, G. (2016). "The Computational Theory of Cognition." In V. C. Müller (ed.), *Fundamental Issues of Artificial Intelligence* (Synthese Library Volume 376) (201–19). Berlin: Springer.

Piccinini, G. (2017a). "Computational Mechanisms." In S. Glennan and P. Illari (eds.), *Routledge Handbook of Philosophy of Mechanisms* (435–46). New York: Routledge.

Piccinini, G. (2017b). "Activities are Manifestations of Causal Powers." In M. Adams, Z. Biener, U. Feest, and J. Sullivan (eds.), *Eppur Si Muove: Doing History and Philosophy of Science with Peter Machamer* (171–82). Berlin: Springer.

Piccinini, G. (2017c). "Access Denied to Zombies." *Topoi* 36(1): 81–93.

Piccinini, G. (forthcoming). "Nonnatural Mental Representation," forthcoming in K. Dolega, T. Schlicht, and J. Smortchkova, eds., *What Are Mental Representations?* Oxford: Oxford University Press.

Piccinini, G. and S. Bahar (2013). "Neural Computation and the Computational Theory of Cognition." *Cognitive Science* 34: 453–88.

Piccinini, G. and C. F. Craver (2011). "Integrating Psychology and Neuroscience: Functional Analyses as Mechanism Sketches." *Synthese* 183(3): 283–311.

Piccinini, G. and C. J. Maley (2014). "The Metaphysics of Mind and the Multiple Sources of Multiple Realizability." In M. Sprevak and J. Kallestrup (eds.) *New Waves in the Philosophy of Mind* (125–52). London: Palgrave Macmillan.

Piccinini, G. and A. Scarantino (2011). "Information Processing, Computation, and Cognition." *Journal of Biological Physics* 37(1): 1–38.

Piccinini, G. and A. W. Schulz (2018). "The Ways of Altruism." *Evolutionary Psychological Science* 5(1): 58–70.

Pikovsky, A., M. Rosenblum, and J. Kurths (2003). *Synchronization: A Universal Concept in Nonlinear Sciences*. Cambridge, MA: Cambridge University Press.

Pilley, J. and A. Reid (2011). "Border Collie Comprehends Object Names as Verbal Referents." *Behavioral Processes* 86: 184–95.

Pineda, D. and A. Vicente (2017). "Shoemaker's Account of Realization: A Review." *Philosophy and Phenomenological Research* 94(1): 97–120.

Pinker, S. (1997). *How the Mind Works.* New York: Norton.

Piper, M. (2012). "You Can't Eat Causal Cake with an Abstract Fork: An Argument Against Computational Theories of Consciousness." *Journal of Consciousness Studies* 19(11–12): 154–90.

Pitowski, I. and O. Shagrir (2003). "Physical Hypercomputation and the Church-Turing Thesis." *Minds and Machines* 13(1): 87–101.

Pitts, W. H. (1942a). "Some Observations on the Simple Neuron Circuit." *Bulletin of Mathematical Biophysics* 4: 121–9.

Pitts, W. H. (1942b). "The Linear Theory of Neuron Networks: The Static Problem." *Bulletin of Mathematical Biophysics* 4: 169–75.

Pitts, W. H. (1943a). "The Linear Theory of Neuron Networks: The Dynamic Problem." *Bulletin of Mathematical Biophysics* 5: 23–31.

Pitts, W. H. (1943b). "A General Theory of Learning and Conditioning: Part I." *Psychometrika* 8(1): 1–18.

Pitts, W. H. (1943c). "A General Theory of Learning and Conditioning: Part II." *Psychometrika* 8(2): 131–40.

Phongphanphanee, P., F. Mizuno, P. H. Lee, Y. Yanagawa, T. Isa, and W. C. Hall (2011). "A Circuit Model for Saccadic Suppression in the Superior Colliculus." *Journal of Neuroscience* 31(6): 1949–54.

Place, U. T. (1956). "Is Consciousness a Brain Process?" *British Journal of Psychology* 47: 44–50.

Poldrack, R. A. (2020). "The Physics of Representation." [Preprint] URL: http://philsci-archive.pitt.edu/id/eprint/16916 (accessed 2020-05-08).

Polger, T. W. (2004). *Natural Minds.* Cambridge, MA: MIT Press.

Polger, T. W. (2007). "Realization and the Metaphysics of Mind." *Australasian Journal of Philosophy* 85(2): 233–59.

Polger, T. W. (2009). "Evaluating the Evidence for Multiple Realization." *Synthese 167*(3): 457–72.

Polger, T. W. and L. A. Shapiro (2008). "Understanding the Dimensions of Realization." *Journal of Philosophy* 105(4): 213–22.

Polger, T. W. and L. A. Shapiro (2016). *The Multiple Realization Book*. Oxford: Oxford University Press.

Polonsky, A., R. Blake, J. Braun, and D. J. Heeger (2000). "Neuronal Activity in Human Primary Visual Cortex Correlates with Perception During Binocular Rivalry." *Nature Neuroscience* 3(11): 1153–9.

Port, R. and T. van Gelder (eds.), (1995). *Mind as Motion: Explorations in the Dynamics of Cognition*. Cambridge, MA: MIT Press.

Posner, M. I. (1976). *Chronometric Explorations of Mind*. Hillsdale, N.J: Lawrence Erlbaum Associates.

Posner, M. I. (2004). *Cognitive Neuroscience of Attention*. New York, NY: Guilford Press.

Posner, M. I. and M. E. Raichle (1994). *Images of Mind*. New York: Scientific American Books.

Potochnik, A. (2017). *Idealization and the Aims of Science*. Chicago: University of Chicago Press.

Poulet, J. F. and B. Hedwig (2003). "Corollary Discharge Inhibition of Ascending Auditory Neurons in the Stridulating Cricket." *Journal of Neuroscience* 23(11): 4717–25.

Poulet, J. F. and B. Hedwig (2006). "The Cellular Basis of a Corollary Discharge." *Science* 311(5760): 518–22.

Pour-El, M. B. (1974). "Abstract Computability and Its Relation to the General Purpose Analog Computer (Some Connections Between Logic, Differential Equations and Analog Computers)." *Transactions of the American Mathematical Society* 199.

Pour-El, M. B. and J. I. Richards (1989). *Computability in Analysis and Physics*. Berlin, Springer Verlag.

Povich, M. (2015). "Mechanisms and Model-Based Functional Magnetic Resonance Imaging." *Philosophy of Science* 82: 1035–46.

Povich, M. (2018). "Minimal Models and the Generalized Ontic Conception of Scientific Explanation." *The British Journal for the Philosophy of Science* 69 (1): 117–37.

Povich, M. (2019). "Model-Based Cognitive Neuroscience: Multifield Mechanistic Integration in Practice." *Theory & Psychology* 29(5). https://doi.org/10.1177/0959354319863880

Prather, J. F., S. Peters, S. Nowicki and R. Mooney (2010). "Persistent Representation of Juvenile Experience in the Adult Songbird Brain." *Journal of Neuroscience* 30(31): 10586–98.

Preston, B. (2013). *A Philosophy of Material Culture: Action, Function, and Mind*. New York: Routledge.

Pruszynski, J. A., M. Omrani and S. H. Scott (2014). "Goal-Dependent Modulation of Fast Feedback Responses in Primary Motor Cortex." *Journal of Neuroscience* 34(13): 4608–17.

Purves, D. (2018). *Neuroscience* (Sixth edition. ed.). New York: Oxford University Press.

Putnam, H. (1960). "Minds and Machines." In S. Hook (ed.), *Dimensions of Mind*. New York: New York University Press.

Putnam, H. (1967a). "Psychological Predicates." In W.H. Capitan & D.D. Merrill (eds.), *Art, Philosophy, and Religion*. Pittsburgh, PA: University of Pittsburgh Press. Reprinted as "The Nature of Mental States" in W. Lycan (ed.) (1999). *Mind and Cognition: An Anthology, Second Edition* (27–34). Malden: Blackwell.

Putnam, H. (1967b). "The Mental Life of Some Machines." In H.-N. Castañeda (ed.), *Intentionality, Minds, and Perception*. Detroit, MI: Wayne State University Press: 177–200.

Putnam, H. (1975a). 'The Meaning of "Meaning."' *Minnesota Studies in the Philosophy of Science* 7: 131–93.

Putnam, H. (1975b). "Philosophy and Our Mental Life." In H. Putnam (ed.), *Mind, Language and Reality: Philosophical Papers, Volume 2* (291–303). Cambridge: Cambridge University Press.

Putnam, H. (1988). *Representation and Reality.* Cambridge, MA: MIT Press.

Pylyshyn, Z. W. (1981). "The Imagery Debate: Analogue Media Versus Tacit Knowledge." *Psychological Review* 88: 16–45.

Pylyshyn, Z. W. (1984). *Computation and Cognition.* Cambridge, MA: MIT Press.

Pylyshyn, Z. W. (2002). "Mental Imagery: In Search of a Theory." *Behavioral and Brain Sciences* 25(2): 157–237.

Pylyshyn, Z. W. (2003). "Return of the Mental Image: Are There Really Pictures in the Head?" *Trends in Cognitive Science* 7(3): 113–18.

Qi, X. L., F. Katsuki, T. Meyer, J. B. Rawley, X. Zhou, K. L. Douglas, C. Constantinidis (2010). "Comparison of Neural Activity Related to Working Memory in Primate Dorsolateral Prefrontal and Posterior Parietal Cortex." *Frontiers in Systems Neuroscience* 4: 12.

Quintana, J., J. Yajeya and J. M. Fuster (1988). "Prefrontal Representation of Stimulus Attributes During Delay Tasks. I. Unit Activity in Cross-temporal Integration of Sensory and Sensory-Motor Information." *Brain Research* 474(2): 211–21.

Raatikainen, P. (2010). "Causation, Exclusion, and the Special Sciences." *Erkenntnis* 73: 349–63.

Raichle, M.E., A. M. MacLeod, A. Z. Snyder, W. J. Powers, D. A. Gusnard and G. L. Shulman (2001). "A Default Mode of Brain Function." *Proceedings of the National Academy of Sciences USA* 92(1): 676–82.

Raichle, M.E. and M. A. Mintun (2006). "Brain Work and Brain Imaging." *Annual Review of Neuroscience* 29: 449–76.

Raja, V. (2017). "A Theory of Resonance: Towards an Ecological Cognitive Architecture." *Minds and Machines* 28: 29–51.

Ramirez, S., X. Liu, P. A. Lin, J. Suh, M. Pignatelli, R. L. Redondo, T. J. Ryan, and S. Tonegawa (2013). "Creating a False Memory in the Hippocampus." *Science* 341 (6144): 387–91.

Ramsey, F. P. (1931). *The Foundations of Mathematics, and Other Logical Essays.* London: Routledge and Kegan Paul.

Ramsey, W. M. (2007). *Representation Reconsidered.* Cambridge: Cambridge University Press.

Ramsey, W. M. (2016). "Untangling Two Questions about Mental Representation." *New Ideas in Psychology* 40: 3–12.

Rapaport, W. J. (1998). "How Minds Can be Computational Systems." *Journal of Experimental and Theoretical Artificial Intelligence* 10: 403–19.

Rashevsky, N. (1938). *Mathematical Biophysics: Physicomathematical Foundations of Biology.* Chicago: University of Chicago Press.

Rashevsky, N. (1940). *Advances and Applications of Mathematical Biology.* Chicago: University of Chicago Press,

Recanzone, G. H., M. M. Merzenich and C. E. Schreiner (1992). "Changes in the Distributed Temporal Response Properties of SI Cortical Neurons Reflect Improvements in Performance on a Temporally Based Tactile Discrimination Task." *Journal of Neurophysiology* 67(5): 1071–91.

Reijmers, L. G., B. L. Perkins, N. Matsuo and M. Mayford (2007). "Localization of a Stable Neural Correlate of Associative Memory." *Science* 317(5842): 1230–3.

Rescorla, M. (2018). "An Interventionist Approach to Psychological Explanation." *Synthese* 195: 1909–40.

Rey, G. (2005). "Mind, Intentionality and Inexistence: An Overview of My Work." *Croatian Journal of Philosophy* 5(15): 389–415.

Richmond, B. J. and R. H. Wurtz (1980). "Vision During Saccadic Eye Movements. II. A Corollary Discharge to Monkey Superior Colliculus." *Journal of Neurophysiology* 43(4): 1156–67.

Rieke, F., D. Warland, R. de Ruyter van Steveninck and W. Bialek (1999). *Spikes: Exploring the Neural Code.* Cambridge MA: MIT Press.

Riley, M. R. and C. Constantinidis (2016). "Role of Prefrontal Persistent Activity in Working Memory." *Frontiers in Systems Neuroscience* 9.

Roberts, T. F., S. M. Gobes, M. Murugan, B. P. Olveczky and R. Mooney (2012). "Motor Circuits are Required to Encode a Sensory Model for Imitative Learning." *Nature Neuroscience* 15(10): 1454–9.

Robertson, T. and P. Atkins (2016). "Essential vs. Accidental Properties." In E. N. Zalta (ed.), *The Stanford Encyclopedia of Philosophy* (Summer 2016 Edition), URL = <https://plato.stanford.edu/archives/sum2016/entries/essential-accidental/>. Stanford, CA: The Metaphysics Research Lab, Centre for the Study of Language and Information, Stanford University.

Robinson, D. L. and R. H. Wurtz (1976). "Use of an Extraretinal Signal by Monkey Superior Colliculus Neurons to Distinguish Real from Self-induced Stimulus Movement." *Journal of Neurophysiology 39*(4): 852–70.

Robinson, Z., C. J. Maley and G. Piccinini (2015). "Is Consciousness a Spandrel?" *Journal of the American Philosophical Association* 1(2): 365–83.

Rockland, K. S and T. Knutson (2000). "Feedback Connections from Area MT of the Squirrel Monkey to Areas V1 and V2." *Journal of Computational Neurology* 425(3): 345–68.

Rodieck, R. W. (1998). *The First Steps in Seeing.* Sunderland, Mass.: Sinauer Associates.

Rogers, H. (1967). *Theory of Recursive Functions and Effective Computability.* New York: McGraw-Hill.

Rohrlich, F. (1990). "Computer Simulation in the Physical Sciences." *PSA: Proceedings of the Biennial Meeting of the Philosophy of Science Association 1990,* no. 2: 507–18.

Rohwer, Y. and C. Rice (2013). "Hypothetical Pattern Idealization and Explanatory Models." *Philosophy of Science* 80: 334–55.

Rosenblatt, F. (1958). "The Perceptron: A Probabilistic Model for Information Storage and Organization in the Brain." *Psychological Review* 65: 386–408.

Rosenblatt, F. (1962). *Principles of Neurodynamics: Perceptrons and the Theory of Brain Mechanisms.* Washington, D.C.: Spartan.

Rosenthal, D. M. (2005). *Consciousness and Mind.* Oxford: Clarendon Press.

Roska, B. and F. Werblin (2001). "Vertical Interactions Across Ten Parallel, Stacked Representations in the Mammalian Retina." *Nature* 410(6828): 583–7.

Roskies, A. (2009). "Brain-Mind and Structure-Function Relationships: A Methodological Response to Coltheart." *Philosophy of Science* 76(5): 927–39.

Ross, L. N (2015). "Dynamical Models and Explanation in Neuroscience." *Philosophy of Science* 81(1): 32–54.

Rossignol, S. and L. Bouyer (2004). "Adaptive Mechanisms of Spinal Locomotion in Cats." *Integr Comp Biol* 44(1): 71–9.

Roth, M. (2005). "Program Execution in Connectionist Networks." *Mind and Language* 20(4): 448–67.

Roy, A., P. N. Steinmetz, S. S. Hsiao, K. O. Johnson and E. Niebur (2007). "Synchrony: A Neural Correlate of Somatosensory Attention." *Journal of Neurophysiology* 98(3): 1645–61.

Rubel, L. A. (1985). "The Brain as an Analog Computer." *Journal of Theoretical Neurobiology* 4.

Rubel, L. A. (1989). "Digital Simulation of Analog Computation and Church's Thesis." *Journal of Symbolic Logic* 54(3): 1011–17.

Rubel, L. A. (1993). "The Extended Analog Computer." *Advances in Applied Mathematics* 14(1): 39–50.

Rubel, L. A. and M. F. Singer (1985). "A Differentially Algebraic Elimination Theorem with Application to Analog Computability in the Calculus of Variations." *Proceedings of the American Mathematical Society* 94(4): 653–8.

Rumelhart, D. E., J. M. McClelland and the PDP Research Group (1986). *Parallel Distributed Processing: Explorations in the Microstructure of Cognition*. Cambridge, MA: MIT Press.

Rupert, R. (2006). "Functionalism, Mental Causation, and the Problem of Metaphysically Necessary Effects." *Noûs* 40: 256–83.

Rusanen, A. M. and O. Lappi (2016). "On Computational Explanation." *Synthese* 193(12): 3931–49.

Rust, N. (2014). "Population-Based Representations: From Implicit to Explicit." In M. Gazzaniga and G. Ronald (eds.), *The Cognitive Neurosciences*, 5th ed. (337–47). Cambridge, MA: MIT Press.

Ryder, D. (2004). "SINBAD Neurosemantics: A Theory of Mental Representation." *Mind & Language* 19(2): 211–40.

Ryder, D. (unpublished). *Models in the Brain*.

Ryle, G. (1949). *The Concept of Mind*. London, New York: Hutchinson's University Library.

Salinas, E. and L. F. Abbott (1995). Transfer of Coded Information from Sensory to Motor Networks. *Journal of Neuroscience* 15(10): 6461–74.

Salmon, W. C. (1984). *Scientific Explanation and the Causal Structure of the World*. Princeton: Princeton University Press.

Salmon, W. C. (2005). *Reality and Rationality*, P. Dowe and M. H. Salmon (eds.), New York: Oxford University Press.

Salzman, C. D., K. H. Britten and W. T. Newsome (1990). "Cortical Microstimulation Influences Perceptual Judgements of Motion Direction." *Nature* 346(6280): 174–7.

Salzman, C. D., C. M. Murasugi, K. H. Britten and W. T. Newsome (1992). "Microstimulation in Visual Area MT: Effects on Direction Discrimination Performance." *Journal of Neuroscience* 12(6): 2331–55.

Samuels, R. (2010). "Classical Computationalism and the Many Problems of Cognitive Relevance." *Studies in History and Philosophy of Science*: 41(3): 280–93.

Sanes, J. R. and R. H. Masland (2015). "The Types of Retinal Ganglion Cells: Current Status and Implications for Neuronal Classification." *Annual Review of Neuroscience* 38: 221–46.

Scarantino, A. and G. Piccinini (2010). "Information Without Truth." *Metaphilosophy* 43(3): 313–30.

Schaffer, J. (2003). "Overdetermining Causes." *Philosophical Studies* 114: 23–45.

Schaffer, J. (2010). "Monism: The Priority of the Whole." *Philosophical Review* 119(1): 31–76.

Schaffner, K. (1993). *Discovery and Explanation in Biology and Medicine.* Chicago: University of Chicago Press.

Schaffner, K. F. (2008). "Theories, Models, and Equations in Biology: The Heuristic Search for Emergent Simplifications in Neurobiology." *Philosophy of Science* 75: 1008–21.

Scheele, M. (2006). "Function and Use of Artefacts: Social Conditions of Function Ascription." *Studies in History and Philosophy of Science* 37(1): 23–36.

Schenk, T. and J. Zihl (1997). "Visual Motion Perception after Brain Damage: I. Deficits in Global Motion Perception." *Neuropsychologia* 35(9): 1289–97.

Scheutz, M. (1999) "When Physical Systems Realize Functions ... ." *Minds and Machines* 9(2).

Scheutz, M. (2001). "Causal Versus Computational Complexity." *Minds and Machines* 11(4).

Scheutz, M. (2004). Comments Presented at the 2004 Pacific APA in Pasadena, CA.

Schlatter, M., and K. Aizawa (2007). "Walter Pitts and 'A Logical Calculus.'" *Synthese* 162: 235–50.

Schlosser, G. (1998). "Self-Re-Production and Functionality: A Systems-Theoretical Approach to Teleological Explanation." *Synthese* 116(3): 303–54.

Schneider, S. (2011). *The Language of Thought: A New Philosophical Direction.* Cambridge, MA: MIT Press.

Schreiner, C. E. and J. A. Winer (2007). "Auditory Cortex Mapmaking: Principles, Projections, and Plasticity." *Neuron* 56(2): 356–65.

Schultz, A. (2018). *Efficient Cognition: The Evolution of Representational Decision Making.* Cambridge, MA: MIT Press.

Schumm, A., W. Rohloff and G. Piccinini (unpublished). "Composition as Trans-Scalar Identity."

Schwartz, E. L. (1990). *Computational Neuroscience.* Cambridge, MA: MIT Press.

Schwartz, P. H. (2007). "Defining Dysfunction: Natural Selection, Design, and Drawing a Line." *Philosophy of Science* 74: 364–85.

Schwartzman, R. J. (1978). "A Behavioral Analysis of Complete Unilateral Section of the Pyramidal Tract at the Medullary Level in Macaca Mulatta." *Annals of Neurology* 4(3): 234–44.

Scott, P. D. (1997). "Crisis? What Crisis? Church's Thesis and the Scope of Cognitive Science." In S. Ó Nualláin, P. Mc Kevitt, and E. Mac Aogáin (eds.), *Two Sciences of Mind: Readings in Cognitive Science and Consciousness* (63–76). Philadelphia: John Benjamins.

Scott, S. H. (2000). Reply to 'One Motor Cortex, Two Different Views'. *Nature Neuroscience* 3(10): 964–5.

Scott, S. H. (2016). "A Functional Taxonomy of Bottom-Up Sensory Feedback Processing for Motor Actions." *Trends in Neuroscience* 39(8): 512–26.

Segal, G. (1991). "Defence of a Reasonable Individualism." *Mind* 100: 485–93.

Segal, G. (2000). *A Slim Book about Narrow Content.* Cambridge, MA: MIT Press.

Sellars, W. (1954). "Physical Realism." *Philosophy and Phenomenological Research* 15(1): 13–32.

Sellars, W. (1956). "Empiricism and the Philosophy of Mind." *Minnesota Studies in the Philosophy of Science* 1: 253–329.

Sereno, M. I. (2014). "Origin of Symbol-using Systems: Speech, but Not Sign, Without the Semantic Urge." *Philosophical Transactions of the Royal Society of London B Biological Sciences* 369(1651): 20130303.

Seth, A. K., Barrett, A. B. and L. Barnett (2015). "Granger Causality Analysis in Neuroscience and Neuroimaging." *Journal of Neuroscience* 35 (8): 3293–7.

Shadlen, M. N. and W. T. Newsome (1998). "The Variable Discharge of Cortical Neurons: Implications for Connectivity, Computation and Information Coding." *Journal of Neuroscience* 18: 3870–96.

Shadmehr, R., J. Brandt and S. Corkin (1998). "Time-Dependent Motor Memory Processes in Amnesic Subjects." *Journal of Neurophysiology* 80(3): 1590–7.
Shadmehr, R. and J. W. Krakauer (2008). "A Computational Neuroanatomy for Motor Control." *Experimental Brain Research* 185(3): 359–81.
Shadmehr, R., M. A. Smith and J. W. Krakauer (2010). "Error Correction, Sensory Prediction, and Adaptation in Motor Control." *Annual Review of Neuroscience* 33: 89–108.
Shagrir, O. (1998). "Multiple Realization, Computation and the Taxonomy of Psychological States." *Synthese* 114(3): 445–61.
Shagrir, O. (2001). "Content, Computation and Externalism." *Mind* 110(438): 369–400.
Shagrir, O. (2005). The Rise and Fall of Computational Functionalism. In Y. Ben-Menahem (ed.), *Hilary Putnam* (220–250). Cambridge: Cambridge University Press.
Shagrir, O. (2006). "Why We View the Brain as a Computer." *Synthese* 153(3): 393–416.
Shagrir, O. (2010a). "Brains as Analog-Model Computers." *Studies in History and Philosophy of Science* 41: 271–9.
Shagrir, O. (2010b). "Marr on Computational-Level Theories." *Philosophy of Science* 77: 477–500.
Shagrir, O. (2012). "Structural Representations and the Brain." *The British Journal for the Philosophy of Science* 63(3): 519–45.
Shagrir, O. (2018). "The Brain as an Input–Output Model of the World." *Minds and Machines* 28: 53–75.
Shagrir, O. and L. Elber-Dorozko (2019). "Integrating Computation into the Mechanistic Hierarchy in the Cognitive and Neural Sciences." *Synthese* https://doi.org/10.1007/s11229-019-02230-9
Shanker, S. G. (1995). "Turing and the Origins of AI." *Philosophia Mathematica* 3: 52–85.
Shannon, C. E. and J. McCarthy (1956). *Automata Studies*. Princeton, NJ: Princeton University Press.
Shannon, C. E. and W. Weaver (1949). *The Mathematical Theory of Communication*. Urbana: University of Illinois Press.
Shenker, O. (unpublished). "Flat Physicalism."
Shapiro, L. A. (1994). "Behavior, ISO Functionalism, and Psychology." *Studies in the History and Philosophy of Science* 25(2): 191–209.
Shapiro, L. A. (2000). "Multiple Realizations." *The Journal of Philosophy* 97(12): 635–54.
Shapiro, L. A. (2004). *The Mind Incarnate*. Cambridge, MA: MIT Press.
Shapiro, L. A. (2016). "Mechanism or Bust? Explanation in Psychology." *British Journal for the Philosophy of Science* 68 (4): 1037–59.
Shapiro, L. A. (2019). "A Tale of Two Explanatory Styles in Cognitive Psychology." *Theory and Psychology*. DOI: 10.1177/0959354319866921
Shapiro, S. C. (1995). "Computationalism." *Minds and Machines* 5(4): 517–24.
Shea N. (2007). Consumers Need Information: Supplementing Teleosemantics with an Input Condition. *Philosophy and Phenomenological Research* 75: 404–35.
Shea, N. (2018). *Representation in Cognitive Science*. Oxford: Oxford University Press.
Shepherd, G. (1999). "Electronic Properties of Axons and Dendrites." In Zigmond, M. J., F. E. Bloom, S. C. Landys, J. L. Roberts, and L. R. Squire (eds.), *Fundamental Neuroscience* (115–17). Amsterdam: Academic Press.
Shepard, R. and S. Chipman (1970). "Second-Order Isomorphism of Internal Representations: Shapes of States." *Cognitive Psychology* 1(1): 1–17.
Sherrington, C. S. (1910). "Flexion-reflex of the Limb, Crossed Extension-reflex, and Reflex Stepping and Standing." *Journal of Physiology* 40(1–2): 28–121.

Sherrington, C. S. (1940). *Man on His Nature.* Cambridge: Cambridge University Press,.

Shoemaker, S. (1980). "Causality and Properties." In P. van Inwagen (ed.), *Time and Cause.* Dordrecht: Reidel.

Shoemaker, S. (1981). "Some Varieties of Functionalism." *Philosophical Topics* 12(1): 93–119.

Shoemaker, S. (2007). *Physical Realization.* Oxford: Oxford University Press.

Sider, T. (2003). "What's So Bad About Overdetermination?" *Philosophy and Phenomenological Research* 67: 719–26.

Sider, T. (2013). "Against Parthood." In K. Bennett and D. Zimmerman (eds.), *Oxford Studies in Metaphysics Volume 8* (237–93). Oxford: Oxford University Press.

Sieg, W. (1994). "Mechanical Procedures and Mathematical Experience." In G. Alexander (ed.) *Mathematics and Mind* (71–117). New York: Oxford University Press.

Sieg, W. (2001). "Calculations by Man and Machine: Conceptual Analysis." In W. Sieg, R. Sommer and C. Talcott (eds.), *Reflections on the Foundations of Mathematics (Essays in Honor of Solomon Feferman).* Urbana, IL: Association for Symbolic Logic. 15: 387–406.

Sieg, W. (2009). "On Computability." In A. Irvine (Ed(ed.), *Philosophy of Mathematics* (Handbook of the Philosophy of Science) (535–630). Amsterdam: North-Holland.

Siegelmann, H. T. (1999). *Neural Networks and Analog Computation: Beyond the Turing Limit.* Boston, MA: Birkhäuser.

Siegelmann, H. T. (2003). "Neural and Super-Turing Computing." *Minds and Machines* 13 (1): 103–14.

Silva, A. J., A. Landreth and J. Bickle (2014). *Engineering the Next Revolution in Neuroscience: The New Science of Experiment Planning.* New York: Oxford University Press.

Skotko, B. G., W. Andrews and G. Einstein (2005). "Language and the Medial Temporal Lobe: Evidence from HM's Spontaneous Discourse." *Journal of Memory and Language* 53(3): 397–415.

Sloman, A. (2001). "The Irrelevance of Turing Machines to AI." In M. Scheutz (ed.), *Computationalism: New Directions.* Cambridge, MA: MIT Press.

Smart, J. J. C. (1959). "Sensations and Brain Processes." *The Philosophical Review* 68(2): 141–56.

Smart, J. J. C. (2007). "The Mind/Brain Identity Theory." In E. N. Zalta (ed.), *The Stanford Encyclopedia of Philosophy (Summer 2007 Edition),* URL = <http://plato.stanford.edu/archives/sum2007/entries/mind-identity/>. Stanford, CA.: The Metaphysics Research Lab, Centre for the Study of Language and Information, Stanford University.

Smith, B. C. (1996). *On the Origin of Objects.* Cambridge, MA: MIT Press.

Smolensky, P. (1988). "On the Proper Treatment of Connectionism." *Behavioral and Brain Sciences* 11(1): 1–23.

Smolensky, P. (1989). "Connectionist Modeling: Neural Computation/Mental Connection." In L. A. Cooper, L. Nadel, P. Culicover, and R. M. Harnish (eds.), *Neural Connections, Mental Computation.* Cambridge, MA: MIT Press.

Smolensky, P. and G. Legendre (2006). *The Harmonic Mind: From Neural Computation to Optimality-Theoretic Grammar. Vol. 1: Cognitive Architecture; Vol. 2: Linguistic and Philosophical Implications.* Cambridge, MA: MIT Press.

Sober, E. (1984). *The Nature of Selection.* Chicago, IL: University of Chicago Press.

Sober, E. (1990). "Putting the Function Back into Functionalism." In W. Lycan (ed.) *Mind and Cognition.* Malden, MA: Blackwell 63–70.

Sober, E. (1999). "The Multiple Realizability Argument Against Reductionism." *Philosophy of Science* 66: 542–64.

Soo, F. S., G. W. Schwartz, K. Sadeghi, and M. J. Berry, 2nd. (2011). "Fine Spatial Information Represented in a Population of Retinal Ganglion Cells." *Journal of Neuroscience* 31(6): 2145–55.

Sparks, D. L., C. Lee and W. H. Rohrer (1990). "Population Coding of the Direction, Amplitude, and Velocity of Saccadic Eye Movements by Neurons in the Superior Colliculus." *Cold Spring Harbor Symposium in Quantum Biology* 55: 805–11.

Spillmann, L. (2014). "Receptive Fields of Visual Neurons: The Early Years." *Perception* 43 (11): 1145–76.

Spivey, M. (2007). *The Continuity of Mind.* Oxford: Oxford University Press.

Sprevak, M. (2010). "Computation, Individuation, and the Received View on Representation." *Studies in History and Philosophy of Science* 41(3): 260–70.

Sprevak, M. (2013). "Fictionalism about Neural Representations." *The Monist* 96: 539–60.

Squire, L. and J. Wixted (2015). "Remembering." *Daedalus, Winter*: 53–66.

Squire, L. R. (2009). "The Legacy of Patient H.M. for Neuroscience." *Neuron* 61(1): 6–9.

Squire, L. R., C. E. Stark, and R. E. Clark (2004). "The Medial Temporal Lobe." *Annual Review of Neuroscience* 27: 279–306.

Staley, K. (1999). "Golden Events and Statistics: What's Wrong with Galison's Image/Logic Distinction." *Perspectives on Science* 7: 196–230.

Stampe, D. (1977). "Toward a Causal Theory of Linguistic Representation." In P. A. French, T. E. Uehling Jr., & H. K. Wettstein (eds.), *Midwest Studies in Philosophy, Vol. 2: Studies in the Philosophy of Language* (81–102). Minneapolis: University of Minnesota Press.

Standing, L. (1973). "Learning 10,000 Pictures." *Quarterly Journal of Experimental Psychology* 25(2): 207–22.

Stangor, C. (2011). *Introduction to Psychology.* Saylor Academy.

Stein, R. (1965). "A Theoretical Analysis of Neuronal Variability." Biophysical Journal 5.

Steinmetz, P. N., A. Roy, P. J. Fitzgerald, S. S. Hsiao, K. O. Johnson, and E. Niebur (2000). "Attention Modulates Synchronized Neuronal Firing in Primate Somatosensory Cortex." *Nature* 404(6774): 187–90.

Stepp, N., A. Chemero and M. T. Turvey (2011). "Philosophy for the Rest of Cognitive Science." *Topics in Cognitive Science* 3: 425–37.

Stich, S. (1983). *From Folk Psychology to Cognitive Science: The Case against Belief.* Cambridge, MA: MIT Press.

Stinson, C. (2016). "Mechanisms in Psychology: Ripping Nature at Its Seams." *Synthese* 193(5): 1585–1614.

Strogatz, S. H. (2003). *Sync: The Emerging Science of Spontaneous Order.* New York: Hyperion.

Strogatz, S. H. (2015). *Nonlinear Dynamics and Chaos*, Second Edition. Boulder, CO: Westview.

Sullivan, J. A. (2008). "Memory Consolidation, Multiple Realizations, and Modest Reductions." *Philosophy of Science* 75(5): 501–13.

Sullivan, J. A. (2009). "The Multiplicity of Experimental Protocols: A Challenge to Reductionist and Non-Reductionist Models of the Unity of Neuroscience." *Synthese* 167: 511–39.

Sullivan, J. A. (2010). "A Role for Representation in Cognitive Neurobiology." *Philosophy of Science* 77(5): 875–87.

Sullivan, J. A. (2016). "Construct Stabilization and the Unity of the Mind-Brain Sciences." *Philosophy of Science* 83: 662–73.

Sussillo, D., M. M. Churchland, M. T. Kaufman, and K. V. Shenoy (2015). "A Neural Network that Finds a Naturalistic Solution for the Production of Muscle Activity." *Nature Neuroscience* 18(7): 1025–33.

Swoyer, C. (1991). "Structural Representation and Surrogative Reasoning." *Synthese* 87(3): 449.

Takeda, K., and Funahashi, S. (2002). "Prefrontal Task-related Activity Representing Visual Cue Location or Saccade Direction in Spatial Working Memory Tasks." *Journal of Neurophysiology* 87(1): 567–88.

Tamburrini, G. (1988). *Reflections on Mechanism*, unpublished Ph.D. dissertation, Columbia University.

Tamburrini, G. (1997). "Mechanistic Theories in Cognitive Science: The Import of Turing's Thesis." In M. L. Dalla Chiara, K. Doets, D. Mundici, and J. van Benthem (eds.), *Logic and Scientific Method* (239–257). Boston: Kluwer.

Tass, P.A., T. Fieseler, J. Dammers, K. Dolan, P. Morosan, M. Majtanik, F. Boers, A. Muren, K. Zilles, and G. R. Fink (2003). "Synchronization Tomography: A Method for Three-Dimensional Localization of Phase Synchronized Neuronal Populations in the Human Brain Using Magnetoencephalography." *Physical Review Letters* 90(8): 088101.

Taube, M. (1961). *Computers and Common Sense: The Myth of Thinking Machines*. New York: Columbia University Press.

Taylor, H. (2017). "Powerful Qualities, the Conceivability Argument and the Nature of the Physical." *Philosophical Studies* 174(8): 1895–1910.

Taylor, H. (2018). "Powerful Qualities, Phenomenal Concepts, and the New Challenge to Physicalism," *Australasian Journal of Philosophy*, 96(1): 53–66.

Thagard, P. (2007). "Coherence, Truth, and the Development of Scientific Knowledge." *Philosophy of Science* 74: 28–47.

Thalos, M. (2013). *Without Hierarchies: The Scale Freedom of the Universe*. Oxford: Oxford University Press.

Thelen, E. and L. Smith (1994). *A Dynamic Systems Approach to the Development of Cognition and Action*. Cambridge, MA: MIT Press.

Thompson, E. (2007). *Mind in Life: Biology, Phenomenology, and the Sciences of Mind*. Cambridge, MA: Harvard University Press.

Thomson, E. E. and W. B. Kristan (2005). "Quantifying Stimulus Discriminability: A Comparison of Information Theory and Ideal Observer Analysis." *Neural Computation* 17(4): 741–78.

Thomson, E. E. and G. Piccinini (2018). "Neural Representation Observed." *Minds and Machines* 28(6): 1–45.

Thorndike, E. L. (1932). *The Fundamentals of Learning*. New York: Teachers College, Columbia University.

Todorov, E. (2000a). "Direct Cortical Control of Muscle Activation in Voluntary Arm Movements: A Model." *Nature Neuroscience* 3(4): 391–8.

Todorov, E. (2000b). "Reply to 'One Motor Cortex, Two Different Views'." *Nature Neuroscience* 3(10): 963–4.

Tong, F., Nakayama, K., Vaughan, J. T., and Kanwisher, N. (1998). "Binocular Rivalry and Visual Awareness in Human Extrastriate Cortex. *Neuron* 21: 753–9.

Tootell, R. B., E. Switkes, M. S. Silverman, and S. L. Hamilton (1988). "Functional Anatomy of Macaque Striate Cortex. II. Retinotopic Organization." *Journal Neuroscience* 8(5): 1531–68.

Trehub, A. (1991). *The Cognitive Brain*. Boston, MA: MIT Press.

Treisman, A. (1996). "The Binding Problem." *Current Opinion in Neurobiology* 6(2): 171–8.

Treisman, A. (2009). "Attention: Theoretical and Psychological Perspectives." In M. Gazzaniga (ed.), *The Cognitive Neurosciences, Fourth Edition* (189–204). Cambridge, MA: MIT Press.

Treisman, A. and G. Gelade (1980). "A Feature Integration Theory of Attention." *Cognitive Psychology* 12: 97–136.

Truccolo, W., G. M. Friehs, J. P. Donoghue, and L. R. Hochberg (2008). "Primary Motor Cortex Tuning to Intended Movement Kinematics in Humans with Tetraplegia." *Journal of Neuroscience* 28(5): 1163–78.

Tucker, Chris (2018). "How to Explain Miscomputation." *Philosophers' Imprint* 18(24): 1–17.

Turing, A. M., (1936–7). "On Computable Numbers, with an Application to the Entscheidungsproblem." *Proceeding of the London Mathematical Society* 42(1): 230–65.

Turing, A. M. (1939). "Systems of Logic Based on Ordinals." *Proceedings of the London Mathematical Society, Ser. 2* 45: 161–228. Reprinted in M. Davis (ed.), *The Undecidable* (1965, 116–54). Hewlett, NY: Raven Press.

Turing, A. M. (1947). "Lecture to the London Mathematical Society on 20 February 1947." In D. Ince (ed.), *Mechanical Intelligence* (87–105). Amsterdam: North-Holland.

Turing, A. M. (1948). "Intelligent Machinery." In D. Ince (ed.), *Mechanical Intelligence* (87–106). Amsterdam, North-Holland.

Turing, A. M. (1950). "Computing Machinery and Intelligence." *Mind* 59: 433–60.

Turner, B. M., Forstmann, B. U., Love, B. C., Palmeri, T. J., and Van Maanen, L. (2017). "Approaches to Analysis in Model-based Cognitive Neuroscience." *Journal of Mathematical Psychology* 76: 65–79.

Tye, M. (1995). *Ten Problems of Consciousness: A Representational Theory of the Phenomenal Mind.* Cambridge, MA: MIT Press.

Uhlhaas, P. J. and W. Singer (2006). "Neural Synchrony in Brain Disorders: Relevance for Cognitive Dysfunctions and Pathophysiology." *Neuron* 52(1): 155–68.

Umiltà, M. A., L. Escola, I. Intskirveli, F. Grammont, M. Rochat, F. Caruana, A. Jezzini, V. Gallese, and G. Rizzolatti (2008). "When Pliers Become Fingers in the Monkey Motor System." *Proceedings of the National Academy of Sciences USA* 105(6): 2209–13.

Ungerleider, L. G. and R. Desimone (1986). "Cortical Connections of Visual Area MT in the Macaque." *J Comp Neurol* 248(2): 190–222.

van Eck, D. (2018). "Rethinking the explanatory power of dynamical models in cognitive science," *Philosophical Psychology*, 31(8): 1131–61.

van Gelder, T. (1995). "What Might Cognition Be, If Not Computation." *The Journal of Philosophy* 92(7): 345–81.

van Gelder, T. (1998). "The Dynamical Hypothesis in Cognitive Science." *Behavioral and Brain Sciences* 21: 615–65.

van Riel, R. and Van Gulick, R. (2019). "Scientific Reduction", *The Stanford Encyclopedia of Philosophy* (Spring 2019 Edition), Edward N. Zalta (ed.), URL = <https://plato.stanford.edu/archives/spr2019/entries/scientific-reduction/>. Stanford, CA: The Metaphysics Research Lab, Centre for the Study of Language and Information, Stanford University.

VanRullen, R., R. Guyonneau, and S. J. Thorpe (2005). "Spike Times Make Sense." *TRENDS in Neuroscience* 28(1): 1–4.

Varela, F. J., E. Thompson, and E. Rosch (1991). *The Embodied Mind: Cognitive Science and Human Experience.* Cambridge, MA: MIT Press.

Varela, F. J., E. Thompson, and E. Rosch (2017). *The Embodied Mind, Revised Edition.* Cambridge, MA: MIT Press.

Vartanian, A. (1973). *Dictionary of the History of Ideas: Studies of Selected Pivotal Ideas, Ed. P.P. Wiener.* New York: Scriners.

Varzi, A. (2016). "Mereology." In E. N Zaltta (ed.), *The Stanford Encyclopedia of Philosophy* (Winter 2016 Edition), URL = <https://plato.stanford.edu/archives/win2016/entries/

mereology/>. Stanford, CA: The Metaphysics Research Lab, Centre for the Study of Language and Information, Stanford University.

Vendler, Z. (1972). *Res Cogitans*. Ithaca, NY: Cornell University Press.

Vera, A. H. and H. A. Simon (1993). "Situated Action: A Symbolic Interpretation." *Cognitive Science* 17: 7–48.

Vernazzani, A. (2019). "The Structure of Sensorimotor Explanation." *Synthese* 196: 4527–4553.

Volkinshtein, D. and R. Meir (2011). "Delayed Feedback Control Requires an Internal Forward Model." *Biological Cybernetics* 105(1): 41–53.

von Neumann, J. (1945). "First Draft of a Report on the EDVAC." Philadelphia, PA, Moore School of Electrical Engineering, University of Pennsylvania.

von Neumann, J. (1951). "The General and Logical Theory of Automata." In L. A. Jeffress (ed.), *Cerebral Mechanisms in Behavior* (1–41). New York: Wiley.

von Neumann, J. (1958). *The Computer and the Brain*. New Haven, Yale University Press.

Wajnerman Paz, A. (2017). "Pluralistic Mechanism." *Theoria* 32(2): 161–75.

Wandell, B. A. (1995). *Foundations of Vision*. Sunderland, MA: Sinauer.

Wang, H. (1974). *From Mathematics to Philosophy*. London: Routledge and Kegan Paul.

Waskan, J. (2006). *Models and Cognition: Prediction and Explanation in Everyday Life and in Science*. Cambridge, MA: The MIT Press.

Watanabe, Y. and S. Funahashi (2004). "Neuronal Activity Throughout the Primate Mediodorsal Nucleus of the Thalamus During Oculomotor Delayed-Responses." I. Cue-, delay-, and Response-period Activity. *Journal of Neurophysiology* 92(3): 1738–55.

Watkins, M. (2002). *Rediscovering Colors: A Study in Pollyanna Realism*. New York: Springer.

Waxman, S. (1972). "Regional Differentiation of the Axon: A Review with Special Reference to the Concept of the Multiplex Neuron." *Brain Research* 47: 269–88.

Webb, J. C. (1980). *Mechanism, Mentalism, and Metamathematics*. Dordrecht: Reidel.

Weber, M. (2005). *Philosophy of Experimental Biology*. Cambridge: Cambridge University Press.

Weber, M. (2008). "Causes without Mechanisms: Experimental Regularities, Physical Laws, and Neuroscientific Explanation." *Philosophy of Science* 75(5): 995–1007.

Weber, M. (2014). "Experiment in Biology." In N. Zalta (ed.), *The Stanford Encyclopedia of Philosophy* (Winter 2014 Edition), URL = <https://plato.stanford.edu/archives/win2014/entries/biology-experiment/>. Stanford, CA: The Metaphysics Research Lab, Centre for the Study of Language and Information, Stanford University.

Weisberg, M. (2013). *Simulation and Similarity: Using Models to Understand the World*. Oxford: Oxford University Press.

Weiskopf, D. A. (2004). "The Place of Time in Cognition." *British Journal for the Philosophy of Science* 55: 87–105.

Weiskopf, D. A. (2011a). "The Functional Unity of Special Science Kinds." *British Journal for the Philosophy of Science* 63: 233–58.

Weiskopf, D. A. (2011b). "Models and Mechanisms in Psychological Explanation." *Synthese* 183(3): 313–38.

Weiskopf, D. A. (2017). "The Explanatory Autonomy of Cognitive Models." In David M. Kaplan (ed.), *Explanation and Integration in Mind and Brain Science* (44–69). Oxford: Oxford University Press.

Wessberg, J., C. R. Stambaugh, J. D. Kralik, P. D. Beck, M. Laubach, J. K. Chapin, J. Kim, S. J. Biggs, M. A. Srinivasan, and M. A. Nicolelis (2000). "Real-time Prediction of Hand Trajectory by Ensembles of Cortical Neurons in Primates." *Nature* 408(6810): 361–5.

Wheeler, J. A. (1982). "The Computer and the Universe." *International Journal of Theoretical Physics* 21(6–7): 557–72.

Wheeler, J. A. (1990). "Information, Physics, Quantum: The Search for Links." In W. H. Zurek (ed.), *Complexity, Entropy, and the Physics of Information*. Redwood City, California: Addison-Wesley.

White, B. J., D. J. Berg, J. Y. Kan, R. A. Marino, L. Itti, and D. P. Munoz (2017). "Superior Colliculus Neurons Encode a Visual Saliency Map During Free Viewing of Natural Dynamic Video." *Nature Communications* 8: 14263.

White, J.A., R. Klink, A. Alonso, and A. R. Kay (1998). "Noise from Voltage-gated Ion Channels May Influence Neuronal Dynamics in the Entorhinal Cortex." *Journal of Neurophysiology* 80(1): 262–9.

Whitlock, J. R., A. J. Heynen, M. G. Shuler, and M. F. Bear (2006). "Learning Induces Long-Term Potentiation in the Hippocampus." *Science* 313(5790): 1093–7.

Wickersham, I. R., D. C. Lyon, R. J. Barnard, T. Mori, S. Finke, K. K. Conzelmann,... E. M. Callaway (2007). "Monosynaptic Restriction of Transsynaptic Tracing from Single, Genetically Targeted Neurons." *Neuron* 53(5): 639–47.

Wiener, N. (1948). *Cybernetics or Control and Communication in the Animal and the Machine*. Cambridge, MA: MIT Press.

Wilkes, K. V. (1982). "Functionalism, Psychology, and the Philosophy of Mind. Mind, Brain, and Function: Essays in the Philosophy of Mind." J. I. Biro and R. W. Shahan. Norman, University of Oklahoma Press: 147–67.

Williams, D. (2018). "Predictive Processing and the Representation Wars." *Minds and Machines* 28: 141–172.

Williams, D. and L. J. Colling (2018). "From Symbols to Icons: The Return of Resemblance in the Cognitive Neuroscience Revolution." *Synthese* 195(5): 1941–67

Wilson, H. R. and J. D. Cowan (1972). "Excitatory and Inhibitory Interactions in Localized Populations of Model Neurons." *Biophysical Journal* 12(1): 1–24.

Wilson, J. (1999). "How Superduper Does a Physicalist Supervenience Need to Be?" *The Philosophical Quarterly* 49(194): 33–52.

Wilson, J. (2010). "Non-Reductive Physicalism and Degrees of Freedom." *British Journal for Philosophy of Science* 61: 279–311.

Wilson, J. (2011). "Non-Reductive Realization and the Power-Based Subset Strategy." *The Monist* 94: 121–54.

Wilson, R. A. (1994). "Wide Computationalism." *Mind* 103(411): 351–72.

Wimmer, K., D. Q. Nykamp, C. Constantinidis, and A. Compte (2014). "Bump Attractor Dynamics in Prefrontal Cortex Explains Behavioral Precision in Spatial Working Memory." *Nat Neurosci* 17(3): 431–9.

Wimsatt, W. (1997). "Aggregativity: Reductive Heuristics for Finding Emergence." *Philosophy of Science* 64(4): 372–84.

Wimsatt, W. C. (1972). "Teleology and the Logical Structure of Function Statements." *Studies in the History and Philosophy of Science* 3: 1–80.

Wimsatt, W. C. (2002). "Functional Organization, Analogy, and Inference." In A. Ariew, R. Cummins, and M. Perlman (eds.), *Functions: New Essays in the Philosophy of Psychology and Biology* (173–221). Oxford: Oxford University Press.

Wimsatt, W. C. (2007). "Re-Engineering Philosophy for Limited Beings." Cambridge, MA: Harvard University Press.

Winning, J, and Bechtel, W. (2018). "Rethinking Causality in Biological and Neural Mechanisms: Constraints and Control." *Minds and Machines* 28(2): 287–310.

Winsberg, E. (2010). *Science in the Age of Computer Simulation*. Chicago: University of Chicago Press.

Wolfram, S. (2002). *A New Kind of Science*. Champaign, IL: Wolfram Media.

Wolpert, D. M. and R. C. Miall (1996). "Forward Models for Physiological Motor Control." *Neural Networks* 9(8): 1265–79.

Wolpert, D. M., R. C. Miall and M. Kawato (1998). "Internal Models in the Cerebellum." *Trends in Cognitive Science* 2(9): 338–47.

Wong, R. K. S., R. D. Traub, and R. Miles (1986). "Cellular Basis of Neural Synchrony in Epilepsy." In A. V. Delgado-Escueta, A. A. Ward, D. M. Woodbury, and R. J. Porter (eds.), *Advances in Neurology*. New York: Raven Press.

Woodward, J. (2003). *Making Things Happen: A Theory of Causal Explanation*. New York: Oxford University Press.

Woodward, J. (2015). "Interventionism and Causal Exclusion." *Philosophy and Phenomenological Research* 91(2): 303–47.

Woodward, J. (2017). "Explanation in Neurobiology: An Interventionist Perspective." In David M. Kaplan (ed.), *Explanation and Integration in Mind and Brain Science* (70–100). Oxford: Oxford University Press.

Wright, C. (1995). "Intuitionists Are Not (Turing) Machines." *Philosophia Mathematica* 3 (3): 86–102.

Wright, L. (1973). "Functions." *Philosophical Review* 82: 139–68.

Yablo, S. (1992). "Mental Causation." *The Philosophical Review* 101(2): 245–80.

Yassin, L., B. L. Benedetti, J. S. Jouhanneau, J. A. Wen, J. F. A. Poulet, and A. L. Barth (2010). "An Embedded Subnetwork of Highly Active Neurons in the Neocortex." *Neuron* 68: 1043–50.

Yizhar, O., L. E. Fenno, T. J. Davidson, M. Mogri, and K. Deisseroth (2011). "Optogenetics in Neural Systems." *Neuron* 71(1): 9–34.

Zangwill, N. (1992). "Variable Realization: Not Proved." *The Philosophical Quarterly* 42 (167): 214–19.

Zednik, C. (2011). "The Nature of Dynamical Explanation." *Philosophy of Science* 78(2): 238–63.

Zednik, C. (2019). "Models and Mechanisms in Network Neuroscience." *Philosophical Psychology* 32(1): 23–51.

Zhang, D. and M. E. Raichle (2010). "Disease and the Brain's Dark Energy." *Nature Reviews Neurology* 6: 15–28.

Zuse, K. (1970). *Calculating Space*. Cambridge, MA: MIT Press.

Zylberberg, J. and B. W. Strowbridge (2017). "Mechanisms of Persistent Activity in Cortical Circuits: Possible Neural Substrates for Working Memory." *Annual Review of Neuroscience* 40: 603–27.

Index

Note: Figures are indicated by an italic "*f*", respectively, following the page number.

For the benefit of digital users, indexed terms that span two pages (e.g., 52–53) may, on occasion, appear on only one of those pages.